U0279227

人类学终身成就奖获奖者风采录
（第2辑）

金　力　王久存　李　辉　主编

上海科学技术出版社

图书在版编目（CIP）数据

人类学终身成就奖获奖者风采录. 第2辑 / 金力，王久存，李辉主编. -- 上海 : 上海科学技术出版社，2024. 11. -- ISBN 978-7-5478-6866-9

Ⅰ. K826.15

中国国家版本馆CIP数据核字第20248YE960号

人类学终身成就奖获奖者风采录（第2辑）

金　力　王久存　李　辉　主编

上海世纪出版(集团)有限公司
上 海 科 学 技 术 出 版 社　出版、发行

(上海市闵行区号景路159弄A座9F-10F)

邮政编码201101　www.sstp.cn

常熟市华顺印刷有限公司印刷

开本 787×1092　1/16　印张 25

字数 390千字

2024年11月第1版　2024年11月第1次印刷

ISBN 978-7-5478-6866-9 / Q·89

定价：128.00元

前　言

人类学是研究人类自身的交叉学科，主要围绕人的生物属性和社会属性展开研究。自胚胎着床的那一刻起，生命的孕育悄然开始，围绕个体的生、长、壮、老、已的生物属性开始形成和演变。从生命诞生到融入社会，个体在心理、认知、思维上发生转变，形成了社会属性。自人类走出非洲那一刻起，因长期受到遗传和环境的交互作用，注定形成了语言不同、肤色各异的群体。于是，当人类学家看到不同的群体，发现不同的遗址，好奇心驱动着他们孜孜不倦地进行田野调查，期待从时空轴上厘清人类的足迹。

人类学萌芽于16世纪，最早属于博物学的四大分支之一。早期的人类学研究是对人类的体质表型进行测量，包括现代人和出土遗骸的测量，形成了体质人类学。对群体间的民俗习惯、宗教信仰和语言表型进行研究，以了解不同群体的社会文化生活，发展形成了考古人类学、文化人类学和语言人类学。20世纪末，人类基因组计划的发起，促进了人类遗传学和测序技术的不断发展，人类学家开始通过对现代人和遗骸进行测序来研究群体关系，形成了生物（分子）人类学。到了今天，随着人类表型组国际大科学计划的发起，开始对自然人类进行跨尺度、多维度、高通量的精密表型测量，从综合人类学的角度精密解析了人类24 000多个表型。从整体上看，人类学从宏观的体质测量到微观的遗传测序，从单一学科研究到多学科交叉融合，俨然发展成了一个兼容并包的学科交叉群。

中国的人类学研究始于20世纪。在人类学研究的背后，我国科学家兢兢业业、不辞辛苦，为中国乃至世界人类学的研究做出了突出贡献。我国是一个多民族国家，从科学角度剖析各民族的起源显得至关重要。我国人类学家在过去一个多世纪以来，解析了汉藏同源，描绘了东亚语系，厘清了民族形成……为了表彰这些人类学家，上海人类学学会从2010年起设立了“人类学终身成就奖”，以表彰在人类学领域艰苦奋斗、勇毅前行的人类学家。这一合辑收录了从2017年到2024年的“人类学终身成就奖”获得者，有溯源发音，促进文理交叉的语言人类学家王士元，奏响了“玉音交振阳关谱”；有解析成分，关联体质与健康的医学人类学家席焕久，测出了“金尺悠鸣杏林春”；有凝聚理论，解释远古社会的考古人类学家陈淳，挖出了“格沙致理万古器”；有优化技术，测量民族身形的体质人类学家郑连斌，踏遍了“称象分石千里邨”；有总结规律，探究汉族成因的历史人类学家徐杰舜，解析了“雪球滚积中华体”；有搜寻线索，挖掘南岛根源的民族人类学家邓晓华，书写了“沧海横流南岛情”；有沉浸山村，刻画群体思维的社会人类学家庄孔韶，勾勒出“山野秘辛绘奇画”；有开创组学，重构东亚语系的语言人类学家潘悟云，道出了“水乡私语书同文”。他们不仅把足迹留在祖国的天南海北，也把论文书写在祖国大地上，进一步筑牢中华民族共同体意识。这一合辑的内容收录了获奖者的研究经历和代表作，他们的研究是人类学领域的知识宝库，他们的研究经历是后辈踔厉奋发的精神财富。希望这一合辑的出版，能够提升人类学事业发展的高度。

王久存
上海人类学学会会长
复旦大学人类遗传学与人类学系系主任，教授
复旦大学现代人类学教育部重点实验室学术委员会委员

目　录

第一章

▼

玉音交振阳关谱

2017年人类学终身成就奖获得者

——王士元

人类学生涯回顾*

王士元1933年8月生于上海，但一个月大时便移居安徽怀远，与祖父母同住。祖母是小时候对他影响最大的人，虽然她是个裹脚、不识字的传统女性，却教导了他许多做人处事的道理。中国古典文学如《水浒传》、《三国演义》里忠孝节义的情节，以及牛郎织女等民间故事，都是在幼年时经由祖母的口进入了他的心灵，使他现在即使早已入了美国籍，仍始终认同自己是中国

与家人摄于上海，左起：妹妹、妈妈、奶奶、爸爸、姑姑

* 本文作者为蔡雅菁。

人，要为振兴中国的学术而努力。1937年七七事变以后，王士元与祖父母从怀远迁到上海的法租界与父母会合，并在上海就读小学，之后也曾入读法国教会学校。后来因父亲在美国加州大学伯克利分校取得经济学硕士学位，并于纽约经商，1948年10月，与祖母（当时祖父已过世）、母亲及妹妹举家赴美，在纽约完成中学教育。

由于中学时代学业成绩优异，哥伦比亚大学提供了他本科四年的奖学金。可是1951年入学后不久，父母亲就和祖母、妹妹先回国了。他一方面不满于无法与家人一起回国而要独自留美，另一方面又因纽约的诱惑太多，进入哥伦比亚大学第一学期的成绩很不理想，甚至被取消了四年的奖学金而退学。之后终于发愤读书，自己靠工读方式完成大学学业。大学期间除了选修语言学的课程，也上了些电子工程方面的课，并开始接触人类学，修读一门讲述东南亚华侨的课，由知名人类学家Morton Fried授课。

语言学的启蒙阶段

有两门语言学的课令他至今难忘，也是促使他日后从事语言学研究的动力。其中一门课名为“Languages of the World”，由意大利裔美籍的语言学家Mario Pei讲授。这门课让他大开眼界，了解到世界语言的多样性。在Pei教授的建议下，他们联名在《纽约时报》上发表了一篇“It’s not all Greek”

与Joseph Greenberg、Vincent Sarich摄于加州大学伯克利分校教职员俱乐部

的文章。这篇短文的内容主要是讲汉语，即使并非是正式的学术论文，却也算是王士元科研生涯中的第一篇文章。另一门课由著名的语言学大师及非洲人类学家Joseph Greenberg所讲授，虽然只是本科生的语言学概论的课，Greenberg却介绍了三四十个非洲语言基本词汇的语料。Greenberg严谨的治学态度给了他非常深刻的启发，也激发了他对语言学的浓厚兴趣。

另外一件激发他对探索语言兴趣的事是，大学时代有一天，他和三个室友一起到学校附近的一家中国饭馆吃饭。当三个美国同学都盼着他替大伙儿点菜时，他却发现自己和餐厅里的侍者沟通不良，和对方讲上海话讲不通，和他说普通话对方也不懂，因为侍者是台山人，最后只得用英文点菜，这让在场的三位同学都调侃他还算是中国人吗。由于自小生活的环境中就接触到多种语言和方言，包括与祖母交谈时所用的怀远话、在上海就学时和童年玩伴所说的上海话，还有在上海租界可以听到的法语、日语等，这些经历都让他体会到中国方言的众多差异和世界语言的丰富多样，并对人类精彩纷呈的语言遗产深深着迷。

早期的学术经历

王士元于1956年和1960年在密歇根大学先后取得硕士、博士学位，硕士研究生时期与Gordon Peterson教授学习声学语音学，并和Kenneth Pike教授学习一般语言学及音系学。后来在Peterson教授指导下取得博士学位，并先后在麻省理工学院电子研究实验中心和IBM Thomas J. Watson Research Center从事博士后研究。1960年，应博士导师Peterson教授的邀请，回到密歇根大学与遗传算法之父John Holland教授合开"Communication Sciences"的课。1961年，赴俄亥俄州立大学任英语系助理教授，随后在该校创办语言学及东亚语言文学两个系，并担任此两系的系主任。

1973年与吕叔湘教授摄于北京

伯克利时期

1966年在赵元任先生的邀请下转到加州大学伯克利分校，任教30年后于1995年

1973 年在北京大学哲学楼讲学

1978 年与林焘、岑麒祥、王力、袁家骅、周祖谟、朱德熙摄于北京大学

退休。任职于伯克利时，在1967年成立语音实验室（Phonology Laboratory），这是美国最早的语音实验室之一；于1973年创办*Journal of Chinese Linguistics*《中国语言学报》并任主编至今，该学报是中国语言学领域第一本国际性刊物。这一年，在北京大学和中国社会科学院的同时邀请下，王士元

第一次有机会回到中国讲学。当他再回到伯克利时，便体会到当时中国已有许多出色的语言本质的研究，而中国语言学的研究传统甚至可溯及两千年前，只可惜这个研究传统却从未和国际上的语言研究有过充分的学术交流。为了将国内学者的研究成果介绍给国际学术界，并让国内学者也能知悉外国学者最新的文献和研究动态，在院长的支持下，《中国语言学报》正式开始运作，1973年创刊的第一期第一篇文章，就刊登了有"非汉语语言学之父"之称的李方桂先生的经典文章《中国的语言和方言》。

除了创办国际性学术刊物，王士元在伯克利期间还有另外两个重要的里程碑：一是建立汉语方言的电子词典，这是当时最早的汉语方言的语料数据库；二是提出语言学界的词汇扩散理论。成立了语音实验室后，他与郑锦全等同事，将《切韵》、《中原音韵》、《汉语方音字汇》等书里的材料都输入计算机里，并加入日语的汉音、吴音及韩语里的汉韩音（Sino-Korean）等材料。根据这套基于汉语方言的计算机词典里的丰富语料，他发现语言变迁在语音上是骤变的，但在词汇上是渐变的，由此提出了词汇扩散理论。享誉国际的遗传学家Luigi Luca Cavalli-Sforza曾经说，词汇扩散理论是"the first thing I heard in linguistics that makes sense"，并在1994年的一篇文章中表示："词汇扩散理论已经公认为是语言演化研究中最重要的创新理论。"已故的哈佛大学教授、人类学家张光直曾在1994年的《亚洲周刊》上这样评论二十世纪中国人文社会学科对国际学术界的贡献和影响："只有在语言学上，中国语言学者如赵元任、李方桂、王士元等先生的著作在普通语言学的书刊里给人引用。"虽然词汇扩散理论刚提出时，曾有些专精于印欧语研究的学者表示质疑，并认为这套理论是基于汉语语料建立的，对欧美语言不一定适用。但如今，词汇扩散理论已被广泛应用在其他语言的研究上，而且不止在语音、声调上，也适用在构词和句法上，包括藏缅语、日语、英语、德语、瑞典语、达罗毗荼语等，甚至有美洲原住民语言尼蒂纳特语（Nitinat）。

王士元在伯克利时期还成立了语言分析实验室（Project on Linguistic Analysis, POLA）。这个实验室让当时鲜有接触的海峡两岸和港澳地区学者，头一次有机会在开放的美国学术环境下进行交流。至今，他所培育出的博士生，在海峡两岸及美国大多成为了知名的语言学教授。他指导过的学生如今任教的单位包括北京大学、台湾新竹清华大学、香港理工大学，以及美国的华盛顿大学、夏威夷大学、马萨诸塞州立大学、加州大学圣迭戈分校等。除了语

2005年与曾志朗教授于台北建国中学演讲

2010年与北京大学副校长张国有摄于名誉教授颁授典礼上

言学家，他也积极和心理学家、遗传学家、统计学家、考古学家等跨领域的学者合作。比如，他和台湾的曾志朗教授开创了汉字认知的神经学研究；他也促进了北京大学语音学实验室的建立；他曾与2014年人类学终身成就奖的得主杜若甫教授合作，研究中国人在遗传演化和语言演化上的一致性。

积极主张并力行跨学科研究

从王士元的学术著作可以看出，他自己身体力行地与不同领域的学者合作发表论文，因为早在1978年，他就以 “The three scales of diachrony” 一文，来强调语言历时研究的三个尺度：宏观、中观与微观。微观的语言研究涉及个人的语言习得与消失，但中观的语言研究则是历史语言学的范畴，可以溯及几千年前的语言，因此必须借助对人群的往来和历史的理解，才能更深入研究语言的变迁。至于宏观的语言研究，则旨在探索语言的涌现和演化，因此更要借助考古学、遗传学等学科。所以在1998年的文章 “Three windows on the past” 中，他特别强调探索过去的三个窗口是考古学、遗传学和语言学，这样才得以追溯并还原几十万年前的远古场景。况且语言的涌现和人类直立行走及大脑演化都有密不可分的关系，因此语言研究的宏观尺度若能结合体质人类学和大脑神经科学的研究成果，也必定会有更重大的发现。

王士元常用盲人摸象的比喻来形容研究人员的划地自限，如果这些学科各行其是，则很难对语言的全貌有透彻的认识，若研究人员能意识到自己的局限而与其他学科的学者合作，取长补短，就能为学术研究开拓新的视野。

田野调查及少数民族语言研究

王士元不只乐于从事理论的探索和实验的证明，也积极投身田野工作。他曾与南开大学石锋教授到广西大瑶山研究瑶族，也曾随北京大学陈保亚教授到云南考察白族、普米族、傈僳族、怒族等少数民族，并曾远赴吉尔吉斯斯坦调查东干人。他走访云南的记录，于2002年由云南民族文化音像出版社发行了名为《走云南：士元讲课》的VCD。这些田野调查的经验，除了让他有机会接触中国西南丰富的语言文化遗产外，也让他对中国的少数民族语言有了第一手的宝贵材料。他与人类学兼民族学家邓晓华教授，合写了多篇有关少数民族语言的文章，如《藏缅语族语言的数理分类及其分析》、《苗瑶语族语言亲缘关系的计量研究——词源统计分析方法》、《壮侗语族语言的数理分类及其时间深度》，这三篇文章都曾获奖。走入田野不仅让他徜徉在丰美的语言天地间，体会到语言多样性的可贵和重要，也让他由衷慨叹语言濒危的可悲和无奈。

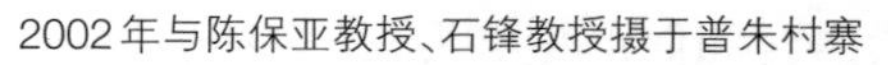

2002 年与陈保亚教授、石锋教授摄于普朱村寨

2003 年与邓晓华教授

移居香港

1997年，王士元在红磡的海逸酒店见证了香港回归，政权的交接也坚定了他重回祖国怀抱的信念。1997—2004年间，他先是在香港城市大学担任语言工程讲座教授，后又应聘为香港中文大学研究教授。在香港中文大学

期间，他把30多年前创办的《中国语言学报》办公室由美国加州迁到香港特区，改由香港中文大学出版社出版这份重要的国际学术期刊。同时担任电子工程系、语言学及现代语言系和东亚研究中心的教授，并偶尔在翻译系授课，讲授自然语言处理。他在电子工程系建立了语言工程实验室，与实验室里理工科背景的学生从事许多语言建模仿真的研究，加深了自己对语言演化的兴趣。基于此，他在2009年于广州召开了第一届演化语言学国际研讨会（International Conference in Evolutionary Linguistics，CIEL）。这一系列的会议后来陆续在天津、上海、北京、香港、厦门、天津成功举办，第八届则是演化语言学会议首次在中国境外举行，由美国印第安纳大学主办，2017年第九届的会议则在八月底在昆明落幕，这一系列会议成为跨学科的国际交流平台，促进了演化语言学的发展。

在香港中文大学期间，另一件令学术界瞩目的盛事是于2012年举办的“A Dialogue on Sound Change: Past, Present and Future”。这是一场由王士元和美国社会语言学权威William Labov探讨关于音变的历史、现状与未来的对话。这两人同样名为Bill，他们的情谊其实可以追溯到20世纪70年代。1971年，Labov就任美国语言学会会长，他在就职演说中率先肯定了词汇扩散理论的贡献，这也逐渐扭转了欧美学者对词汇扩散理论的看法。Labov教授著有三卷经典的语言学巨著*Principles of Language Change*，他在卷首的致谢中写道：“本书的许多研究、调查、论述都是对于王士元在竞争性语音演变、词汇扩散、方言混合方面极具远见卓识的开创性研究工作的响应。……虽然我自己的团队在这些问题上的研究跟他们不都一样，但是他们理论观点的冲击力和与之密切相关的语言资料对我影响至深。”

离开香港中文大学后，2015年王士元又获聘为香港理工大学语言与认知科学讲座教授，因为此时，他的研究兴趣已由单纯的语言研究转向认知退化的探索。目前，语言习得领域对婴儿如何习得母语已经有了丰硕的研究文献与可观的研究成果，但对于老年阶段语言如何退化消失，所知却远远不足。有鉴于人口老龄化已是全球不得不面对的严峻挑战，随着医疗进步和环境卫生改善，保持健康身体的同时该如何维护认知能力的活跃，是高龄化社会必定要应对的问题。王士元在香港理工大学积极开展研究，冀望能借由测量脑电波来对认知障碍进行早期的诊断评估，以期能防患于未然，在大脑开始呈现失智退化的迹象前，就能及早发现并采取相应措施来减缓这个趋势。

王士元曾说："不同学科之间的边界犹如画在沙滩上的线条，随着每一次先进知识波涛的到来，这边界就会发生变化，甚至完全消失。人类的知识，特别是研究语言的知识，应该是彼此相连的，并且最终是相互贯通的。"虽然他自幼赴美，心里却始终记挂着自己是中国人，要为中国人争光，因此每次他到中国各地讲学，都一定会告诉在场听众：19世纪语言学发迹于欧洲，20世纪语言学兴盛于美国，但我们要让21世纪的中国成为语言学的重心。他这番语重心长的话，不仅是针对所有中国语言学研究者的衷心期许，也是他一辈子兢兢业业、念兹在兹的奋斗目标。期盼中国的学术界能避免闭门造车、故步自封的保守心态，培养国际视野并时刻关注最前沿的研究，多多提倡跨学科、多领域的研究方法，共同为推进中国语言学的迈步向前而努力！

2004年与Jone Holland教授摄于香港城市大学

2004年与杨振宁家人摄于清华大学，左二起：杨振平之子Mason、杨振宁、杨振宁之弟杨振平及夫人史美

2005年与Tom Schoemann教授摄于美国新墨西哥州圣菲学院

2009年与Sydney Lamb、邹嘉彦摄于香港城市大学

2007 年与 Paul Kay 教授摄于香港中文大学办公室

2010 年在香港中文大学接受叶新燕、Stephen Matthews 访问

2010 年与沈钟伟教授摄于天津

2011 年与 Bernard Comrie 教授摄于香港中文大学

2011 年与 2009 年诺贝尔物理学奖得主高锟摄于香港中文大学

2011 年摄于北京语言大学荣誉教授颁授典礼

2011 年与郑锦全教授摄于台北

2012 年与 William Labov 教授对话于香港中文大学

2012 年与丁邦新夫妇摄于香港

2013 年与谭力海教授摄于香港中文大学

2013 年与洪兰教授摄于香港中文大学

2013 年与潘悟云教授摄于香港

2013年与石锋、彭刚摄于香港沙田的80华诞晚宴

2013年在台湾认知神经科学学会成立大会上发言

2017年与Salikoko Mufwene教授及张梦翰摄于第九届演化语言学研讨会

与Murray Gell-Mann教授摄于香港城市大学

与Charles Fillmore教授摄于香港城市大学

与夫人蔡雅菁骑双人协力车

个人简历

学历

1956—1960　密歇根大学博士

1955—1956　密歇根大学硕士

1951—1955　哥伦比亚大学学士

工作经历

2015—2018　语言与认知科学讲座教授，香港理工大学

2013—2015　香港中文大学-北京大学-台湾联合大学系统语言与人类复杂系统联合研究中心主任，香港中文大学

2004—2015　电子工程系研究教授（2009—2012期间为伟伦研究教授），香港中文大学

1996—2004　语言工程讲座教授，香港城市大学

1973—　创刊编辑及主编，《中国语言学报》

1966—1995　语言学教授，加州大学伯克利分校

1963—1965　新创立的语言学及东亚语言文学两系之副教授兼系主任，俄亥俄州立大学

1961—1963　英语系助理教授，俄亥俄州立大学

1960—1961　沟通科学讲师，密歇根大学

荣誉与奖项

2017—2020　香港城市大学翻译及语言学系荣誉教授

2017—　华侨大学荣誉教授

2015—2018　香港中文大学文学院荣誉教授

2013—　南京师范大学荣誉教授

2012—2015　大连民族大学荣誉教授

2011—　北京语言大学荣誉教授

2010—　北京大学荣誉教授

2010　台湾科技大学金语言奖（Golden Language Award）

2006—2007　日本京都国际高等研究所奖助

1992—	台北“中央研究院”院士
1992	国际中国语言学会首届会长
1991	意大利贝拉乔（Bellagio）高等研究中心奖助
1983—1984	斯坦福行为科学高等研究中心奖助
1971—1972	福布莱特研究奖助及瑞典国家教授（National Professor）
1969—1970	斯坦福行为科学高等研究中心奖助

代表性论著

[1] Wang W S-Y, Peterson G E. Segment inventory for speech synthesis[J]. Journal of the Acoustical Society of America, 1958, 30(8): 743–746.

[2] Wang W S-Y, Peterson G E. Segment inventory for speech synthesis[R]. The University of Michigan Speech Research Laboratory Report No. 1, 1958: 19–28.

[3] Shen Y, Wang W S-Y, Yotsukura S. Some suprasegmental features in English, Japanese and Mandarin Chinese[J]. The English Teachers Magazine, 1958, 7(9): 463–472.

[4] Wang W S-Y. Transition and release as perceptual cues for final plosives[J]. Journal of Speech and Hearing Research, 1959, 2(1): 66–73.

[5] Wang W S-Y, Crawford J. Frequency studies of English consonants[J]. Language and Speech, 1960, 3(3): 131–139.

[6] Wang W S-Y, Fillmore C J. Intrinsic cues and consonant perception[J]. Journal of Speech and Hearing Research, 1961, 4(2): 130–136.

[7] Reeds J A, Wang W S-Y. The perception of stops after s[J]. Phonetica, 1961, 6(1): 78–81.

[8] Wang W S-Y. Stress in English[J]. Language Learning, 1962, 12(1): 69–77.

[9] Meyers L F, Wang W S-Y. Tree representations in linguistics[R]. Project on Linguistic Analysis Report 3, Ohio State University Research Foundation, 1963: 54–111.

[10] Wang W S-Y. Mandarin phonology[R]. Project on Linguistic Analysis Report 6, Ohio State University Research Foundation, 1963: 1–6.

[11] Wang W S-Y. Some syntactic rules for Mandarin[C]//Proceedings of the Ninth International Congress of Linguists. Mouton, 1964: 191–202.

[12] Wang W S-Y. Two aspect markers in Mandarin[J]. Language, 1965, 41(3): 457–470.

[13] Wang W S-Y. 句法分析的原则[J]. 中国语文月刊，1966，108：2–16.

[14] Wang W S-Y. Conjoining and deletion in Mandarin syntax[J]. Monumenta Serica, 1967, 26: 224–236.

[15] Wang W S-Y. The measurement of functional load[J]. Phonetica, 1967, 16: 36–54.

[16] Wang W S-Y. Phonological features of tone[J]. International Journal of American Linguistics, 1967, 33(2): 93–105.

[17] Wang W S-Y. Bibliography of Chinese Linguistics [C]//Sebeok T A, ed. Current Trends in Linguistics. Mouton, 1967: 188–499.
[18] Wang W S-Y, Li K P. Tone 3 in Pekinese [J]. Journal of Speech and Hearing Research, 1967, 10(3): 629–636.
[19] Wang W S-Y. Vowel features, paired variables and the English vowel shift [J]. Language, 1968, 44(4): 695–708.
[20] Mohr B, Wang W S-Y. Perceptual distance and the specification of phonological features [J]. Phonetica, 1968, 18: 31–45.
[21] Wang W S-Y. Competing changes as a cause of residue [J]. Language, 1969, 45: 9–25.
[22] Wang W S-Y. Project DOC: its methodological basis [J]. Journal of American Oriental Society, 1970, 90(1): 57–66.
[23] Wang W S-Y, Cheng C C. Implementation of phonological change: the Shuang-Feng Chinese case [C]//Papers from Sixth Regional Meeting. Chicago Linguistic Society, 1970: 552–559.
[24] Cheng C C, Wang W S-Y. Phonological change of Middle Chinese initials [J]. Tsing Hua Journal of Chinese Studies, 1971, 9(1): 216–270.
[25] Wang W S-Y. The basis of speech [C]//Reed C E, ed. The Learning of Language. New York: Kenkyusha Publishers, 1971: 267–306.
[26] Wang W S-Y. Approaches to phonology [C]//Sebeok T A, ed. Current Trends in Linguistics. Mouton, 1972: 101–121.
[27] Wang W S-Y. The many uses of F0 [C]//Valdman A, ed. Papers in Linguistics and Phonetics to the Memory of Pierre Delattre. The Hague: Mouton, 1972: 487–503.
[28] Wang W S-Y. The Chinese language [J]. Scientific American, 1973, 228: 53–62.
[29] Cheng C C, Wang W S-Y. Tone change in Chaozhou Chinese: a study in lexical diffusion [C]//Kachru B, et al, eds. Issues in Linguistics: Papers in Honor of Henry and Renee Kahane. University of Illinois Press, 1973: 99–113.
[30] Wang W S-Y, Chan S W, T'sou B K. Chinese linguistics and the computer [J]. Linguistics, 1973, 118: 89–117.
[31] Wang W S-Y. Why and how do we study the sounds of speech [C]//Mitchell J L, ed. Computers in the Humanities. Edinburgh University Press, 1974: 39–53.
[32] Wang W S-Y. 语言研究讲话 [J]. Journal of Chinese Linguistics, 1974, 2(1): 1–24.
[33] Chen M, Wang W S-Y. Sound change: actuation and implementation [J]. Language, 1975, 51(1): 255–281.
[34] Wang W S-Y. Language change [J]. Annals of the New York Academy of Science, 1976, 280: 61–72.
[35] Lehiste I, Wang W S-Y. Perception of sentence boundaries with and without semantic information [C]//Phonologica 1976. Innsbrucker Beiträge zur Sprachwissenschaft, 1976, 19: 277–283.
[36] Hardyck C, Tzeng O, Wang W S-Y. Cerebral lateralization effects in visual half-field

experiments [J]. Nature, 1977, 269: 705–707.

[37] Tzeng O J L, Hung D L, Wang W S-Y. Speech recoding in reading Chinese characters [J]. Journal of Experimental Psychology: Human Learning and Memory, 1977, 3: 621–630.

[38] Chuang C K, Wang W S-Y. Use of optical distance sensing to track tongue motion [J]. Journal of Speech and Hearing Research, 1978, 21: 482–496.

[39] Hardyck C, Tzeng O J L, Wang W S-Y. Lateralization of function and bilingual decision processes: is thinking lateralized? [J]. Brain and Language, 1978, 5: 56–71.

[40] Chuang C K, Wang W S-Y. Psychophysical pitch biases related to vowel quality, intensity difference, and sequential order [J]. Journal of the Acoustical Society of America, 1978, 64(4): 1004–1014.

[41] Wang W S-Y. The three scales of diachrony [C]//Kachru B B, ed. Linguistics in the Seventies: Directions and Prospects. University of Illinois, 1978: 63–75.

[42] Fillmore C J, Kempler D, Wang W S-Y. Introductory chapter to Individual Differences in Language Ability and Language Behavior [M]. Academic Press, 1979: 1–10.

[43] Wang W S-Y. Language change: a lexical perspective [J]. Annual Review of Anthropology, 1979, 8: 353–371.

[44] Tzeng O J L, Hung D, Cotton B, Wang W S-Y. Visual lateralization effect in reading Chinese characters [J]. Nature, 1979, 282: 499–501.

[45] Wang W S-Y. Language structure and optimal orthography [C]//Tzeng O J L, Singer H, eds. The Perception of Print: Reading Research in Experimental Psychology. Lawrence Erlbaum Associates, 1981: 223–236.

[46] Wang W S-Y. Variation and selection in language change [J]. Bulletin of the Institute of History and Philology, 1982, 53: 495–519.

[47] Wang W S-Y. Speech and script relations in some Asian languages [C]//Chu-Chang M, ed. Asian- and Pacific-American Perspectives in Bilingual Education: Comparative Research. New York: Teachers College Press, 1983: 56–72.

[48] Tzeng O J L, Wang W S-Y. The first two R's [J]. American Scientist, 1983, 71: 238–243.

[49] Polich J M, McCarthy G, Wang W S-Y, Donchin E. When words collide: orthographic and phonological interference during word processing [J]. Biological Psychology, 1983, 16: 155–180.

[50] Tzeng O J L, Wang W S-Y. Search for a common neuro-cognitive mechanism for language and movements [J]. American Journal of Physiology, 1984, 246: R904–R911.

[51] Wang W S-Y, Xu Tongqiang. Conversation on historical linguistics [J].语言学论丛，1984，13：250–258.

[52] Lin T, Wang W S-Y.声调感知问题[J].中国语言学报，1984，2：59–69.

[53] Wang W S-Y, Cheng C C. Middle Chinese tones in modern dialects [C]//Channon R, Shockey L, eds. In Honor of Ilse Lehiste. Foris Publishers, 1987: 513–523.

[54] Cavalli-Sforza L L, Wang W S-Y. Spatial distance and lexical replacement[J]. Language, 1986, 62: 38–55.
[55] Wang W S-Y. Representing language relationships[C]//Hoenigswald H, Wiener L, eds. Biological Metaphor and Cladistic Classification. University of Pennsylvania Press, 1987: 243–256.
[56] Wang W S-Y. A note on tone development[C]//The Chinese Language Society of Hong Kong, eds. Wang Li Memorial Volumes. Hong Kong: Joint Publishing Company, 1987: 435–443.
[57] Shen Z W, Wooters C, Wang W S-Y. Closure duration in the classification of stops: a statistical analysis[C]//Studies Presented to Ilse Lehiste. Ohio State University Working Papers in Linguistics, 1987, 35: 197–209.
[58] Wang W S-Y. 计算机在语言学里的运用[C]//Proceedings of ROCLING I. Taipei: R.O.C. Computational Linguistics Workshop I, 1988: 257–287.
[59] Wang W S-Y. The migrations of the Chinese people and the settlement of Taiwan[C]// Kuang-chou Li, ed. Anthropological Studies of the Taiwan Area: Accomplishments and Prospects. Taipei: Department of Anthropology, Taiwan University, 1989: 15–36.
[60] Wang W S-Y. Language prefabs and habitual thought[C]//TESOL Summer Institute. Forum Lectures, San Francisco State University, 1989.
[61] Wang W S-Y. Language in China: a chapter in the history of linguistics[J]. Journal of Chinese Linguistics, 1989, 17(2): 183–222.
[62] Ogura M, Wang W S-Y. A phonetic study of Japanese stop consonants[C]//Grosser W, et al, eds. Phonophilia: Festschrift for Professor Zaic. Salzburg University Press, 1989: 115–128.
[63] Wang W S-Y. 汉语教学三点看法[C]//Proceedings of the Second World Chinese Language Conference. Volume 1. Taipei, 1990: 13–23.
[64] Wang W S-Y. Theoretical issues in studying Chinese dialects[J]. Journal of the Chinese Language Teachers Association, 1990, 25: 1–34.
[65] Wang W S-Y, Shen Z W. 词汇扩散的动态描写[J]. 语言研究, 1991, 20: 15–33.
[66] Ogura M, Wang W S-Y. The development of the Indo-European languages: from a perspective of dynamic dialectology[C]//Studies in Dravidian and General Linguistics: a Festschrift for Bh. Krishnamurti. Hyderabad: Osmania University Press, 1991: 280–337.
[67] Ogura M, Wang W S-Y. Isoglosses: artifactual or real[C]//Brogyanyi B, ed. Prehistory, History and Historiography of Language, Speech and Linguistic Theory. Amsterdam: John Benjamins, 1991: 153–181.
[68] Wang W S-Y, 张文轩. 汉语语言学发展的历史回顾[J]. 兰州学刊, 1991, 59(2): 79–89, 104.
[69] Husmann L E, Wang W S-Y. Ethnolinguistic notes on the Dungan[J]. Sino-Platonic Papers, 1991, 27: 71–84.

[70] Ogura M, Wang W S-Y, Cavalli-Sforza L L. The development of Middle English /i/ in England: a Study in Dynamic Dialectology[C]//Eckert P, ed. New Ways of Analyzing Sound Change. Academic Press, 1991: 63–106.

[71] Wang W S-Y, Shen Z W. 方言关系的计量表述[J]. 中国语文, 1992(2): 81–92.

[72] Mountain J L, Wang W S-Y, Du R F, Yuan Y D, Cavalli-Sforza L L. Congruence of genetic and linguistic evolution in China[J]. Journal of Chinese Linguistics, 1992, 20: 315–331.

[73] Wang W S-Y, Lien C F. Bidirectional diffusion in sound change[C]//Jones C, ed. Historical Linguistics: Problems and Perspectives. Essex: Longman, 1993: 345–400.

[74] Wang W S-Y. 语言变化的机理[J]. 中国境内语言暨语言学, 1994, 2: 1–20.

[75] Wang W S-Y. 汉语方言研究的理论断想[J]. 张文轩, 译, 刘汉城, 校. 青海民族学院学报(社会科学版), 1994, 3: 57–61, 80.

[76] Shen Z W, Wang W S-Y. 吴语浊塞音的研究——统计上的分析和理论上的考虑[C]//Xu Yunyang 徐云扬, ed. 吴语研究. Hong Kong: Chinese University Press, 1995: 219–238.

[77] Ogura M, Wang W S-Y. Lexical diffusion in semantic change: with special reference to universal changes[J]. Folia Linguistica Historica, 1995, XVI(1–2): 29–73.

[78] Wang W S-Y, Asher R E. Chinese linguistic tradition[C]//Koerner E F K, Asher R E, eds. Concise History of the Language Sciences: From the Sumerians to the Cognitivists. New York: Pergamon Press, 1995: 41–44.

[79] Wang W S-Y. Lexical diffusion: retrospect and prospect[J]. Contemporary Chinese Linguistics, 1996, 1(1).

[80] Wang W S-Y. Linguistic diversity and language relationships[C]//Huang C T J, Li Y H A, eds. New Horizons in Chinese Linguistics. Dordrecht: Kluwer Academic Publishers, 1996: 235–267.

[81] Wang W S-Y. Genes, dates and the writing system[J]. International Review of Chinese Linguistics, 1996, 1(1): 45–47.

[82] Freedman D A, Wang W S-Y. Language polygenesis: a probabilistic model[J]. Anthropological Science, 1996, 104(2): 131–138.

[83] Liu G K F, Wang W S-Y. 白马非马: a case of folk etymology[J]. Journal of Chinese Linguistics, 1996, 24: 128–137.

[84] Ogura M, Wang W S-Y. Snowball effect in lexical diffusion: the development of -s in the third person singular present indicative in English[C]//Britton D, ed. English Historical Linguistics 1994. Amsterdam/Philadelphia: John Benjamins, 1996: 119–141.

[85] Ogura M, Wang W S-Y. Lexical diffusion and evolution theory[C]//Hickey R, Puppel S, eds. Language History and Linguistic Modelling: A Festschrift for Jacek Fisiak on his 60th Birthday. Berlin: Mouton de Gruyter, 1996: 1083–1098.

[86] Ogura M, Wang W S-Y. Explorations in the origins of the Japanese language[C]//Interdisciplinary Perspectives on the Origins of the Japanese. International Research Center for Japanese Studies, 1996: 309–334.

[87] Wang W S-Y. A quantitative study of Zhuang-Dong languages[C]//Yue A O, Endo M, eds. In Memory of Mantaro J. Hashimoto桥本万太郎纪念中国语学论集.Tokyo: Uchiyama Books内山书店,1997：81–96.

[88] Wang W S-Y. Languages or dialects?[J]. CUHK Journal of Humanities, 1997, 1: 54–62.

[89] Wang W S-Y. Language and the evolution of modern humans[C]//Omoto K, Tobias P V, eds. Recent Advances in Human Biology, vol. 3, The Origins and Past of Modern Humans- Towards Reconciliation. Singapore: World Scientific, 1998: 247–262.

[90] Wang W S-Y. Three windows on the past[C]//Mair V H, ed. The Bronze Age and Early Iron Age Peoples of Eastern Central Asia. Philadelphia: University of Pennsylvania Museum Publications, 1998: 508–534.

[91] Wang W S-Y. Representing relationships among linguistic elements[C]//T'sou B K, et al, eds. Quantitative and Computational Studies on the Chinese Language. Hong Kong: City University of Hong Kong, 1998: 1–14.

[92] Ogura M, Wang W S-Y. Evolution theory and lexical diffusion[C]//Fisiak J, Krygier M, eds. Advances in English Historical Linguistics 1996. Berlin/New York: Mouton de Gruyter, 1998: 315–344.

[93] Qiao S Z, Minett J, Wang W S-Y. Evaluating phylogenetic trees by matrix decomposition[J]. Anthropological Science, 1998, 106(1): 1–22.

[94] Ogura M, Wang W S-Y, Cavalli-Sforza L L. The development of Middle English I in English: a study in dynamic dialectology[C]//Fisiak J, Oizumi A, eds. English Historical Linguistics and Philology in Japan. Berlin: Mouton de Gruyter, 1998: 237–286.

[95] Wang W S-Y. Languages and peoples of China[C]//Yin Y, Yang I L, Chan H C, eds. 中国境内语言暨语言学.第五辑.语言中的互动[Chinese Languages and Linguistics V: Interactions in Language]. Taipei: Institute of Linguistics, 1999: 1–26.

[96] Wang W S-Y. Language emergence and transmission[C]//Peyraube A, Sun C F, eds. In Honor of Mei Tsu-Lin: Studies on Chinese Historical Syntax and Morphology. Paris: École de Hautes Études en Sciences Sociales, Centre de Recherches Linquistiques sur l'Asie Orientale, 1999: 247–257.

[97] Schoenemann P T, Budinger T F, Sarich V M, Wang W S-Y. Brain size does not predict general cognitive ability within families[J]. Proceedings of the National Academy of Sciences USA, 2000, 97(9): 4932–4937.

[98] Wang W S-Y. The joys of research: Mendel, Jones, and Human Origins[C]// Symposium on Broadening Research Frontiers at the City University of Hong Kong.

[99] Wang W S-Y. Human diversity and language diversity[C]//Jin L, Seielstad M, Xiao C, eds. Genetic, Linguistic and Archeological Perspectives on Human Diversity in Southeast Asia. Singapore: World Scientific, 2001: 17–33.

[100] Wang W S-Y.门德尔与琼斯,道不同不相为谋? [J].科学中国人,2001,11：28–31.

[101] Wang W S-Y, Ke J Y.语言的起源及建模仿真初探[J].中国语文,2001,282：195–

200.
[102] Karafet T, Xu L, Du R, Wang W S-Y, et al. Paternal population history of East Asia: sources, patterns, and microevolutionary processes[J]. American Journal of Human Genetics, 2001, 69: 615–628.
[103] Wang W S-Y. 语言是云南的文化宝藏[J]. 科学人, 2002, 10: 58–59.
[104] Ke J Y, Minett J, Au C P, Wang W S-Y. Self-organization and selection in the emergence of vocabulary[J]. Complexity, 2002, 7(3): 41–54.
[105] Wang W S-Y. Yunnan and her cultural treasures[J]. International Association of Chinese Linguistics Newsletter, 2003, 11(2): 3–5.
[106] Minett J W, Wang W S-Y. On detecting borrowing: distance-based and character-based approaches[J]. Diachronica, 2003, 20(2): 289–330.
[107] 王士元, 李文雄, 王明珂. 中国民族的起源与形成过程[R]. "中央研究院" 主题研究计划, 2003.
[108] 邓晓华, 王士元. 苗瑶语族语言的亲缘关系的计量研究——词源统计分析方法[J]. 中国语文, 2003(3): 253–263.
[109] 邓晓华, 王士元. 藏缅族语言的数理分类及其形成过程的分析[J]. 民族语文, 2003(3): 1–12.
[110] 邓晓华, 王士元. 古闽、客方言的来源以及历史层次问题[J]. 古汉语研究, 2003(2): 1–10.
[111] Ke J Y, Ogura M, Wang W S-Y. Optimization models of sound systems using genetic algorithms[J]. Computational Linguistics, 2003, 29(1): 1–18.
[112] Wang W S-Y, Ke J Y, Minett J W. Computational studies of language evolution[C]//Huang C R, Lenders W, eds. Computational Linguistics and Beyond, Frontiers in Linguistics 1, Language and Linguistics Monograph Series B. Taipei: Institute of Linguistics, 2004: 65–108.
[113] Ogura M, Wang W S-Y. Dynamic dialectology and complex adaptive system[C]//Dossena M, Lass R, eds. Methods and Data in English Historical Dialectology. Bern: Peter Lang Publisher, 2004: 137–170.
[114] Peng G, Wang W S-Y. An innovative prosody modeling method for Chinese speech recognition[J]. International Journal of Speech Technology, 2004, 7: 129–140.
[115] Wang F, Wang W S-Y. Basic words and language evolution[J]. Language and Linguistics, 2004, 5(3): 643–662.
[116] Wang W S-Y, Minett J W. The invasion of language: emergence, change, and death[J]. Trends in Ecology and Evolution, 2005, 20(5): 263–269.
[117] Whitehouse P, Usher T, Ruhlen M, Wang W S-Y. Kusunda: an Indo-Pacific language in Nepal[J]. Proceedings of the National Academy of Sciences USA, 2004, 101: 5692–5695.
[118] Gong T, Ke J, Minett J W, Wang W S-Y. A computational framework to simulate the coevolution of language and social structure[C]//Pollack J, Bedau M A, Husbands

P, Ikegami T, et al, eds. Artificial Life IX: Proceedings of the 9th International Conference on the Simulation and Synthesis of Living Systems. Cambridge, MA: MIT Press, 2004: 214–219.

[119] Gong T, Wang W S-Y. Computational modeling on language emergence: a coevolution model of lexicon, syntax and social structure[J]. Language and Linguistics, 2005, 6(1): 1–42.

[120] Gong T, Ke J, Minett J W, Holland J H, Wang W S-Y. Coevolution of lexicon and syntax from a simulation perspective[J]. Complexity, 2005, 10(6): 50–62.

[121] Wang W S-Y, Minett J W. The invasion of language: emergence, change, and death [J]. Trends in Ecology and Evolution, 2005, 20(5): 263–269.

[122] Wang W S-Y, Minett J W. Vertical and horizontal transmission in language evolution [J]. Transactions of the Philological Society, 2005, 103(2): 121–146.

[123] 汪锋，王士元. 语义创新与方言的亲缘关系[J]. 方言，2005，2：157–167.

[124] Wang W S-Y. 语言学的回顾与前瞻[J]. 辅仁外语学报，2005，2：1–18.

[125] 汪锋，王士元. 基本词汇与语言演变[J]. 谷峰，译. 语言学论丛，2006，33：340–358.

[126] Gong T, Minett J W, Wang W S-Y. Computational simulation on the co-evolution of compositionality and regularity[C]//Cangelosi A, Smith A D M, Smith K, eds. The Evolution of Language: The Proceedings of the 6th International Conference on Language Evolution. London: World Scientific Publishing Co. Pte. Ltd., 2006: 99–106.

[127] Minett J W, Gong T, Wang W S-Y. A language emergence model predicts word order bias[C]//Cangelosi A, Smith A D M, Smith K, eds. The Evolution of Language: Proceedings of the 6th International Conference. World Scientific, 2006: 206–213.

[128] Ogura M, Wang W S-Y. Ambiguity and language evolution: evolution of homophones and syllable number of words[C]//Fisiak J, eds. Studia Angelica Posnaniensia. Uniwersytet Im. Adama Mickiewicza W Poznaniu, 2006, 42: 3–30.

[129] 王士元. 演化语言学中的计算机建模[J]. 北京大学学报（哲学社会科学版），2006，43：17–22.

[130] 王士元. 语言是一个复杂适应系统[J]. 清华大学学报（哲学社会科学版），2006，21：5–13.

[131] 王士元. 语言演化的探索[C]//钟荣富，刘显亲，胥嘉陵，何大安，eds. 门内日与月：郑锦全先生七秩寿庆论文集. 语言暨语言学专刊外编之七. Taipei: Institute of Linguistics, 2006: 9–32.

[132] Wang W S-Y. 索绪尔与雅柯布森：现代语言学历史略谈[C]//刘翠溶，ed. 四分溪论学集下册：庆祝李远哲先生七十寿辰论文集. 台北：允晨文化，2006：669–686.

[133] Wang W S-Y. Hooked on language[C]//何大安，张洪年，潘悟云，吴福祥，eds. 山高水长：丁邦新先生七秩寿庆论文集. 语言暨语言学专刊外编之六. Taipei: Institute of Linguistics, 2006: 1–20.

[134] Wang W S-Y. 浅谈索绪尔与雅柯布森对现代语言学的贡献[J]. 辅仁外语学报，2006，3：1–21.

[135] Wong F, Minett J W, Wang W S-Y. Reassessing combinatorial productivity exhibited by simple recurrent networks in language acquisition[C]//Proceedings of the 20th International Joint Conference on Neural Networks. Vancouver, Canada, July 2006: 1596–1603.

[136] 王士元.语言习得与关键年龄的问题[J].华文世界,2007,99: 64–68.

[137] Wong F C K, Wang W S-Y. Combinatorial productivity through the emergence of categories in connectionist networks[J]. Dynamics of Continuous, Discrete & Impulsive Systems (DCDIS) A Supplement, 2007, 14(S1): 650–657.

[138] Deng X H, Wang W S-Y.壮侗语族语言的数理分类及其时间深度[J].中国语文, 2007,6(321): 536–548.

[139] Wang W S-Y. The language mosaic and its biological bases[J]. Journal of Bio-Education, 2007, 2(1): 8–16.

[140] Ke J Y, Gong T, Wang W S-Y. Language change and social networks[J]. Communications in Computational Physics, 2008, 3: 935–949.

[141] Minett J W, Wang W S-Y. Modeling endangered languages: the effects of bilingualism and social structure[J]. Lingua, 2008, 118: 19–45.

[142] 王士元.宏观语音学[J].中国语音学,2008,1: 1–9.

[143] Gong T, Puglisi A, Loreto V, Wang W S-Y. Conventionalization of linguistic knowledge under communicative constraints[J]. Biological Theory, 2008, 3: 154–163.

[144] Wang W S-Y. Recent advances in evolutionary linguistics[C]//Hsiao Y E, Hsu H-C, Wee L-H, et al, eds. Interfaces in Chinese Phonology: Festschrift in Honor of Matthew Y. Chen on His 70th Birthday. Language and Linguistics Monograph Series W-8. Taipei, 2008: 279–294.

[145] Gong T, Wang W S-Y. The reserve role of the naming game in social structure [C]//Smith A D M, Smith K, Ferrer i Cancho R, eds. The Evolution of Language: Proceedings of the 7th International Conference (EvoLang7). Singapore: World Scientific, 2008: 139–146.

[146] Gong T, Minett J W, Wang W S-Y. Exploring social structure effect on language evolution based on a computational model[J]. Connection Science, 2008, 20(2–3): 135–153.

[147] Gong T, Minett J W, Wang W S-Y. Computational Simulation Study on Coevolution of Compositionality and Regularity[M]//The Evolution of Language. Singpore: World Scientific, 2006: 99–106.

[148] Gong T, Minett J W, Wang W S-Y. The role of cultural transmission in individual's language-related ability[C]//Smith A D M, Smith K, Ferreri Cancho R, eds. The Evolution of Language: Proceedings of the 7th International Conference (EvoLang7). Singapore: World Scientific, 2008: 139–146.

[149] Peng G, Minett J W, Wang W S-Y. The networks of syllables and characters in Chinese

[J]. Journal of Quantitative Linguistics, 2008, 15: 243–255.

[150] 王士元.演化论与中国语言学[J].南开语言学刊,2008,2: 1–15.

[151] Wang F, Tsai Y, Wang W S-Y. Chinese literacy[C]//Olson D, Torrance N, eds. Cambridge Handbook on Literacy. Cambridge University Press, 2009: 386–417.

[152] Gong T, Minett J W, Wang W S-Y. A simulation study on word order bias[J]. Interaction Studies, 2009, 10: 51–76.

[153] Gong T, Minett J W, Wang W S-Y. A framework triggering displacement in human language[C]//Ways to Protolanguage Conference. Torun, Poland. Sept 21–23, 2009.

[154] Siok W T, Kay P, Wang W S-Y, Chan A H D, Chen L, Luke K-K, Tan L H. Language regions of brain are operative in color perception[J]. Proceedings of the National Academy of Sciences, 2009, 106: 8140–8145.

[155] 王士元.中国的语言与民族[J].茶马古道研究集刊,2010,1: 1–16.

[156] Gong T, Minett J W, Wang W S-Y. A simulation study exploring the role of cultural transmission in language evolution[J]. Connection Science, 2010, 22: 69–85.

[157] Peng G, Minett J W, Wang W S-Y. Cultural Background Influences the Liminal Perception of Chinese Characters: An ERP Study[J]. Journal of Neurolinguistics, 2010, 23: 416–426.

[158] Peng G, Zheng H-Y, Gong T, Yang R-X, Kong J-P, Wang W S-Y. The influence of language experience on categorical perception of pitch contours[J]. Journal of Phonetics, 2010, 38: 616–624.

[159] 王士元.演化语言学的演化[J].当代语言学,2011,13(1): 1–21.

[160] Peng G, Wang W S-Y. Hemisphere lateralization is influenced by bilingual status and composition of words[J]. Neuropsychologia, 2011, 49: 1981–1986.

[161] Wang W S-Y, Tsai Y. The alphabet and the sinogram: Setting the stage for a look across orthographies[M]//McCardle P, Lee J R, Miller B, et al, eds. Dyslexia Across Languages: Orthography and the Brain-Gene-Behavior Link. Baltimore: Brookes Publishing, 2011: 1–16.

[162] Wang W S-Y. Ambiguity in language[J]. Korean Journal of Chinese Language and Literature, 2011, 1: 3–20.

[163] Ogura M, Wang W S-Y. The global organization of the English lexicon and its evolution[M]//Sauer H, Waxenberger G, eds. English Historical Linguistics 2008, Volume II: Words, Texts and Genres. Amsterdam: John Benjamins Publishing Company, 2012: 65–83.

[164] Peng G, Zhang C C, Zheng H Y, Minett J W, Wang W S-Y. The effect of intertalker variations on acoustic-perceptual mappings in Cantonese and Mandarin tone systems[J]. Journal of Speech, Language, and Hearing Research, 2012, 55: 579–595.

[165] Minett J W, Zheng H Y, Fong M C-M, Zhou L, Peng G, Wang W S-Y. A Chinese text input brain-computer interface based on the P300 speller[J]. International Journal of Human-Computer Interaction, 2012, 28: 472–483.

[166] Zheng H Y, Minett J, Peng G, Wang W S-Y. The impact of tone systems on the categorical perception of lexical tones: An event-related potentials study[J]. Language and Cognitive Processes, 2012, 27(2): 184–209.

[167] Zhang C C, Peng G, Wang W S-Y. Unequal effects of speech and nonspeech contexts on the perceptual normalization of Cantonese level tones[J]. Journal of the Acoustical Society of America, 2012, 132(2): 1088–1099.

[168] Zhou L, Peng G, Zheng H Y, Su I-F, Wang W S-Y. Sub-lexical phonological and semantic processing of semantic radicals: A primed naming study[J]. Reading and Writing, 2012. DOI: 10.1007/s11145-012-9402-7.

[169] Wang W S-Y. Language learning and the brain: An evolutionary perspective[J]// Cao G, Chappell H, Djamouri R, Wiebusch T, editors. Breaking Down the Barriers: Interdisciplinary Studies in Chinese Linguistics and Beyond. Taipei: Institute of Linguistics, 2013: 21–48.

[170] 王士元.语言演化的三个尺度[J].科学中国人,2013,1: 16–20.

[171] Zhang C C, Peng G, Wang W S-Y. Achieving constancy in spoken word identification: Time course of talker normalization[J]. Brain & Language, 2013, 126: 193–202.

[172] Peng G, Yang R X, Wang W S-Y. Lateralized Stroop interference effect with Chinese characters[J].实验语言学,2013,2(1): 1–8.

[173] Peng G, Deutsch D, Henthorn T, Su D, Wang W S-Y. Language experience influences non-linguistic pitch perception[J]. Journal of Chinese Linguistics, 2013, 41(2): 447–467.

[174] 王士元.谁是中国人?[J]科学中国人,2013,8: 38–43.

[175] 龚涛,帅兰,王士元.用计算器仿真研究语言演化[J].语言科学,2013,12(1): 82–100.

[176] 拉波夫,王士元.语音变化前沿问题演讲录[J].语言教学与研究,2014,1: 1–12.

[177] Ogura M, Inakazu T, Wang W S-Y. Evolution of tense and aspect[C]//Cartmill E A, Roberts S, Lyn H, Cornish H, editors. Proceedings of the 10th International Conference. Vienna, Austria, 14–17 April 2014: 213–220. Singapore: World Scientific.

[178] Zhou L, Fong M C-M, Minett J W, Peng G, Wang W S-Y. Pre-lexical phonological processing in reading Chinese characters: An ERP study[J]. Journal of Neurolinguistics, 2014, 30: 14–26.

[179] Zheng H Y, Peng G, Chen J-Y, Zhang C C, Minett J W, Wang W S-Y. The influence of tone inventory on ERP without focal attention: A cross-language study[J]. Computational and Mathematical Methods in Medicine, 2014: 1–7.

[180] Wang W S-Y. Ancestry of languages and of peoples(语言及人类的祖先)[J].语言学论丛,2014,50: 1–28.

[181] 王士元.语言演变中的相变[J].张妍,蔡雅菁,译.南开语言学刊,2014,2: 1–13.

[182] Wang W S-Y. The peoples and languages of China: Evolutionary background[M]// Wang W S-Y, Sun C F, eds. Oxford Handbook of Chinese Linguistics. Oxford: Oxford

University Press, 2015: 19–33.
[183] Wang W S-Y, Sun C F. Introduction[M]//Wang W S-Y, Sun C F, editors. Oxford Handbook of Chinese Linguistics. Oxford: Oxford University Press, 2015: 3–18.
[184] 王士元.我们的两门远房亲戚[J].科学人,2015,3：74–77.
[185] Wang W S-Y. Some issues in the study of language evolution（语言演化研究的几个议题）[J].语言研究[Studies in Language and Linguistics], 2015, 35(3): 1–11.
[186] Zhang C C, Pugh K R, Mencl W E, Molfese P J, Frost S J, Magnuson J S, Peng G, Wang W S-Y. Functionally integrated neural processing of linguistic and talker information: An event-related fMRI and ERP study[J]. NeuroImage, 2015, 124: 536–549.
[187] Mai G, Minett J, Wang W S-Y. Delta, theta, beta, and gamma brain oscillations index levels of auditory sentence processing[J]. NeuroImage, 2016, 133: 516–528.
[188] Wang W S-Y. Chinese linguistics[M]//Chan S-w, editor. The Routledge Encyclopedia of the Chinese Language. London/New York: Routledge, 2016: 152–183.
[189] 王士元.达尔文、华莱士与人类演化[J].科学中国人,2017,1：30–37.
[190] Zhang C C, Peng G, Shao J, Wang W S-Y. Neural bases of congenital amusia in tonal language speakers[J]. Neuropsychologia, 2017, 97: 18–28.
[191] 王士元.人类起源、语言的形成及其演化问题[M]//吴璐，曹鹏鹏整理.汉语史与汉藏语研究,2017,1：1–23.
[192] 王士元.复杂系统与音节语言的形成[M]//汪锋，林幼菁.语言与人类复杂系统.昆明：云南大学出版社,2017：19–49.
[193] Wang W S-Y.变换律语法理论[M].Hong Kong: Hong Kong University Press, 1965: 124.
[194] Wang W S-Y, Lyovin A. CLIBOC: Chinese Linguistics Bibliography on Computer [M]. Cambridge: Cambridge University Press, 1970.
[195] Wang W S-Y. The Lexicon in Phonological Change[M]. The Hague: Mouton, 1977.
[196] Fillmore C J, Kempler D, Wang W S-Y. Individual Differences in Language Ability and Language Behavior[M]. New York: Academic Press, 1979.
[197] Wang W S-Y. Human Communication: Language and Its Psychobiological Bases [M]. San Francisco: W. H. Freeman and Company, 1982.
[198] Wang W S-Y. Explorations in Language Evolution[M]. India: Osmania University Press, 1982.
[199] Wang W S-Y.实验语音学讲座[M].语言学论丛第十一辑.北京：商务印书馆，1983.
[200] Wang W S-Y.现代语言学专题讲座[M].成都：四川师范大学,1985.
[201] Wang W S-Y. Language, Writing and the Computer[M]. San Francisco: W. H. Freeman and Company, 1986.
[202] Wang W S-Y, Shen Z W.方法，理论与方言研究[M].Hong Kong: Hong Kong Linguistic Society, 1988.

[203] Wang W S-Y. The Basis of Speech [M]. 东京：研究社印刷株式会社，1989.

[204] Wang W S-Y. Languages and Dialects of China [M]. Berkeley: Project on Linguistic Analysis, 1991.

[205] Wang W S-Y. The Emergence of Language: Development and Evolution. Readings from Scientific American Magazine [M]. San Francisco: W. H. Freeman, 1991.

[206] Wang W S-Y. Explorations in Language [M]. Taibei: Pyramid Press, 1991.

[207] Wang W S-Y, editor. The Ancestry of the Chinese Language [M]. Hong Kong: Journal of Chinese Linguistics, 1995.

[208] T'sou B K, Lai T B Y, Chan S W K, Wang W S-Y, editors. Quantitative and Computational Studies on the Chinese Language (汉语计量与计算研究) [M]. Hong Kong: Language Information Sciences Research Centre (LISRC), City University of Hong Kong, 1998.

[209] 王士元. 语言的探索：王士元语言学论文选译 [M]. 石锋，等译. 北京：语言文化大学出版社，2000.

[210] 王士元. 王士元语言学论文集 [M]. 北京：商务印书馆，2002.

[211] Minett J W, Wang W S-Y, editors. Language Acquisition, Change and Emergence: Essays in Evolutionary Linguistics [M]. Hong Kong: City University of Hong Kong Press, 2005: xii+538.

[212] 王士元，Peng G. 语言，语音与技术 [M]. 上海：上海教育出版社，2006.

[213] Wang W S-Y, Peng G. 语言，语音与技术 [M].Hong Kong: City University of Hong Kong Press, 2007.

[214] Minett J W, Wang W S-Y, editors. Language, Evolution, and the Brain [M]. Hong Kong: City University of Hong Kong Press, 2009.

[215] 邓晓华，王士元. 中国的语言及方言的分类 [M]. 北京/香港：中华书局，2009.

[216] 王士元. 王士元语音学论文集 [M]. 北京：世界图书出版公司，2010.

[217] 王士元. 语言、演化与大脑 [M]. 北京：商务印书馆，2011.

[218] 王士元. 语言、演化与大脑 [M]. 北京：高等教育出版社，2014.

[219] Wang W S-Y. Love and War in Ancient China: The Voices of Shijing [M]. Hong Kong: City University of Hong Kong Press, 2013.

[220] 王士元. 演化语言学论集 [M]. 北京：商务印书馆，2013.

[221] Wang W S-Y, Sun C F, editors. Oxford Handbook of Chinese Linguistics [M]. Oxford: Oxford University Press, 2015.

[222] Wang W S-Y. Section editor for Linguistics, International Encyclopedia of Social and Behavioral Sciences, 2nd ed [M]. Amsterdam: Elsevier, 2015.

[223] Wang W S-Y, Tikofsky R S. Speech//McGraw-Hill Encyclopedia of Science and Technology. 1st ed [M]. New York: McGraw-Hill, 1960: 593–599.

[224] Wang W S-Y. The Chinese language [M]//The Academic American Encyclopedia. 1980.

[225] Wang W S-Y. Origins of language [M]//Frawley W. Oxford International

Encyclopaedia of Linguistics. Oxford: Oxford University Press, 1992: 139–141.

[226] Wang W S-Y, Chao Y R. Encyclopedia of Language and Linguistics[M]. Oxford: Pergamon Press, 1994: 504.

[227] Wang W S-Y. Glottochronology, lexicostatistics, and other numerical methods[M]// Asher R E, editor. The Encyclopedia of Language and Linguistics. Vol. 3. New York: Pergamon Press, 1994: 1445–1450.

[228] Wang W S-Y, Asher R E. Chinese linguistic tradition[M]//The Encyclopedia of Language and Linguistics. Amsterdam: Elsevier, 1994: 524–527.

[229] Wang W S-Y. Sino-Tibetan[M]//The Encyclopedia of Language and Linguistics. Oxford: Pergamon Press, 1994: 3951–3953.

[230] Wang W S-Y. Chinese[Mandarin][M]//Encyclopedia of Linguistics. Fitzroy Dearborn Publishers, 2004.

[231] Wang W S-Y. Tone languages[M]//Malmkjaer K, editor. The Linguistics Encyclopedia. 2nd ed. London: Routledge & Kegan, 2004: 552–558.

代表作

FOUR PHASE TRANSITIONS IN LANGUAGE EVOLUTION*

William S-Y. Wang
(*The Chinese University of Hong Kong*)

ABSTRACT

How language emerged uniquely in our species is a central issue toward understanding the basis of our humanity. Giving the issue a name, such as 'language organ', and attributing it vaguely to some genetic mutation is not productive. Rather, the issue should be examined from the perspective of evolution theory. Here I suggest that the first phase transition, the trajectory toward language, started when we first assumed bipedal posture. This first phase transition occurred with the Australopithecine over 3,000,000 years ago. The second phase transition occurred with the emergence of our genus Homo over 2,000,000 years ago, when our ancestors exhibited symbolic behavior by producing and maintaining a variety of stone tools. The third occurred when primary communication changed from gestures and prosodies to sequences of syllables made up of vowels and consonants, which provided an efficient signal space; this occurred some 20,000 years ago with the emergence of our species Homo sapiens. The fourth phase transition was the invention of writing some 6,000 years ago, with numerous far reaching consequences.

---

* 原载于 Journal of Chinese Linguistics Monograph Series, 2016(26): 1–20.

KEY WORDS

Language evolution

"I think you should be more explicit here in step two."

When I saw the above cartoon in New Yorker magazine some years ago, the issue of language evolution immediately came to mind. Many linguists were then mesmerized by the belief that languages are homogeneous systems that can be explained in simple, algebra-like rules. However, the more language data one looked at, the more the rules became complicated, abstract, and often implausible. Terms like 'language organ', 'language instinct', and 'language bioprogram' were used, much as 'step two' depicted in the cartoon.

Instead of exploring in depth the abilities which enable infants to master language so effortlessly, terms like 'language acquisition device' and 'universal grammar' were advocated, as if giving the problem a rhetoric can substitute for solving it. When the late Roger Brown pioneered empirical methods with corpora and experiments to study how children actually construct their language from the sparse data in their environment, his project at Harvard was irresponsibly and rudely criticized as "the biggest waste of time in the history of science".[1]

The difficulty with the 'language organ' approach was remarked upon in a recent book by the psychologist Corballis:"The idea that the basis for language emerged in a single step in a single individual is remarkable, and smacks of the miraculous." (2011: 24) Several years earlier, the neuroscientist Ramachandran made the same point about this approach; he further noted a similarity of this approach to the dilemma that famously divided the two discoverers of the theory of evolution, i.e., Charles Darwin and Alfred Russell Wallace. As Ramachandran (2004) put it:

But how could an extraordinarily complex mechanism like language with so many interlocking components have evolved through the blind workings of chance — through natural selection. ... Alfred Russell Wallace said the mechanism is so complicated it couldn't have evolved through natural selection at all and must have resulted from divine intervention. ... Chomsky said something quite similar, although he didn't invoke God. ... He almost says it's a miracle. Unfortunately, neither Wallace's nor Chomsky's theory can be tested.

The critical ingredient missing in the 'language organ' approach is the dimension of evolutionary ***time***. Going back to Corballis, the salient observation is that evolution is gradual and takes time on a grand scale:

> ... language can be understood to have evolved gradually, rather than having emerged suddenly in some comparatively recent individual on the family tree, called Prometheus. (2011: 34)

Prometheus existed only in Greek mythology, and unfortunately is not here to make 'step two' in the cartoon more explicit for us. Several decades earlier, the linguist Hockett also linked the 'language organ' approach to Greek mythology. He titled his 1978 discussion "In Search of Jove's Brow", recalling the myth of Minerva suddenly emerging from Jove. Following in Hockett's tracks several years later, I contrasted such a Minerva Theory with what I called a Mosaic Theory, which I characterized in [1982] 1991 as follows:

> [Language] evolved in a **mosaic** fashion, with the emergence of semantics, phonology, morphology and syntax all at different times and according to different schedules ... language is regarded as a kind of 'interface' among a variety of **more basic abilities**. These abilities underlie nonlinguistic processes as well, and involve the perception of patterns in the frequency and temporal domains, the coding and storage of events and objects at different levels of memory, the manipulation of various hierarchical mental structures.

Continuing the metaphor in Greek mythology in 1984, I specifically compared the 'language organ' to the gods that would suddenly spring out of nowhere in classical Greek tragedies to resolve impossible dilemmas. The literary device is called *Deus ex machina* in Latin, where the machine in the phrase refers to some mechanical contraption to help bring a god onto the stage. Although the emergence of language is a very hard problem, resorting to miracles or to such literary devices is no help. Rather, we need to bring together all pieces of relevant information that may eventually contribute to its solution. To my mind, the following thoughts from Ramachandran are consistent with the Mosaic theory and point in the right direction:

> Early hominids were very good at tool use, especially what is known as the sub-assembly technique ... There is a close operational analogy between this function and the embedding of noun clauses within longer sentences. So perhaps what originally evolved for tool use in the hand area is now **exapted** and assimilated in the Broca's area to be used in aspects of syntax such as hierarchical embedding.
>
> Each of these effects is a small bias, but acting in conjunction they may have paved the way for the emergence of sophisticated language. This is very different from Steve Pinker's idea that language is a specific adaptation which evolved step by step for the sole purpose of communication. I suggest, instead, that it is the fortuitous synergistic combination of a number of mechanisms which evolved for other purposes initially that later became assimilated into the mechanism that we call language.

I have highlighted the word 'exapted' above because of its importance for the theme of the discussion here. This new term was introduced by the biologists Gould and Vrba in 1982 to refer to new functions served by old structures, which is something often observed in the life sciences. The concept is an elaboration of what the geneticist Jacob advocated in his famous 1977 paper, Evolution and Tinkering. Nature hardly ever introduces organs *de novo*, but typically makes use of old structures to serve new functions. As Jacob put it:

> Living organisms are historical structures, literally creations of history. They represent not a perfect product of engineering, but a patchwork of odd sets pieced together when and where opportunities arose.

In the context of linguistics, exaptation or tinkering is a concept that was known much earlier. In his brilliant book on language, 1921, Sapir used the word 'overlaid' to refer to the same phenomenon:

> Physiologically, speech is an **overlaid** function, or, to be more precise, a group of overlaid functions. It gets what service it can out of organs and functions, nervous and muscular, that have come into being and are maintained for very different ends than its own. (p.9)

For producing speech, the situation is easy to see. We cannot speak without an outgoing stream of air provided by the more fundamental function of respiration. However, we should note that many respiratory refinements have additionally evolved over past millennia for much more precise control of this outgoing air stream so that it can synchronize with the varying phrases and stresses in the speech. Even more importantly, syllables are overlaid on the fundamental function of mastication. The rhythmic movements of opening and closing the jaw when we chew food, and the minute movements of the tongue as it transports the food to be processed by different dental structures, provided the sensori-motor skills exapted for producing our vowels, consonants, and syllables.

Going back to Ramachandran's observation quoted above, what he called 'the sub-assembly technique' lies at the heart of hierarchical structures, which we find ubiquitously in the cognitive behaviors of higher animals. The utility of hierarchic patterns and processes was charmingly explained by Herbert Simon in his classic paper of 1962 on the architecture of complexity; he gave a parable of two watch-makers, the one using sub-assembly winning over the one who does not. The idea of hierarchical structures has been expressed by a variety of notations, including nested parentheses linearly as well as tree diagrams.

In linguistics, the idea of sub-assembly or hierarchic processes has long been used when sentences were parsed in language textbooks. In mid-20th century, the idea was discussed in depth by Wells (1947) for syntax and by Pike & Pike (1947) for phonology, under the label 'immediate constituents' or IC; the idea was further extended to 'discontinuous constituents' when a constituent is interrupted, such as the underlined morphemes of the auxiliary separated by the main verb in 'he *was dreaming*.' The terminology of 'immediate constituents' later changed to 'constituent structure' and 'phrase structure', though the basic idea remains largely intact. Indeed, it is the iterative and recurrent use of linguistic materials that underlies the open ended nature of language, both in phonology and in syntax.

Another observation Sapir made in his book on the age of language is also worth quoting in the present context:

> The universality and the diversity of speech lead to a significant inference. We are forced to believe that language is an **immensely ancient** heritage of the human race, whether or not all forms of speech are the historical outgrowth of **a single pristine form**. It is doubtful if any other cultural asset of man, be it the art of drilling for fire or of chipping stone, may lay claim to a greater age. I am inclined to believe that it antedated even the lowliest developments of material culture, that these developments, in fact, were not strictly possible until language, the tool of significant expression, had itself taken shape. (p.23)

Let us first comment briefly on his remark on whether all languages come from a single source — '*a single pristine form*', which is the hypothesis of monogenesis. Reasoning from probability theory and basing on demographic data available for prehistoric times, Freedman and Wang (1996) gave various reasons to argue for the opposite hypothesis of polygenesis, namely that the languages of the world descend from many sources. Polygenesis admittedly makes for a messier evolutionary scenario; but it is more realistic.

Sapir's main conjecture in the quote was that language emerged earlier than '*the lowliest form of material culture*.' The basis for this conjecture is that language was the mental tool that our ancestors used to build a rudimentary culture some millions of years ago that was beyond the capacity of all other species at that time. As the diagram below suggests, the launching pad for the trajectory toward language and culture was the transition to erect posture and bipedal movement. This took place with a pre-*Homo* genus called *Australopithecus* over 3 million years ago.

Over 2 million years ago, the genus *Homo* appeared, which systematically made the first stone tools. Sapir's conjecture was that some form of primitive language must have been in place in their minds which enabled them to achieve this rudimentary culture. As discussed above, Ramachandran similarly linked tool use to the evolution of language. Whereas other species evolved primarily by ***biological evolution***, through transmission of genes from generation to generation, the evolution of *Homo* was driven in addition by ***cultural evolution***, through much more complex modes of transmission of teaching, learning, and numerous forms of social interaction. It is this powerful combination of ***dual evolution***, the biological plus the cultural that has created the unique endowments that characterize our species, *Homo sapiens*.

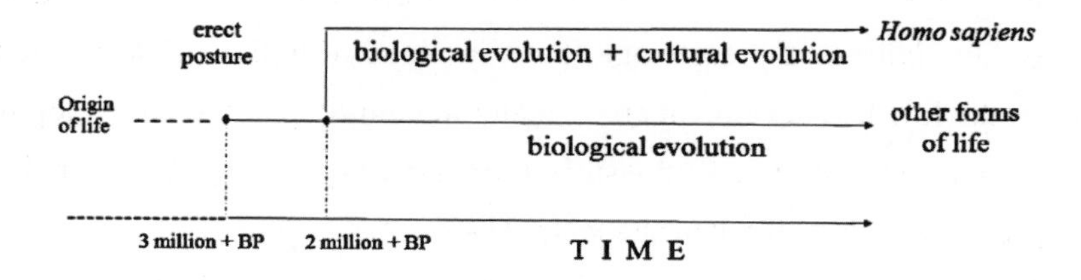

Over the millions of years, many species have become extinct. Our closest relatives today are the chimpanzees, whose lineage diverged from ours some 6 million years ago. Sapir, writing in 1921, was not in a position to go the next step and compare our culture with that of the chimpanzee, since most of what we know about the chimpanzee was learned in recent decades, as de Waal recently summarized (2005). Call and Tomasello (2008) is a more advanced discussion of the chimpanzee's mental world; Herrmanm et al (2007) report on cognitive differences among the chimpanzee, the orang utan, and the human child (Senghas Kita and Ozyurek 2004).

Nor could Sapir know much about the earliest material culture of our ancestors; he passed away in 1939, just about when Louis Leakey was discovering the stone tools at Olduvai Gorge in Tanzania. Now we have much more data for such comparisons, including in addition better knowledge of the biological equipment that makes language possible.

The remainder of this essay will attempt an update on these issues, almost a century after Sapir's conjecture. The term 'phase transition' in the title of this essay is borrowed from physics, where it refers to qualitative changes brought about by nonlinear development. A common example is the gradual addition of a common amount of heat to H_2O, changing it successively from solid ice to liquid water and to gaseous steam. We will discuss four such phase transitions in the evolution of language.

PHASE TRANSITION 1: BIPEDAL POSTURE

The first phase transition took place over 3 million years ago. 1974 was a good year for paleoanthropology, for the discovery of one of the most famous set of fossils in human evolution in Ethiopia. The remarkably complete set was the remains of a young girl, known to science as Lucy, a member of a species named *Australopithecus afarensis*. The importance of these fossils for language has been discussed by Johanson, one of their discoverers (Johanson and Blake 2006). Detailed anatomical studies of Lucy and of many other fossils discovered later in nearby regions of Ethiopia have convinced scientists that these creatures walked upright.

Unlike all other primates, Lucy was bipedal, standing erect with her head balanced vertically upon her vertebral column. The other great apes, i.e., chimpanzees, gorillas and orangutans, may walk bipedally occasionally, even carrying things in their arms over short distances sometimes; they may also use 'knuckle walking' complementing the two legs by leaning on an extended hand. Nonetheless, Lucy's species was the first species in the primate order that assumed an erect stance much as humans do now.

This major 'tinkering' that evolution has done with our bodies has significantly restructured our anatomy, especially the head and neck areas. As a reminder of our quadrupedal past, the nerve pathway between the brain and the larynx has been greatly lengthened because it needs to descend below the aorta and then double back up again to control the laryngeal muscles. The larynx itself descends as well during early infancy, creating a vertical volume in the throat approximately perpendicular to the mouth. The resulting anatomy, resembling a bent tube, is shared by food and drink, as well as by the air we breathe and speak with. While such an arrangement allows us to speak a greater diversity of sounds, the negative side is that it makes us more vulnerable to choking, as noted early by Darwin:

> The strange fact that every particle of food and drink which we swallow has to pass over the orifice of the trachea with some risk of falling into the lungs, notwithstanding the beautiful contrivance by which the glottis is closed. (1859: 191)

Below is a very informative diagram from Vallender et al (2008) which shows some of the vital differences between us and our primate relatives.

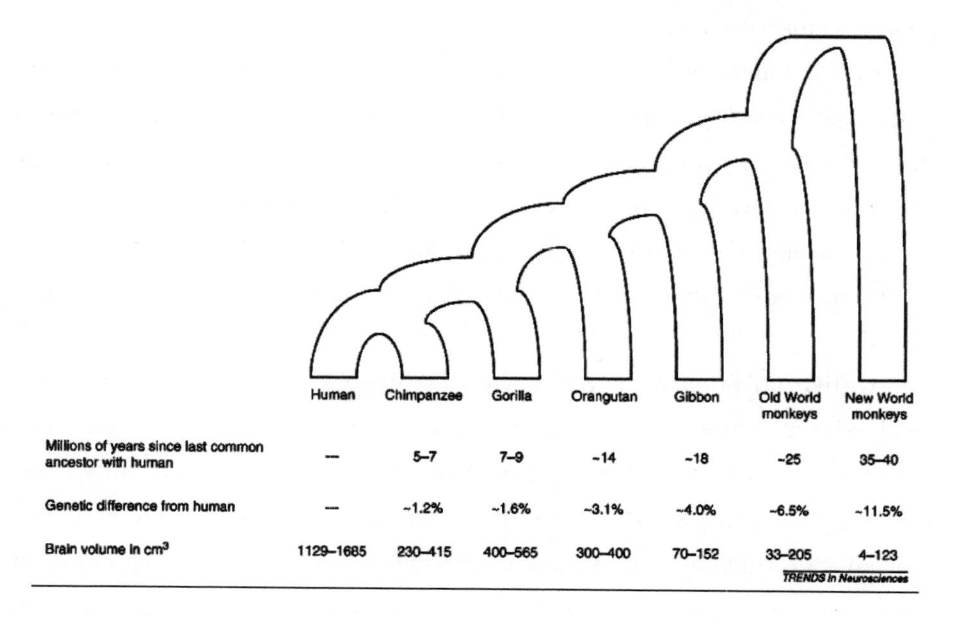

	Human	Chimpanzee	Gorilla	Orangutan	Gibbon	Old World monkeys	New World monkeys
Millions of years since last common ancestor with human	—	5–7	7–9	~14	~18	~25	35–40
Genetic difference from human	—	~1.2%	~1.6%	~3.1%	~4.0%	~6.5%	~11.5%
Brain volume in cm^3	1129–1685	230–415	400–565	300–400	70–152	33–205	4–123

Comparing humans with the chimpanzee is particularly instructive. Note that while we are only some 1.2% different genetically, our brain is more than 4 times larger than the chimpanzee brain. Measurement of the many skulls of *Australopithecus afarensis* that have been unearthed since Lucy shows a volume similar to that of the other great apes. Such an explosive growth of the brain over these 3 million years — from a mean of 450 cc to 1350 cc - is remarkable in biological evolution.

Nonetheless, there is still much cognitive that we share with the chimpanzee due to our long common lineage. The chimpanzees also have socio-political structure living in small groups, and their cultures vary from region to region. They are the only primate other than man that has been shown to have self-recognition. They use twigs to fish for termites. They use stones to crack nuts; mothers have even been observed to shape the hands of young ones in teaching them to do this. They coordinate with each other when trapping monkeys for food. While there is still quite a gap between our cognition and theirs, a good case can be made that we lie on the same continuum of development.

On the other hand, efforts to teach the other great apes language or language-like systems have stopped at a relatively low ceiling. The most sustained studies so far are those of Kanzi the chimpanzee (Savage-Rumbaugh and Lewin 1994), and of Koko the gorilla (Patterson and Matevia 2001). Whether the medium of instruction is the American Sign Language, or plastic chips on a magnetic board, or lexigrams on a computer screen, the language attained by the ape students has been compared to human five-year-olds at best. Herrmann et al (2007) report interesting experiments comparing human infants, chimpanzees and orangutans, proposing that humans have a greater advantage in 'cultural intelligence'. Call and Tomasello (2008) provide insightful discussions on the mental differences between us and the chimpanzee with respect to 'theory of mind', i.e., the ability to understand and/or empathize the thoughts and/or feelings of another individual.

An obvious consequence of bipedal posture is the freeing of the hands from the heavy task of pounding the earth during walking and running, allowing them to develop finer and finer skills. While the primate hand in general has achieved some flexible use from climbing trees and various simple chores, it was in the human lineage that the hand developed the greatest dexterity. This remarkable achievement reaches its peak in our times in musical performances, as exemplified in the rapid and precise movements of the pianist's fingers as they fly across the keyboard. A similar kind of finger dexterity is displayed as modern teenagers play various video games on their cell phones.

Increasing demands on the hand select for more brain tissues for controlling its many parts; witness the large areas in the sensory and motor cortices devoted to the thumb and the fingers. Conversely, finer neural control from the brain enables more diverse activities for the hands to execute. In general, control and execution reinforce each other mutually, as is often the case of co-evolution of structure and function. This interaction between hand and brain, plus all the associated infrastructure of vision, kinesthesia, etc., must have been one of the major engines pushing the explosive growth of the brain.

PHASE TRANSITION 2: SYMBOLIZATION

Bipedal posture provided the foundation for many aspects of behavior that are uniquely human, including linguistic behavior but not exclusively so. The second phase transition upon which language was launched was the concept of symbolization — that one thing could represent something else. Deacon (1997) has aptly named us *The Symbolic Species*. It was the mental manipulation of geometric relations that led the genus *Homo* to initiate making a variety of stone tools, for pounding, for cutting, for drilling, and so on. These mark the '*lowliest developments of material culture*' that Sapir spoke of, and the technique of sub-assembly mentioned by Ramachandran, quoted earlier. Bipedal posture provided us with a biology with immense potential. Symbolization marked a new phase when cultural evolution began to replace biological evolution as the major engine of change.

The earliest stone tools were made in Africa over 2 million years ago. Shortly after that date, wave after wave of early *Homo* left Africa for many parts of Asia. For instance, a recent report in *Science* magazine describes a rare complete skull found in Dmanisi in the Republic of Georgia, dated to 1.8 million years ago (Lordkipanidze et al., 2013). The more famous representatives of our genus include: Peking Man, whose fossils were found in the outskirts of modern Beijing and whose dates center around 500,000 years ago; the Neanderthals that ranged over Europe and Western Asia, whose remains date to 600,000 through 30,000 years ago; and the diminutive *Homo floresiensis* of Indonesia, who may have survived until as recently as 12,000 years ago.

A recent overview of the origin of our species is Stringer (2012), who is an articulate advocate of the single origin version of the Out-of-African hypothesis. In contrast, many scholars prefer a multiregional hypothesis, which posits significant interbreeding of *Homo sapiens* with the earlier *Homo* species. There is as yet no consensus on the details of our evolutionary past, giving a consistent account that accommodates all the data from paleoanthropology, population genetics, and evolutionary linguistics. Relevant data are constantly emerging in the form of unearthed fossils or newly discovered genetic materials. There can be no doubt that impressive advances have been made in recent decades since the various disciplines have begun to work together integrating each other's results toward a coherent consensus.

Taking clue from the modern behavior of the other primates, early *Homo* must have communicated with a combination of gestures and prosodies. Gestures require less shared information to communicate intent — extending the palm to beg, raising the fist to strike, baring the teeth to bite. It should not take much to learn when your neighbor bares his teeth that he means harm. Prosodies in the form of growls, grunts, hoots, screams, etc. to express pleasure, anger, and other emotions easily become conventionalized as signals, and typically may co-occur with gestures. (Baring the teeth is a lot more threatening when paired with a growl!) A famous case is the use of prosody for alarm calls in vervet monkeys, as reported by Seyfarth et al (1980).

The coupling of gestures with vocalization continues to this day. Our facial expressions change and we motion with our hands and shoulders even when our addressee is not within sight, such as when we are speaking on a phone. McNeill (2005) and Goldin-Meadow et al (2008) discuss gesture and thought from a modern perspective; Corballis (2011) connects gesture with vocalization; and Arbib (2013) integrates these concepts with the exciting discoveries of the mirror neuron system.

PHASE TRANSITION 3: SEGMENTAL PHONOLOGY

Gestures and prosodies persist in modern languages though in roles which are much reduced in importance; see Corballis (2002). They vary in form and content according to the cultural context. Movements of the head to show agreement or disagreement, of the hand for beckoning, protrusion of the tongue as a sign of respect or disgust, etc. all depend critically on social norms. Similarly, prosodies in the form of intonations and contrasting stress vary from language to language. On the whole, however, these two communicative devices are limited in their semantic range, and do not combine sequentially with each other easily. They function on continua in analog fashion rather than digitally as in segmental phonology.

The expressive power of language was greatly enhanced by the introduction of phonetic segments, i.e., consonants and vowels, as the basic building blocks of speech, organized around the syllable, stress group, and breath group according to the sub-assembly technique. Segments require much less energy to produce, when compared with bodily gestures or complete intonations, and take a great deal less time — in the order of tenths of a second. As a simple example, a phonology of 5 vowels and 20 consonants can generative 100 distinct syllables of the shape CV, and many more syllables of the shape CVC. When these syllables are combined they generate ample phonetic space for a vocabulary of many thousands of words.

The generative power is further enhanced at the level of syntax, where the words are combined, again making use of the sub-assembly technique (Schoenemann and Wang, 1996; Schoenemann, 2005; Vogt, 2005). Given the time efficiency of segmental phonology, a working memory with a chunk size of several seconds can therefore easily take in sentences of moderate length for multiple analyses. It is likely that the third phase transition occurred shortly after modern *Homo sapiens* appeared in Africa, and the new form of language played a critical role in enabling our ancestors to settle in all parts of the world. Recent research in genetics has greatly increased our understanding of the biological bases of these linguistic abilities; see Chow (2005), Lieberman (2013), Sia, Clem and Huganir (2013).

PHASE TRANSITION 4: WRITING

Writing was invented much more recently, the earliest samples dating back some 6,000 years ago in some civilizations. Daniels and Bright (1996) give a comprehensive survey of the history of various systems of writing; Wang and Tsai (2011) compared alphabetic writing, derived from the Phoenician script, with sinograms, invented in Ancient China.

Writing first emerged as an instrument of privileged groups, such as royalty or priests. Mass literacy among the common people is a very recent phenomenon, within the last century or so; see Olson and Torrance (2009). Because of its relative recency in human evolution, our brain has not yet co-adapted with writing as it has co-adapted with speech; see Wang (2012). As a consequence, the acquisition of written language does not come effortlessly to all children as spoken language does. Instead, there is a minority of children in all cultures who have difficulty in learning to read and write; Dehaene (2009) is a general discussion of such issues from the viewpoint of cognitive neuroscience.

Whereas spoken language evaporates the instant it is produced, written language allows for information to be transmitted across space as well as to accumulate across time. The sharing and accumulation of information, made possible by writing, has grown exponentially, mostly in the form of books during past centuries. Now this immense amount of information, and still growing ever faster, is in the form of electronic files, accessible on the Internet and other resources.

Before the advent of language information about how to do a particular task could only be transmitted by actual demonstration by the teacher in the presence of the student. With the emergence of phonology, this information can be presented in spoken form. With the invention of writing, the information can be externalized as graphic patterns on paper or as polarizations on various magnetic or optic media. In all these cases, the information to be transmitted originates from the sender's brain. The processes of externalization and transmission have certain constraints as well as possibilities of error.

It would be foolhardy to predict at this point what the fifth phase transition for language will be. But extrapolating from the line of reasoning given above, it is not impossible that the time will come when we can monitor the sender's brain waves with enough sensitivity and fidelity, and transmit these directly to the brains of the receivers, without any intervening media. Science fiction writers have already depicted such scenarios, sometimes suggesting that information can be transplanted into the receivers' brain by implanting chips containing integrated circuits. Fiction has been known to predict the future, so perhaps the fifth phase transition will embody some of these scenarios.

Let us now recapitulate the four phase transitions that led to language which we have reviewed above. [1] The first phase transition occurred when *Australopithecus* changed from quadrupedat[2] posture to a bipedal one, over 3 million years ago, causing a fundamental re-structuring of our anatomy. This freed the hands for forming gestures, and provided the mouth and throat with greater phonetic possibilities. [2] The second phase transition occurred with the emergence of our genus *Homo*, over 2 million years ago. The production and maintenance of a variety of stone tools give indirect evidence that these individuals had achieved symbolic behavior. [3] The third phase transition occurred when phonology emerged, with the appearance of our species *Homo sapiens* over 100,000 years ago, built on hierarchies of vowels, consonants, syllables and prosody. Information of much greater complexity can be transmitted within the narrow time window of working memory than is possible with gestures and prosody. [4] The fourth phase transition occurred with the invention of writing, over 5,000 years ago, when words were represented by graphic symbols. This overcame the limitations of time and space, and ushered in a much more powerful mode of thinking.

On this last point, it is useful to ponder a response from the physicist Richard Feynman to historian Charles Weiner, reported in Gleick (1993: 409). Clark (2008: xxv) retold it in this way:

> Weiner, encountering with a historian's glee a batch of Feynman's original notes and sketches, remarked that the materials represented 'a record of [Feynman's] day-to-day work'. But instead of simply acknowledging this historic value, Feynman reacted with unexpected sharpness:
>
> "I actually did the work on the paper," he said.
>
> "Well," Weiner said, "the work was done in your head, but the record of

it is still here."

"No, it's not a record, not really. It's working. You have to work on paper and this is the paper. Okay?"

As I understand this famous exchange, Feynman was insisting on the point that writing it down is itself an important part of the creative process. Indeed, although language is our heads in its entirety, but the creation of a complex piece of work, whether it be in science or in literature, is not simply downloading something from mind to paper, or to a computer file. Rather the process is very much an interactive process going back and forth between mind and paper.

Here is an example to illustrate the point. A sentence with one degree of center embedding like "*The black cat his dog chased killed the big rat*" would normally present no problem for comprehension. On the other hand, a sentence with three degrees of center embedding like "*The black cat his dog the horse my uncle bought kicked chased killed the big rat*"[3] would not be intelligible without writing it down, even though its constituent sentences are all simple. The situation is not unlike not being able to multiply large numbers in our head, even when we are perfectly well versed in the rules of multiplication. Neither very complex sentences nor multiplication of large numbers would have been possible without writing. Thus writing does much more than *record* language for messages to be sent across space and preserved across time; rather it *extends* language by enabling it to form ever more complex and intricate messages.

NOTES

1. Quote attributed to Chomsky by S. Pinker, *Cognition* 1998: 206. For detailed critiques of Chomsky's work, see Postal 2009 and references therein.
2. *Quadruped* is to be distinguished from *Tetrapod*. The latter is a taxonomic unit in evolutionary biology which includes all vertebrates with quadrupedal ancestors, including mammals, reptiles, amphibians, and birds.
3. The five constituent sentences are: *The black cat killed the big rate; his dog chased the black cat; the horse kicked the dog*; and *my uncle bought the horse*.

REFERENCES

ARBIB, Michael A. 2013. *How the Brain got Language: The mirror System Hypothesis*. New York: Oxford University Press.

CALL, Josep and Michael Tomasello. 2008. Does the chimpanzee have a theory of mind? 30 years later. *Trends in Cognitive Sciences* 12: 187–192.

CHOW, King L. 2005. Speech and language — a human trait defined by molecular genetics. In *Language Acquisition, Change and Emergence: Essays in Evolutionary Linguistics*, ed. J. W. Minett and W.S.-Y. Wang, 21–45. Hong Kong: City University of Hong Kong Press.

CLARK, Andy. 2008. *Supersizing the Mind: Embodiment, Action, and Cognitive Extension*. Oxford: Oxford University Press.

CORBALLIS, Michael C. 2002. *From Hand to Mouth: The Origins of Language*. Princeton: Princeton University Press.

________. 2011. *The Recursive Mind: The Origins of Human Language, Thought, and Civilization*. Princeton: Princeton University Press.

DANIELS, Peter T. and William Bright, eds. 1996. *The World's Writing Systems*. New York: Oxford University Press.

DARWIN, Charles. 1859. *On the Origin of Species by Means of Natural Selection or The preservation of Favored Races in the Struggle for Life*. London: John Murray.

De WAAL, Frans B. M. 2005. A century of getting to know the chimpanzee. *Nature* 437: 56–59.

DEACON, Terrence W. 1997. *The Symbolic Species: the Co-evolution of Language and the Brain*. New York: W.W. Norton.

DEHAENE, Stanislas. 2009. *Reading in the Brain: The Science and Evolution of a Human Invention*. New York: Viking.

FREEDMAN, D.A. and W.S-Y. Wang. 1996. Language polygenesis: A probabilistic model. *Anthropological Science* 104.2: 131–138.

GLEICK, James. 1993. *Genius: The Life and Science of Richard Feynman*. New York: Vintage.

GOLDIN-MEADOW, Susan, Wing Chee So, Asli Ozyurek and Carolyn Mylander. 2008. The natural order of events: How speakers of different languages represent events nonverbally. *PNAS* 105: 9163–9168.

GOULD, Stephen Jay and Elizabeth S. Vrba. 1982. Exaptation — a missing term in the science of form. *Paleobiology* 8(1): 4–15.

HERRMANN, E., J. Call, M.V. Hernández-Lloreda, B. Hare and M.Tomasello. 2007. Humans have evolved specialized skills of social cognition: The cultural intelligence hypothesis. *Science* 317: 1360–1366.

HOCKETT, C.F. 1978. In search of Jove's brow. *American Speech* 53: 243–313.

JACOB, Franrçois. 1977. Evolution and tinkering. *Science* 196: 1161–1166.

JOHANSON, Donald and Edgar Blake. 2006. *From Lucy to Language*. New York: Simon & Schuster.

LIEBERMAN, Philip. 2013. Synapses, Language, and Being Human. *Science* 342: 944–945.

LORDKIPANIDZE, David, Marcia S. Ponce de León, Ann Margvelashvili, Yoel Rak, G. Philip Rightmire, Abesalom Vekua, and Christoph P. E. Zollikofer. 2013. A complete skull from Dmanisi, Georgia, and the evolutionary biology of early *homo*. *Science* 342: 326–331.

McNEILL, David. 2005. *Gesture and Thought*. Chicago: University of Chicago Press.

MINETT, J.W. and W.S-Y. Wang, eds. 2005. *Language Acquisition, Change and Emergence: Essays in Evolutionary Linguistics*. Hong Kong: City University of Hong Kong Press.

OLSON, David and N. Torrance, eds. 2009. *Cambridge Handbook on Literacy*. Cambridge, N.Y.: Cambridge University Press.

PATTERSON, F. G. P. and M. L. Matevia. 2001. Twenty-seven Years of Project Koko and Michael. In *All Apes Great and Small*. Vol. 1 *African Apes*, ed. B.M.F. Galdikas, N.E. Briggs, L.K. Sheeran, G.L. Shapiro and J. Goodall, 165–76. New York: Kluwer Academic Publishers.

PIKE, Kenneth L. and Eunice Victoria Pike. 1947. Immediate Constituents of Mazateco Syllables. *International Journal of American Linguistics* 13: 78–91.

PINKER, Steven. 1998. Obituary: Roger Brown. *Cognition* 66: 199–213.

POSTAL, Paul M. 2004. *Skeptical Linguistic Essays*. New York: Oxford University Press.

______. 2009. The incoherence of Chomsky's 'Biolinguistic' ontology. *Biolinguistics* 3: 104–123.

RAMACHANDRAN, V.S. 2004. *A Brief Tour of Human Consciousness: From Impostor Poodles to Purple Numbers*. New York: Pi Press.

SAPIR, Edward. 1921. *Language: An Introduction to the Study of Speech*. New York: Harcourt.

SAVAGE-RUMBAUGH, Sue and Roger Lewin. 1994. *Kanzi: The Ape at the Brink of the Human Mind*. New York: John Wiley & Sons.

SCHOENEMANN, P. Thomas. 2005. Conceptual complexity and the brain: Understanding language origins. In *Language acquisition, change and emergence: Essays in evolutionary linguistics*, ed. J. W. Minett and W.S.Y. Wang, 47–94. Hong Kong: City University of Hong Kong Press.

SCHOENEMANN, P.T. and W. S-Y. Wang. 1996. Evolutionary principles and the emergence of syntax- Commentary on Müller: Innateness, autonomy, universality. *Behavioral and Brain Sciences* 19(4): 646–647.

SENGHAS, Ann, Sotaro Kita and Aslı Ozyurek. 2004. Children creating core properties of language: Evidence from an emerging sign language in Nicaragua. *Science* 305: 1779–1782.

SEYFARTH, R. M., D. L. Cheney and P. Marler. 1980. Monkey responses to three different alarm calls: Evidence of predator classification and semantic communication. *Science* 210: 801–803.

SIA, G. M., R. L. Clem and R. L. Huganir. 2013. The Human language-associated gene SRPX2 regulates synapse formation and vocalization in mice. *Science* 342: 987–991.

SIMON, Herbert. 1962. The architecture of complexity. *Proc. Amer. Philos. Soc.* 106: 467–482.

STRINGER, Chris B. 2012. *The Origin of Our Species*. London: Penguin.

V ALLENDER, Eric J., Nitzan Mekel-Bobrov and Bruce T. Lahn. 2008. Genetic basis of human brain evolution. *Trends in Neurosciences* 31(12): 637–644.

VOGT, Paul. 2005. On the acquisition and evolution of compositional languages: Sparse input and the productive creativity of children. *Adaptive Behavior* 13: 325–46.

WANG, W.S-Y. [1982]1991. *Explorations in Language Evolution*. Osmania papers in linguistics. Supplement; v. 8 Hyderabad: Osmania University Press. Reprint, Taipei: Pyramid Press, in *Explorations in Language*, 105–131.

______. 1984. Organum ex machina. *Behavioral and Brain Sciences* 7(2): 210–1.

______. 2012. Language learning and the brain: An evolutionary perspective. In *Breaking Down the Barriers: Interdisciplinary Studies in Chinese Linguistics and Beyond*, ed. G. Cao, H. Chappell, R. Djamouri and T. Wiebusch, 21–48. Taipei: Institute of Linguistics.

WANG, William S-Y. and Yaching Tsai. 2011. The alphabet and the sinogram: Setting the stage for a look across orthographies. In *Dyslexia Across Languages: Orthography and the Brain-Gene-Behavior Link*, ed. P. McCardle, J.R. Lee, B. Miller and O. Tzeng, 1–16. Baltimore, Md.: Paul H. Brookes Pub. Co.

WELLS, Rulon S. 1947. Immediate constituents. *Language* 23: 81–117.

语言演化的四级相变

王士元

（香港中文大学）

提要

要想理解人之所以为人的基础何在，那么探索语言如何独一无二地在人类涌现是个核心议题。把这个议题取个像「语言器官」(language organ)一类的名字，并把它含糊地归因于某种基因突变，是无济于事的作法。这个议题理应从演化论的观点加以检视。在此我主张，语言涌现的轨迹，始于我们首次采取双脚直立的姿势，这比科学分类上我们「人属」(*Homo*)的出现还要早。第二个相变出现在当我们的祖先制造各类石器工具展现了象征行为时。促成语言涌现的第三个相变，出现在当口语沟通从手势和韵律的模式转变为主要依靠元音、辅音构成音节串时，这种转变为人类提供了有效的信号空间。第四个相变则是文字的发明，也造成了若干深远的影响。

关键词

语言演化

第二章

▼

金尺悠鸣杏林春

2018年人类学终身成就奖获得者

——席焕久

人类学生涯回顾

普通的农家子弟

1945年腊月十八，我出生在辽宁省绥中县六股河畔普通农民家里，家境贫寒，父辈几乎都是文盲。我记事时常跟着大人干活，目睹父辈们的辛劳与节俭。给我留下深刻印象的是爷爷在碾米时把掉在地上的米一粒一粒地捡起来。五六岁时，大人收土豆我就用筐往屋里运，由于过劳，一干重活就咳嗽，鼻子出血，每次都吃中药调治，因而从小就对中医有较深刻的印象。

1954年我在当地上小学，经历了“大跃进”和大炼钢铁，学校经常停课劳动，修水渠，挖铁砂，我也经常因劳累而犯病。

父亲只读了两年书，倍感文盲之苦，希望儿子好好读书，所以，只要孩子说看书，一切工作都可以免除，让孩子学习。

1960年，我以优异的成绩考取了县里的实验中学——绥中一中。开始，每天走读，上学时常常遇到学校的教导主任——王景元老师，他走路很快，我们跟不上，只好在后边跟着小跑，三九天汗从棉袜子往下淌，虽然有些苦，但这种锻炼却练就了我一双铁脚板。后来住校，由于三年自然灾害，国家处于困难时期，粮食不够吃，学校组织我们到绥中西北沟的山里捡橡树的果实和叶子，碾成粉末补充口粮，苦涩难吃，导致便秘。由于肚子饿，冬天晚自习到食堂后边捡来白菜尾巴烤着吃。读高中时有了自行车，为了备考，高三又住校。

初中时代

读初中时，每天早晨还要替母亲到生产队上工，收工后吃早饭，然后走10多里路上学，晚上回来还要挖野菜喂猪，秋天要拾柴火。母亲从不休息，当妇女队长多年，年年是县乡劳动模范，那时她常常保护无辜的群众，这是母亲去世时来吊唁的乡里人说的。父母对爷爷奶奶十分孝顺。

由于身在农村，经常参加劳动，所有的农活我都会干，所以学校放假留些学生在校义务劳动，我每次都被选中参加劳动。施肥、浇水、除草，什么活都干，尤其是掏粪，又脏又累，近一个月的工作，衣服都熏得很臭。班级每次选干部，我不是被选为劳动委员就是学习委员，因为我不仅劳动好，学习也好。

我的学习时间很多时候都在晚上和路上，路上背课文、公式，晚上在煤油灯下看书到深夜，眼睛常被熏成黑圈。母亲在灯下为全家做针线活陪着我学习，夜深了，因次日还要起早给我做饭，她睡了，但我还在学习。

从童年到青少年，父辈们的身教培养了我的勤劳、节俭、不怕困难的品质，对塑造我后来的性格起了很大作用。不论后来的外语学习、研究生备考，还是人类学的采样及实验室研究，不管多难我都能克服，这与小时候父辈的身教有直接关系。

不寻常的决定

1965年高中毕业前夕，我们开始填写高考志愿，普通院校与重点院校各12个志愿。我填得很快，24个志愿除一个医学专业外，其余全部是理工科，普通院校的志愿是随意填的，因为我有充分信心能考上一流大学。结果没考上重点院校，别人祝贺我考上锦州医学院，我却高兴不起来，我怎么会到锦州医学院？不会考得这么差呀！于是我决定第二年再考。父亲听说后劝我："去吧，学医不错"。最后我听了父亲的话。没想到，这个意见造就了我的一生。没有这个决定，就不会有我的今天，因为第二年"文化大革命"开始，不招生了。

入校后，我才知道医学要学33门课，其中解剖课是重要的基础课，当时不知道解剖学是怎么回事，一上课才知道要接触尸体，我后悔极了。上解剖学课时我总是精神溜号，想我学的数理化知识白费了，太可惜！由于害怕尸体也不敢去实验室。整整煎熬了一个学期。但毕竟是木已成舟，只好准备新学期好好从头学习。但不久，"文化大革命"开始了。当时我做团支部书记，当上了"文化革命"小组长，一切都按上级精神办的，一下子却成了保守派，很不理解。

大串联开始了。我高中的同学来锦州串联，问我："席焕久，你知道你为什么来锦州医学院？"我说："不知道。""高三时你是否写过一篇作文——我们的好老师？这个老师是特务嫌疑。”我说:“我一个学生，哪知道这些！”因为此事我被鉴定为“是非不清”。在当时这是很要命的。后来听招生的老师说，第四志愿是锦州医学院，通常是不会录取的，因为分数较高才被录取。我这才恍然大悟。

1970年我毕业留校做外科医生。后来，国家要加强基础课，我服从学校的需要到解剖教研室做教师，后又改做行政工作直到1979年。

偶然的机会

1978年国家开始招收研究生，第二年正是我行政生涯“结束”的一年，在研究生报名截止的前一天我报了名。因为过去带过解剖实习，就报了解剖学专业。考试前一天我还在听组织胚胎课。说心里话，没有条件报考，因为10年没有接触业务工作了，当年学的是光镜结构，而现在是电镜结构。有的老师认为我是“癞蛤蟆想吃天鹅肉，不知天高地厚”。家里人也认为“瞎子看戏白打工”。然而，“瞎猫碰上了死耗子”，我考取了西安医科大学。

入学后最大的困难是英语听力难度太大，过去初中、高中、大学一年级学的都是俄语，1978年我才开始自学英语，读音完全按汉语拼音字母读的，几乎

与导师张怀韬教授(前排右2)合影

每个单词都读错，一直影响到今天。读研时我们班级同学英语都很棒，可我就是听不懂，我用95%时间听录音，耳朵听得发聋，屁股坐出了老茧，但终于取得优异成绩，为出国深造打下了基础。

公共课程学完后，开始进教研室做课题。这时开始接触人类学，过去连什么是人类学都不知道。我的导师是西安医科大学解剖学教研室的奠基人——张怀韬教授，他让我做西安市儿童青少年生长发育的研究。教研室有一台X光机，有一套马丁测量设备。为了做好这项研究，我事先在原单位学习摄片、洗片、观片技术，跟教研室老师学习测量技术。后来，教研室给我安排一个小女孩作助手，帮助记录，导师借给我一辆自行车。我跑遍西安市各高校（临潼除外），选择研究对象进行体质测量，拍X光片，做骨的显微切片，用计算机做多元统计分析。今天，这些看起来很平常、简单，但在1982年，人类学与解剖学领域应用计算机和多元统计分析算是开创性的。我到西安市各高校做体质测量时，有的名牌大学的教务长不同意，要我们开具省委介绍信；有的校长不同意接待。难度很大，最终我冲破了一个一个阻力，完成了2 612例，我们教研组的老师还惊奇地发现，西安医科大学后勤还帮助我出大客车，拉学生来校做体质测量。

在读研时，我偶然碰见了吴新智教授，他去陕西蓝田考察猿人后路经西安，来校指导工作，导师让我为吴老师服务，其实没有什么服务，就是打打开水，引领去食堂吃饭，老师自己拿饭票。我趁机请教一些问题，收获颇丰。后来每当我在北京等待转车回锦州时，我都到他那里请教几个预先准备好的问题。我从电视剧《霍元甲》中得到启发，他有两个徒弟——陈真和大安，陈真武艺较高强。除个人努力外，陈真还有个老师是独臂老人，他善于从两位老师那里学习，而大安只有一位老师。于是我利用在北京一天的等车时间，带着准备好的问题去请教北京医院李果珍教授、积水潭医院王云钊教授、中国科学院吴汝康教授，后来还有天津医科大学的吴恩惠

与吴新智院士合影

教授，他们的指点大大丰富了我的研究论文，扩展了我的思路。

出国深造

1985年省委派中国医科大学刘述舜同志来校任院长，他在中国医科大学被称为四大台柱子之一，重视人才，团结师生，有远见，有魄力。他看中我，想让我做院长助理，我说："刘院长，我先出国，等我回来后，听候组织安排"。这时美国Fels研究所、哥伦比亚大学和加拿大麦吉尔大学都同意接收我做访问学者。我决定去美国，先到Fels研究所，这个研究所是用Fels基金于1929年建立的，是世界上人类学研究数据最多，研究对象年龄跨度最大、追踪时间最长，信息最丰富的研究所。所长是A. F. Roche教授，澳大利亚人，曾任美国生物人类学会主席，发表过很多文章与著作，其中有关于人体测量方法和手腕部、膝部骨龄的书。他给我出了三个题目，让我自选一个。经过夜以继日的工作，这三个题目我全部都完成了，为我后来编写测量骨龄手册打下了基础。当时的人体成分研究主要是用水下测量仪器测骨密度。我很快写完第一篇文章——手腕部与膝部骨龄的差异，发表在美国人类学杂志上，引起了强烈的反响，12个国家20多所大学的专家学者或索取论

与导师Roche教授

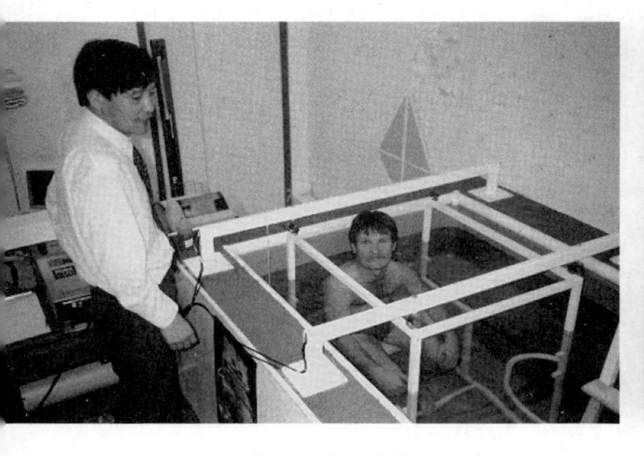

在Fels研究所测水下身体比重

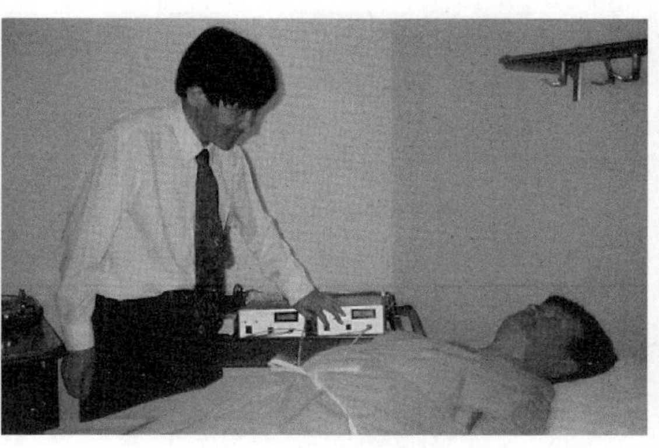

在Fels研究所测骨密度

文，或希望合作。

Roche教授对我的工作很满意，他决定资助20万美元与我合作在中国开展工作。为训练我穿刺，他让研究所副教授以上的人员每人让我做一次穿刺。我很是感动，也很紧张。他写的文章草稿也让我看，每次我都提出一些意见，他采纳了不少。后来他的材料都让我先看，提提意见。我回国后，他的论著一直都寄给我。

出国前，我患了阑尾炎，本该做手术，但急于出国没来得及做，我知道美国医疗费用很贵，临行前我带好了抗生素、注射工具等以防不测。后来在美国发病时我用针灸和按摩治好了，现在也没犯。所里的秘书知道了这件事，后来她感冒鼻塞，我按摩一下就通了，她问我是怎么回事，我告诉她这是中医治疗。她告诉旁边的家庭诊所，说中国的中医很有效，家庭诊所只要有重要的病人来就诊都让我看看。这使我对民族医学产生了新的认识，为后来研究医学人类学埋下了伏笔。

后来我去哥伦比亚大学学习医学人类学。一次我去听课，老师拿出一本书，指着书上的照片说，你认识这位教授吗？我说认识，他是我的老师吴新智教授。这时候我才知道吴教授在那时就很有名了。说心里话，我对医学人类学很生疏，当年来美时我曾经征求吴新智老师的意见，去美国应学点什么。他说你有医学的背景，又懂人类学，学医学人类学，回国以后推动我国的医学人类学研究。我遵照吴老师的意见才开始走进这个领域。

1989年我学习期满，回到了母校锦州医学院，开始向全国介绍医学人类学，出版了全国第一本医学人类学著作。

访问哥伦比亚大学，与王志勉教授合影

后来，我还先后到美国哥伦比亚大学体成分研究室考察，到美国亚利桑那大学访问，参观宾夕法尼亚大学博物馆，考察加州大学洛杉矶分校人类学系。还在英国牛津大学、澳大利亚新南威尔士大学、新加坡国立大学学习和考察，取得不少收获。

访问加州大学洛杉矶分校人类学系

访问美国亚利桑那大学，与 T. G. Lohman 教授（右）和 S. B. Going 教授

走进人类学

从西安医科大学毕业回来，我真想大干一场，把学到的东西贡献给母校，但当时的环境难以发挥作用，我正式向学校提出调到大连医学院工作。时任党委副书记刘国富同志亲自到家劝我留下，学校也坚决不放。当时的基础部主任，教我化学的卢树德教授也帮助我分析，我无奈留下。后来我到辽宁省铁岭地区开原县进行人类学的研究。到了开原县，在当地地方病研究所王辉亚同志的全力帮助下，先进行体质测量。用了大半年的时间，我踏遍了开原的山山水水，测量了 2 046 例。教研室主任李泽山教授，对我很支持，给我安排了新参加工作的谷学静老师当助手。这个地区也是地方性克汀病高发区，我们还开始了病理人类学研究，把疾病的表现和体质特征与文化、社会、经济结合起来，分析病因，制订防治措施。

参加第五届中国生态健康论坛期间与王如松院士合影

2001 年，我应邀参加中山大学“21 世纪都市可持续发展暨纪念人类学百年国际学术研讨会”并作报告，结识了中国科学院生态研究中心王如松院士。我们经常交流，后来他也常邀请我参加全国生态学会议并作报告，有的大城市的生态规划也让我参与。从人类学角度回答生态健康问题，大大丰富了我

的人类学知识，我也从生态学角度研究人类学问题。

分子生物学的深入发展促进了个性化医疗，进化医学的发展需要有进化论的理论，这些都涉及人类学的根本问题——人的差异。我注意从形态、机能、代谢、免疫、体能等多方面研究人的差异。人的生长发育问题与成长、衰老都有关系，骨的健康与民族/种族、年龄、性别、地域都有关系。2005年，我利用去美国的机会查阅大量文献，结合自己的研究，在国家科学技术学术著作出版基金的支持下编写出版了《生物医学人类学》。

20世纪80年代的人类学界，除古人类学外，多侧重在体质测量方面尤其是在骨测量方面的研究，吴新智院士提出要研究体成分，迎接人类学的第二个春天。2013年10月12—13日，在中央民族大学主持召开的“二十一世纪中国人类学发展高峰论坛”上，国内一些著名专家和美国亚利桑那大学的教授都发表了演讲。为了做好体成分的研究工作，在中国科协基金支持下，我对全国的体成分研究工作者进行培训，推动了体成分的研究。除西安市和开原县的体质调查之外，还有藏族人类学的研究，在国家科学技术学术著作出版基金的支持下出版了专著并获中国出版政府奖提名奖。还参加全国汉族体质人类学的研究、中国各民族体质人类学表型特征的调查。我们的团队多次去西藏、新疆及广大城乡进行采样调查。

主持“二十一世纪中国人类学发展高峰论坛”

在辽宁新宾满族自治县红庙子乡采样

出席内蒙古自治区自然科学年会并作学术报告

赴西藏采样的团队

在新疆调研

首届体成分培训班

在辽宁阜新作人类学调查

指导研究生体质测量

给医学本科生讲医学人类学课

在中学作科普报告

在美国亚利桑那大学作学术讲座

在国际人类学与民族学联合会第十六届大会上与外国学者交流

2005 年出席国际学术会议作会议主持人

在研究的同时，我还开展了人类学的教育、学术交流及科普工作。率先在全国为本科生开设“医学人类学”选修课，为研究生开设“人类学与医学”课程；在解剖学教学中，强调人的个性，讲人的差异，谈进化医学，为精准医学和个性化医疗打下基础。

在全国的学术会上，介绍这一新兴学科；在《健康报》上发表人的差异与健康之关系的文章；受聘于中国科协成为全国人类学首席科学传播专家，为中学生和大众作科普报告。

作为中国解剖学会人类学专业委员会主任，每年都组织人类学专业委员会开年会进行学术交流。在美国亚利桑那大学作学术讲座，介绍中国学者的工作；作为国际人类学与民族学联合会第十六届大会分会场主席，参与筹办国际学术会议，被国家民委、云南省政府评为筹备工作先进个人。我在国内外交流的题目有：Anthropological study on Tibetan in Tibet、人的差异、生活方式演变与健康、生态学与健康、人类学的研究与发展、民族与人种问题的讨论等。

其他工作

从1986年开始，我先后任锦州医学院基础部副主任、主任，1995年任副院长，1998年任院长、党委书记。任职期间开展教学改革，获国家教学成果奖。在任期间学校形成教学改革、科技产业和医牧结合三个办学特色。2017年，获全国医学教育杰出贡献奖。1990年开始先后被选为古塔区人民代表，锦州市人民代表和人大常委，辽宁省九、十届人大代表并作为主席团成员和全国九届人大代表。任职期间，写意见建议上百件，很多都被政府采纳，如高校扩招、出国人员的文明教育和新疆生产建设兵团有关政策等。

古人云，莫道桑榆晚，为霞尚满天。我已入古稀之年，但仍进行科学研究等工作。先后主持或参与国家科技部项目、国家自然科学基金项目、老科学家采集工程项目等。2018年刚刚获批“中国解剖学学科史”研究项目，一部新作刚刚出版发行，体成分的三篇论文正在陆续发表。在科研工作的同时还学习游泳，弹钢琴，练书法，缓解工作的疲劳。

结语

1970年我毕业于锦州医学院，先做了一段时间的外科医生，后任解剖学

前教育部长周济到锦州医科大学视察

出席全国九届人大会议

与吴新智院士一起接受CCTV采访

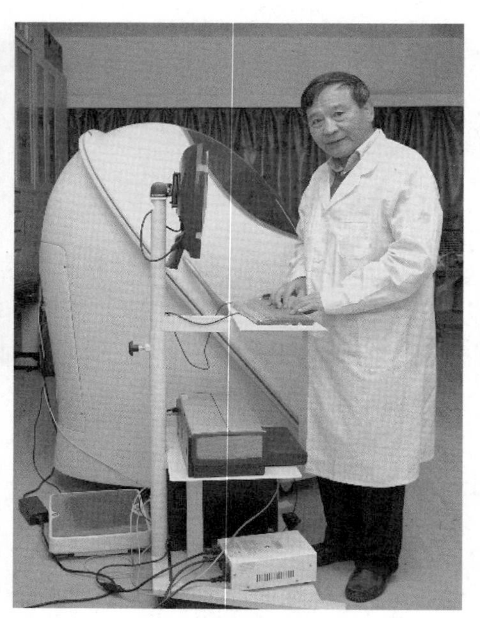

用BOD POD测体成分

教师，开始接触人类学。但真正学习人类学是从研究生开始，出国深造则进一步强化了人类学领域的知识。在我的人类学生涯中，有三位老师起着重要作用。首先是我的启蒙老师——张怀韬教授，他把我引进门，使我开始认识人类学。第二位是Roche教授，在他的指导下学习了骨骼年龄的评价、体成分和生长发育研究。第三位老师是吴新智院士，从确定出国学习医学人类学，到培养我当博士生导师，再到做人类学专业委员会的工作，他都细心地帮助我。

2020年将迎来我从教50周年。回顾这近50年历程，对人类学由不知到略知，由了解到熟悉，由一般性研究到深入探讨，得益于学校的培养和教育，得益于老师的帮助与指导，得益于我的学生们的支持和促进，得益于家庭的支持与理解，我衷心地感谢他们。我虽到了古稀之年，已从教师岗位退休，但人类学的道路还没有走完，我要力所能及地继续发光发热，推动人类学的发展。

主要论著

[1] Xi H J, Roche A F. Differences between the hand-wrist and the knee in assigned skeletal ages[J]. Am J Phys Anthropol, 1990, 83: 95–102.

[2] Xi H J, Roche A F, Baumgartner R N. Association of adipose tissue distribution with relative skeletal age in boys: The Fels longitudinal study[J]. Am J Hum Biol, 1989, 1(5): 589–597.

[3] Xi H J, Roche A F, Gao M. Sibling correlations for skeletal age assessments by Fels method[J]. Am J Hum Biol, 1989, 1(5): 613–621.

[4] Ren F, Li C Y, Xi H J, et al. Estimation of human age according to telomere shortening in peripheral blood leukocytes of Tibetan[J]. Am J Forensic Med Pathol, 2009, 30(3): 252–255.

[5] Xi H J, Zhang L P, Guo Z Y, et al. Serum leptin concentration and its effect on puberty in Naqu Tibetan adolescents[J]. J Phys Anthropol, 2011, 30(3): 111–117.

[6] Lei Z, Ma H J, Xu N, et al. The evaluation of anti-angiogenic treatment effects for implanted rabbit VX2 breast tumors using functional multi-slice spiral computed tomography (f-MSCT)[J]. European J Radilogy, 2011, 78: 277–281.

[7] Liu Y X, Xi H J, Xing W W, et al. Aquaporin changes in compound 48/80 induced inflammatory sublaryngeal edema in rat[J]. Journal of Voice, 2012, 26(6): 815, e17–815, e23.

[8] He X, Zhang J F, Li Z X, et al. The traits of five types of tongue movement in Han of Shannxi, China[J]. Anat Sci Int, 2012, 87(4): 181–186.

[9] Xi H J, Li C Y, Ren F, et al. Telomere, aging and age-related diseases[J]. Aging Clin Exp Res, 2013, 25: 139–146.

[10] Xi H J, Li M, Fan Y, et al. A comparison of measurement methods and sexual dimorphism for digit ratio (2D: 4D) in Han ethnicity[J]. Arch Sex Behav, 2014, 43(2): 329–333.

[11] Li Y L, Zheng L B, Xi H J, et al. Body weights in Han Chinese population[J]. Chin Sci Bull, 2014, 59(35): 5096–5101.

[12] Li Y L, Zheng L B, Xi H J, et al. Stature of Han Chinese dialect groups: a most recent survey[J]. Sci Bull, 2015, 60(5): 565–569.

[13] Zheng L B, Li Y L, Xi H J, et al. Stature in Han populations of China[J]. Sci China

Life Sci, 2015, 58(2): 215–217.

[14] Zhang H L, Fu Q, Li W H, et al. Gender differences and age-related changes in body fat mass in Tibetan children and teenagers: an analysis by the bioelectrical impedance method[J]. J Pediatr Endocrinol Metab, 2014, 28(1–2): 87–92.

[15] Xi H J, Chen Z, Li W, et al. Chest circumference and sitting height among children and adolescents from Lhasa, Tibet compared to other high altitude populations[J]. Am J Hum Biol, 2015 Aug 7.doi:10.1002/adjhb.22772.

[16] C J J, Xi H J. Vascular endothelial growth factor 165-transfected adipose-derived mesenchymal stem cells promote vascularition-assisted fat transplantation[J]. Artif Cells Nanomed Biotechnol, 2015, 26: 1–9.

[17] Zhang F, Sun D, Chen J, et al. Simvastatin attenuates angiotensin II-induced inflammation and oxidative stress in human mesangial cells[J]. Mol Med Rep, 2015, 11(2): 1246–1251.

[18] Wang B, Wu N, Liang F, et al. 7, 8–dihydroxyflavone, a small-molecule tropomyosin-related kinase B(TrkB)agonist, attenuates cerebral ischemia and reperfusion injury in rats[J]. J Mol Histol, 2014, 45(2): 129–140.

[19] Guo J P, Xi H J, Ren F, et al. Analysis of polymorphism at sites-597 and-572 of interleukin-6 promoterin Tibetan population from Tibet autonomous region[J]. Journal of Clinical Rehabilitative Tissue Engineering Research, 2007, 11(34): 6912–6914.

[20] Xi H J, Li W H, Wen Y F, et al. Analysis of environmental factors influencing on growth and development of Plateau children and adolescents in China[J].解剖学报, 2018,50(6): 769–781.

[21] 席焕久，李文慧，温有锋，等.人体成分研究概览[J].人类学学报，2018，37(2): 241–252.

[22] 席焕久，李文慧，温有锋，等.海拔对儿童和青少年生长发育的影响[J].人类学学报，2016，35(2): 267–281.

[23] 李文慧，席焕久，侯续伟，等.西藏那曲藏族成人体成分分析[J].解剖学杂志，2017，40(1): 63–66.

[24] 王健，席焕久，李文慧，等.拉萨藏族成人体成分现状[J].解剖学杂志，2017，40(2): 192–196.

[25] 李文慧，席焕久，候续伟，等.那曲藏族成人身体各部肌肉量分析[J].解剖学杂志，2017，40(3): 326–329.

[26] 李文慧，席焕久，侯续伟.拉萨藏族儿童青少年跟骨骨强度的年龄变化及其影响因素[J].解剖学杂志，2017，40(5): 599–602.

[27] 刘详君，张兴，席焕久，等.中国汉族乡村成年人的身高与体质量的地理学分布[J].人类学学报，2018，37(3): 484–495.

[28] 席焕久，李文慧，赵宏.藏族生物人类学研究回顾[J].人类学学报，2015，34(2): 260–266.

[29] 李咏兰，郑连斌，席焕久，等.中国北方、南方汉族体重差异[J].解剖学报，2015，46(2): 270–274.

[30] 席焕久.关于解剖变异的辨析与思考[J].解剖学杂志,2014,37(6): 717-718.
[31] 海向军,何烨,何进全,等.甘肃农村汉族成人Heath-Carter体型[J].解剖学杂志,2013,36(6): 1105-1110.
[32] 席焕久,吕坡,李文慧,等.国际大赛获奖运动员的克托莱指数分析[J].解剖学杂志,2014,37(2): 229-237.
[33] 席焕久.藏族的高原适应——西藏藏族生物人类学研究回顾[J].人类学学报,2013,32(3): 247-255.
[34] 席焕久,张海龙,李文慧,等.高原地区居民的体成分与形态学变化[J].解剖科学进展,2013,19(2): 178-181,183.
[35] 李文慧,席焕久,吕坡,等.辽宁汉族6项不对称行为特征[J].解剖学报,2013,44(3): 413-418.
[36] 顾平,席焕久,李文慧.辽宁盘锦地区体校篮球运动员体型特征[J].解剖学报,2013,44(3): 423-426.
[37] 张文化,席焕久.辽宁青少年投掷运动员投掷能力与指长比的相关性[J].解剖学杂志,2013,36(1): 113-117.
[38] 吕坡,范杰,席焕久.EGLN1基因两个位点多态性与藏族人群高原低氧适应的关系[J].解剖学报,2013,44(3): 419-422.
[39] 张海龙,席焕久,付强,等.利用生物电阻抗法分析西藏藏族青少年肌肉发育特点[J].解剖学报,2013,44(2): 292-296.
[40] 席焕久,任甫.人的体质千差万别[N].健康报,2013-1-11.
[41] 张海龙,席焕久,李文慧,等.利用生物电阻抗法分析辽宁汉族成人脂肪分布特点[J].解剖学报,2012,43(6): 850-854.
[42] 敖哈拉主编,席焕久、夏桂兰、吴伟康、赖世隆副主编.中国老年医学[M].沈阳: 辽宁民族出版社,1992.
[43] 席焕久,张美芝,张海龙.谈谈人的年龄[J].解剖科学进展,2012,18(1): 87-90.
[44] 徐国昌,杨雷,席焕久,等.河南汉族成人皮褶厚度和人体成分分析[J].解剖学杂志,2012,35(2): 229-232.
[45] 徐国昌,杨雷,席焕久,等.河南汉族人群头部7项长度指标与身高的相关性[J].解剖学报,2012,43(4): 553-558.
[46] 席焕久,赵红,李文慧,等.对指长比研究现状的初步分析[J].解剖学报,2012,43(4): 569-573.
[47] 张美芝,席焕久,温有锋,等.健康绝经妇女体成分差异对跟骨密度的影响[J].解剖科学进展,2012,18(2): 168-169.
[48] 李超,席焕久,张海龙,等.辽宁地区1127例汉族成年人跟骨密度分析[J].解剖科学进展,2012,18(3): 219-221,226.
[49] 裴林国,张圆圆,刘荣志,等.辽宁农村汉族成人皮褶厚度的年龄变化[J].解剖学杂志,2011,34(5): 684-686,690.
[50] 张海龙,席焕久,闫文柱,等.辽宁城市汉族成人的Heath-Cater法体型研究[J].解剖学报,2011,42(6): 846-850.

[51] 姜东，闫文柱，刘素伟，等.辽宁农村汉族青壮年身高与指距的关系[J].解剖学报，2011，42（6）：851-853.
[52] 席焕久，李超，张美芝，等.500位中国高校教学名师奖获得者探析与启示[J].中华医学教育杂志，2011，31（1）：74-77.
[53] 李丽，席焕久，牛志民，等.辽宁省男犯攻击行为心理探析[J].现代预防医学，2011，38（12）：450-456.
[54] 李丽，席焕久，牛志民，等.汉族大学生指长比与人格关系探讨[J].中华行为医学与脑科学杂志，2011，20（3）：261-263.
[55] 李丽，席焕久，牛志民，等.大学男生指长比与攻击行为的相关分析[J].中国卫生学校，2011，32（1）：4-5.
[56] 姜东，闫文柱，刘素伟，等.辽宁农村汉族成人手长、足长与身高的关系[J].解剖学报，2011，42（2）：249-252.
[57] 李明，王立军，范英南，等.根据拉萨藏族儿童青少年掌指骨参数进行性别判定[J].解剖学杂志，2011，34（1）：98-102.
[58] 席焕久，李文慧，张美芝，等.人的差异及其影响因素[J].解剖科学进展，2011，17（5）：478-483，486.
[59] 裴林国，贺生，张园园，等.拉萨藏族人群线粒体DNA多态性分析[J].解剖学杂志，2011，34（4）：539-543.
[60] 李明，王立军，范英南，等.拉萨藏族儿童青少年掌骨皮脂厚度的X线测量[J].解剖学杂志，2010，33（6）：798-801.
[61] 闫文柱，姜东，刘素伟，等.辽宁农村汉族成人头面部特征[J].解剖学杂志，2010，33（6）：811-815.
[62] 席焕久.对人体解剖学教育的思考[J].局解手术学杂志，2010，19（1）：1-2.
[63] 席焕久，胡荣，温有锋，等.辽宁汉族指长及指长比特点[J].解剖学报，2010，41（3）：477-479.
[64] 席焕久，裴林国，程鹏，等.北京奥运会运动成绩的人种差异[J].体育科学，2010，30（6）：81-84.
[65] 阎文柱，姜东，刘素伟，等.辽宁农村汉族成人体质特征分析[J].解剖学报，2010，41（5）：756-760.
[66] 李明，王立军，席焕久，等.拉萨藏族儿童青少年掌骨长与身高关系[J].解剖学杂志，2010，33（3）：394-397.
[67] 王红梅，王平，席焕久.藏族青少年血清钙磷及钙调节激素水平分析[J].中国学校卫生，2009，30（4）：351-352.
[68] 李明，席焕久，任甫，等.拉萨藏族青少年左手指长比和指长/身高的特点[J].解剖学杂志，2009，32（2）：260-262.
[69] 温有锋，叶丽平，席焕久，等.西藏藏族青少年体型[J].人类学学报，2009，28（1）：64-72.
[70] 刘永，温有锋，席焕久.西藏藏族mtDNA D-Loop区HVR-Ⅱ序列多态性[J].解剖学杂志，2008，31（3）：409-411.

[71] 李宁，苏玉虹，席焕久，等．西藏藏族人群15个短串联重复序列基因座的遗传多态性[J]．遗传，2008，30（7）：851-856.
[72] 梁艳明，席焕久，李明，等．人类指长比的研究进展[J]．解剖科学进展，2008，14（4）：449-452.
[73] 肖艳杰，席焕久，李志强．西藏地区藏族女生月经初潮状况调查[J]．现代预防医学，2008，35（22）：4355-4357.
[74] 肖艳杰，席焕久．藏族中小学生体表面积现况调查[J]．中国学校卫生，2008，29（5）：393-397.
[75] 肖艳杰，席焕久．西藏地区7～18岁藏族学生体格发育状况调查[J]．中国学校卫生，2007.28（11）：968-970.
[76] 文普帅，温有锋，席焕久，等．西藏藏族DYS287位点多态性研究[J]．武警医学，2007，18（2）：89-90.
[77] 刘永，温有锋，席焕久．西藏藏族mtDNA COⅡ/tRNAlys基因间区9-bp缺失多态性[J]．武警医学，2007，18（2）：100-102.
[78] 张路萍，席焕久，朱永林，等．藏族男性青少年血清瘦素水平及其影响因素的研究[J]．武警医学，2007，18（3）：170-172.
[79] 任甫，李长勇，席焕久，等．人外周血白细胞端粒DNA长度与年龄的定量关系[J]．中国法医学杂志，2007，22（3）：160-162.
[80] 李宁，苏玉红，席焕久，等．拉萨市藏族人群15个STR基因多态性的研究[J]．遗传，2006，28（11）：1361-1364.
[81] 席焕久，任甫，郭景鹏．再谈加强人类学的研究，促进人类学的发展[J]．解剖学杂志，2006，29（4）：401-402.
[82] 温有锋，席焕久，叶丽平，等．那曲地区藏族青少年体型Heath-Carter法研究[J]．解剖学杂志，2006，29（4）：417-420.
[83] 冯强，温有锋，任甫，等．中国西藏拉萨藏族群体mtDNA环区HVR-Ⅰ序列多态性研究[J]．武警医学，2006，17（4）：258-260.
[84] 张路萍，席焕久，张曼，等．藏族女性青少年血清瘦素与青春发育[J]．中国医学检验杂志，2006，7（1）：10-12.
[85] 冯强，席焕久，温有锋，等．那曲地区藏族mtDNA D-环区高变区Ⅰ序列多态性[J]．解剖学杂志，2006，29（2）：150-152.
[86] 任甫，温有锋，席焕久，等．拉萨市藏族青少年体型的Heath-Carter法研究[J]．解剖学杂志，2005，28（2）：215-218.
[87] 席焕久，张鹏飞，罗俊生．海绵窦的解剖学概念及相关结构的观察[J]．解剖学报，2005，36（4）：395-399.
[88] 任甫，李长勇，席焕久，等．拉萨藏族男女青少年手腕部和膝部长骨干骺融合时间与新疆哈萨克族及辽宁省开原市农村青少年的对照[J]．中国临床康复，2005，9（23）：18-20.
[89] 李春山，李长勇，席焕久，等．拉萨藏族青少年手腕部骨龄发育评价[J]．中国临床康复，2005，9（23）：36-38.

[90] 席焕久.生态学与人类健康,生态健康与科学发展观[C].首届中国生态学健康论坛文集,2004：178-187.
[91] 席焕久.加强人类学的研究　促进人类学的发展[J].解剖学杂志,2005,28(4)：369-370.
[92] 席焕久,任甫.对肋骨进行组织形态测量推断年龄[J].人类学学报,2002,(2)：126-133.
[93] 王剑,任甫,席焕久.基因水平骨骼性别判定研究进展[J].解剖科学进展,2001,7(4)：323-325.
[94] 任甫,席焕久.用长骨密质骨推断年龄研究进展[J].解剖科学进展,2000,6(1)：31-34.
[95] 席焕久,李永新,任甫.颅中窝径路内听道手术的CT辅助定位[J].解剖学杂志,1999,22(5)：408-413.
[96] 席焕久.医学人类学——一个崭新的领域.健康报,1997,第二版.
[97] 崔世英,席焕久.铁蛋白在兔肾近端小管中的穿细胞运输[J].解剖学报,1996,27(4)：437-441.
[98] 席焕久.一部医学与人类文化的专著诞生了[J].人类学学报,1996,15(1)：92.
[99] 席焕久,李锦平,谷学静,等.开原县农村青少年膝部长骨干骺融合的研究[J].人类学学报,1995,14(2)：157-161.
[100] 席焕久,李泽山,王辉亚,等.地方性克汀病病理人类学的研究[J].人类学学报,1992,11(2)：176-184.
[101] 席焕久,谷学静,李泽山,等.月经初潮年龄的研究[J].人类学学报,1987,6(3)：213-222.
[102] 席焕久,王志君,夏桂兰.国人锁骨的测量[J].解剖学杂志,1986,9(3)：212-214.
[103] 席焕久.西安市学生中从膝部长骨干骺融合推算身高与年龄的回归方程[J].人类学学报,1985,9(3)：212-214.
[104] 席焕久.膝部长骨干骺融合的X线观察与分析——西安市学生膝部长骨干骺融合的研究[J].人类学学报,1984,3(2)：107-113.
[105] 席焕久.西安市学生膝部长骨的干骺融合时间与身高等发育指标的关系[J].人类学学报,1984,3(4)：341-348.
[106] 席焕久,曲巍,张慧,等.试论对医学生进行人的差异的教育[J].中华医学教育杂志,2013,33(5)：648-650.
[107] 席焕久,邵帅,文普帅,等.医学文化与医学教育改革[J].中华医学教育杂志,2010,30(1)：1-4.
[108] 席焕久,李明.试论医学人类学作为医学生课程的必要性[J].中华医学教育杂志,2008,28(3)：4-10.
[109] 席焕久.生物医学人类学[M].北京：科学出版社,2018.
[110] 郑连斌,李泳兰,席焕久,等.中国汉族体质人类学研究[M].北京：科学出版社,2017.
[111] 席焕久,柏树令.中华医学百科全书·人体解剖学[M].北京：中国协和医科大学

出版社,2015.
[112] 席焕久.人类学发展现状与展望[M]//中国科学技术协会主编.人体解剖学与组织胚胎学学科发展报告.北京：中国科学技术出版社,2014.
[113] 席焕久.心胸外科手术解剖彩色图谱[M].北京：人民军医出版社,2014.
[114] 席焕久(参编).人体解剖学名词(第二版)[M].人体解剖学与组织胚胎学名词审定委员会.北京：科学出版社,2014.
[115] 席焕久,刘武,陈昭.21世纪中国人类学发展[M].北京：知识产权出版社,2014.
[116] 席焕久.体质人类学(英文)[M].北京：知识产权出版社,2012.
[117] 席焕久(总主编).腹部外科手术解剖色彩图谱[M].北京：人民军医出版社,2011.
[118] 席焕久(总主编).妇产科手术解剖色彩图谱[M].北京：人民军医出版社,2012.
[119] 席焕久.人类学发展的九十年[M]//中国解剖学会编.中国解剖学会90年历程.西安：第四军医大学出版社,2010.
[120] 席焕久,陈昭.人体测量方法[M].北京：科学出版社,2010.
[121] 席焕久.西藏藏族人类学研究[M].北京：北京科学技术出版社,2009.
[122] 席焕久,文普帅.生活方式的演变与健康[M]//桑国卫,陈家兴主编.生态健康与生态文明建设.北京：中国医药科技出版社,223-226.
[123] 席焕久.新编人体解剖学[M].北京：人民军医出版社,2008.
[124] 席焕久.新编人体解剖学学习指导与习题册[M].北京：人民军医出版社,2008.
[125] 吴新智,刘武主编.席焕久参编.百年中国——人类学民族学文库[M].北京：知识产权出版社,2008.
[126] 席焕久.人体解剖学[M].北京：人民卫生出版社,2007.
[127] 席焕久.人体解剖学学习指导与习题集[M].北京：人民卫生出版社,2007.
[128] 席焕久参编.中国大百科全书.第二版[M].北京：中国大百科全书出版社,2007.
[129] 席焕久.现代医学基础(第一、二、三、四、五、六册)[M].北京：人民军医出版社,2005.
[130] 席焕久.医学人类学[M].北京：人民卫生出版社,2004.
[131] 席焕久.局部解剖学学习指导[M].北京：人民军医出版社,2004.
[132] 席焕久.中国人类学的发展[M]//周大鸣主编.21世纪人类学.北京：民族出版社,2003.
[133] 席焕久.人体解剖学[M].北京：高等教育出版社,2003.
[134] 黄瀛主编,吴晋宝、党瑞山、许家军、张为龙、张朝佑、席焕久等副主编.中国人解剖学数值[M].北京：人民卫生出版社,2002.
[135] 席焕久.人体解剖学学习指导[M].北京：人民卫生出版社,2002.
[136] 吴汝康主编,席焕久参编.南京直立人[M].江苏：江苏科学技术出版社,2002.
[137] 席焕久.人类学的研究与发展[M]//中国科学技术协会,国家自然科学基金委员会主编.学科发展蓝皮书.北京：中国科学出版社,2002.
[138] 席焕久.体育人类学[M].北京：北京体育大学出版社,2001.
[139] 金宏义主编,王临虹、邓静云、席焕久副主编.重点人群保健[M].北京：人民卫生出版社,2001.

［140］ 席焕久. 人体解剖学［M］. 北京：人民卫生出版社，2001.
［141］ 席焕久. 中国人类学的发展［M］// 中国解剖学会编. 中国解剖学会八十年. 北京：中国科学技术出版社，2000.
［142］ 孙荣鑫主编，席焕久参编. 人体解剖学［M］. 北京：人民卫生出版社，2000.
［143］ 席焕久. 新编老年医学［M］. 北京：人民卫生出版社，2000.
［144］ 黄瀛主编，席焕久参编. 中国人体质调查第三集［M］. 上海：第二军医大学出版社，1999.
［145］ 孙宝志主编，刘国良、富成志、席焕久、唐建武、张君邦副主编. 临床医学导论［M］. 北京：高等教育出版社，1999.
［146］ 席焕久. 人的骨骼年龄［M］. 沈阳：辽宁民族出版社，1997.
［147］ 席焕久，戴生富. 人体解剖学［M］. 重庆：中国人事出版社，1997.
［148］ 席焕久. 365日谈——家庭预防保健［M］. 北京：科技文献出版社，1995.
［149］ 席焕久. 医学人类学［M］. 沈阳：辽宁大学出版社，1994.

获奖情况

［1］《西藏藏族人类学研究》2010年获第二届中国出版政府奖图书奖提名奖。
［2］《颅底中央部显微解剖与手术入路的研究及临床应用》2011年获辽宁省科技进步一等奖。
［3］《经颅中窝入路的内听道手术CT辅助定位》2002年获辽宁省科技进步一等奖。
［4］《国家级新药——小牛血清去蛋白注射液》2002年获辽宁省科技进步二等奖。
［5］《颅底中央前方手术入路的显微解剖学研究及临床应用》2006年获辽宁省科技进步二等奖。
［6］《用肋骨的组织结构特征推断死者年龄》2006年获辽宁省科技进步三等奖。
［7］《医学人类学的研究》2006年获辽宁省自然科学学术成果二等奖。
［8］ 2004年获锦州市科学技术特别贡献奖。
［9］ 2009年获国家第五届高等学校教学名师奖及辽宁省名师奖。
［10］ 2017年获中华医学会医学教育分会医学教育杰出贡献奖。
［11］《以"器官系统为中心"医学课程模式的研究与成功实践》2005年获国家教学成果二等奖和辽宁省教学成果一等奖。
［12］《"以器官系统为中心"的医学专科综合型课程体系改革的研究》2001年获国家教学成果二等奖和辽宁省教学成果一等奖。
［13］《高校内部管理体制改革中存在的问题分析》2002年获国家人事部科研成果三等奖。
［14］《高校内部管理体制改革中存在的问题分析》2002年获辽宁省机构编制委员会办公室，人事厅科研成果一等奖。
［15］ 1993年被国家教委、国家人事部授予"在社会主义建设中做出突出贡献的回国留学人员"称号。
［16］ 2005年被中国科学技术协会评为全国优秀科技工作者。
［17］ 1992年获国务院政府特殊津贴。

[18] 2005年被辽宁省总工会授予辽宁五一劳动奖章。
[19] 2005年被中共辽宁省委、辽宁省人民政府授予辽宁省优秀专家称号。
[20] 1998年被辽宁省委宣传部、辽宁省精神文明建设活动办公室授予学雷锋先进个人称号。
[21] 2008年被评为辽宁省攀登学者。
[22] 2009年被评为中华人民共和国国家民族事务委员会，云南省人民政府国际人类学与民族学联合会第十六届大会筹办工作先进个人。

学术任职

[1] 先后任中国解剖学会常务理事、副理事长、名誉理事长；中国解剖学会人类学专业委员会委员、副主任、主任。
[2] 辽宁省解剖学会副理事长。
[3] 中国生态学会生态健康与人类健康专业委员会副主任。
[4] American Anthropological Association会员。
[5] 《解剖学杂志》副主编，《人类学学报》、《解剖学报》、《解剖学科学进展》、《局部手术学杂志》编委，《中国医师能力评价》常务副主编。
[6] International Journal of Clinical and Experimental Anatomy: Membership of Scientific Advisory Board.
[7] 中国科学史学会科学家研究分会创始理事。
[8] 国家医学考试中心：教育测量与医学教育评价专家。
[9] 中华医学会医学教育学会委员，全国高等教育学会医学教育分会常务理事。
[10] 中华医学会辽宁省分会副理事长。
[11] 中国高等教育协会，中国教育报编辑部，中国教育协会，北京高等教育部老干部协会：中国西部教育顾问。
[12] 中国科协首席科学传播专家。

代表作

CHEST CIRCUMFERENCE AND SITTING HEIGHT AMONG CHILDREN AND ADOLESCENTS FROM LHASA, TIBET COMPARED TO OTHER HIGH ALTITUDE POPULATIONS*

HUANJIU XI,[1*] ZHAO CHEN,[2] WENHUI LI,[1] YOUFENG WEN,[1]
HAILONG ZHANG,[1] YANJIE XIAO,[1] SUWEI LIU,[1] LINGUO PEI,[3]
MEIZHI ZHANG,[1] PO LV,[1] FU REN,[1] KEQIANG HUANG,[4]
LIPING YE,[1] CHUNSHAN LI,[4] AND LIGUANG ZHAO[1]

([1]*Institute of Anthropology, Liaoning Medical University, Jinzhou 121001, China*

[2] *Mel and Enid Zuckerman College of Public Health, University of Arizona, Tucson 85724–5211*

[3] *Department of Basic Medicine, Nanyang Medical College, Nanyang 473061, China*

[4] *Department of Stomatology, Liaoning Medical University, Jinzhou 121001, China*)

Objectives: The adaptation of human beings to a high altitude environment during growth has been reported in several populations but is less known for Tibetans. The objective of this study was to investigate similarities and differences of Tibetans in patterns and characteristics of physical growth and development in comparison to other high altitude populations.

Methods: We measured the stature, weight, chest circumference and sitting height of 2,813 healthy children and adolescents aged 6- to 21-year-old living at 3,658–4,500 m in Tibet, China, and compared them with published data from other high altitude populations. Eligible participants must have been born and raised in Tibet, and both their parents' families have to be Tibetan for at least the past three generations.

---

* 原文发表于 American Journal of Human Biology, 2016, 28: 197–202.

Results: The physical growth and development of children and adolescents in Tibet and the Andes followed similar patterns, such as delayed growth, short stature and sitting height, and large chest dimensions. Relative to stature, Tibetan sitting heights are similar to Andeans, but chest circumferences are smaller.

Conclusions: Findings from this study reinforce the conclusion that Tibetan and Andean populations have adapted differently to high altitude hypoxia. The physical features of each population may result from unique adaptation to hypoxia, as well as socio-ecological factors, such as poor nutrition. Am. J. Hum. Biol. 28: 197–202, 2016.

Human biologists have reported that people living at high altitudes have significant adaptations to hypoxia in their respiratory, cardiovascular, and hematological systems (Bingham et al., 2010). An enlarged chest is a commonly observed trait in response to high-altitude hypoxia (Weinstein, 2007). The Qinghai-Tibetan Plateau and Andes are the two major high altitude regions in the world where humans have lived for millennia. Tibet is located on the highest plateau of the world, where 86% of the areas is above 4,000 m, with various natural environments and climate.

Most studies of high-altitude growth have concentrated on overall body size or chest dimensions, but a number of studies note that indigenous high-altitude Andeans are characterized by relatively long trunks and short legs (Dittmar, 1997; Palomino et al., 1979; Stinson et al., 1978). Sitting height and chest circumference serve as morphological indicators of thorax growth at high altitude, and both are affected by hypoxia and nutrition during development (Brutsaert et al., 2004; Stinson, 2009). Frisancho (1969) found both sitting height and sternum length were correlated with vital capacity in high-altitude Andeans, though those variables were not as strong predictors of lung volume as was chest circumference at maximum inspiration. Since there is a high correlation between chest size measurements and lung volume in high altitude Andean natives (Frisancho, 1969; Mueller et al., 1978), the large chest characteristic of Andean high-landers may be a morphological trait associated with pulmonary structural adaptations to high altitude hypoxia (Beall, 1982).

Several scholars (Beall, 1982; Pawson, 1976; Malik and Singh, 1978; Miklashevskaia et al., 1972; Tripathy et al., 2007; Weitz et al., 2004) have studied morphological growth among children and adolescents from Tibet. However, a systematic comparison of different patterns and characteristics of physical growth and development in sitting height and chest circumference among different Tibetan and Andean populations has not been undertaken. This study was designed to compare measures of body and chest size among children and adolescents living on the Tibetan Plateau with published data from other high altitude populations to determine if differences in the pattern of growth exist.

SUBJECTS AND METHODS

During 2004–2010, a total of 2813 healthy children and adolescents (males = 1,417; females = 1,396) aged 6–21 years old were recruited, using cluster random sampling, from Lhasa and Negev schools in central Tibet. Lhasa at an altitude of 3,658 m is the capital of the Tibet Autonomous Region, China. The Engle Coefficient of urban residents (the proportion of income devoted to food) was 45.2%. Children and adolescents were from families with high-middle income. However, Negev is located in northern Tibet Plateau at an altitude of 4,500 m. The Engle Coefficient of pastoral area herdsman was 58.0% (Tibet Autonomous Region Bureau of Statistics, 2007). Participants in Negev come from many herding families with middle-low income. The parents of each participant signed informed consent for their children. To be eligible for this study, participants had to have been born and grown up in Tibet, and both their paternal and maternal sides of the family had to be Tibetan for at least the previous three generations. To establish this, each participant was asked to identify their ethnicity and birthplaces of both parents and grandparents. Statements regarding the ethnicity of both parents and grandparents were verified by teachers and health care workers who were familiar with the participants and their families. Birthplaces and birth dates of participants were determined from school or university records or other official forms of identification.

Stature, weight, chest circumference (during normal breathing) and sitting height were measured using an anthropometer, weighting balance scale, and highly flexible inelastic tape with a precision error of less than 0.1 cm/kg, respectively following a standard protocol by Wu et al. (1984) and Lohman et al. (1988). The subjects were barefoot and wore thin socks and minimal clothing. The measurements were made by two well-trained observers, working independently. The inter/intra-observer correlation coefficients were more than 0.96. The comparison data on growth and development of children and adolescents in Qinghai, India, Peru, and Chile were abstracted from published studies (Dittmar, 1997; Frisancho, 1969; Leatherman et al., 1995; Tripathy et al., 2007; Weitz et al., 2004; Zhang et al., 2010).

RESULTS

We compared chest circumference and sitting height relative to stature. Table 1 presents the means and standard deviations of height, weight, chest circumference and sitting height for children and adolescents from Tibet. Table 2 presents means and standard deviations for height, weight, chest circumference and sitting height of children and adolescents derived from five other studies.

Chest circumferences and sitting heights of children and adolescents in most of Tibet age groups were different from other regions and the differences were significant ($P<0.05$ or 0.01).

Chest circumference

Overall, Andean chest circumferences were larger than those of Tibetans (either from Lhasa, Qinghai or from Ladakh in India), despite the fact that Andeans were shorter than Tibetans. Figure 1 shows that the chest circumferences relative to stature of Lhasa Tibetans in this study are very much like children and adolescents from Qinghai, while Tibetan children and adolescents from India show somewhat larger chest circumferences relative to compared to children and adolescents in Tibet. However, the Peruvian data indicate that Andeans have much larger chest circumferences relative to stature than any of the Tibetan populations.

Sitting height

Figure 2 compares data on sitting height relative to stature from this study with previous studies on populations at altitude. Even though our cross-sectional comparison showed that sitting height with age of children and adolescents for

TABLE 1 Stature, weigh, chest circumference and sitting height of children and adolescents in Tibet, PRC ($\bar{X} \pm SD$)

Age (years)	Tibet		Stature (cm)		Weight (kg)		Chest circumference (cm)		Sitting height (cm)	
	Males	Females	Males	Females	Males	Females	Males	Females	Males	Females
6	9	22	118.9 ± 3.5	117.9 ± 3.7	21.8 ± 3.5	21.3 ± 2.7	57.3 ± 4.0	56.3 ± 3.2	65.2 ± 3.3	64.5 ± 1.9
7	112	102	122.9 ± 6.5	120.5 ± 5.6	23.1 ± 3.9	21.6 ± 3.3	58.9 ± 4.3	57.2 ± 4.1	66.5 ± 3.1	65.2 ± 3.0
8	106	105	128.0 ± 6.2	127.1 ± 6.5	26.1 ± 4.2	24.2 ± 5.2	61.0 ± 4.5	58.8 ± 4.7	68.9 ± 2.9	68.1 ± 4.0
9	105	136	129.9 ± 5.1	130.9 ± 7.2	26.7 ± 3.9	26.2 ± 6.4	61.9 ± 4.5	60.5 ± 6.3	69.8 ± 2.6	69.7 ± 3.3
10	102	111	133.8 ± 8.5	136.7 ± 6.3	28.9 ± 4.6	30.1 ± 5.5	63.6 ± 3.9	63.3 ± 5.3	71.9 ± 4.5	72.7 ± 3.5
11	107	97	141.2 ± 7.3	143.4 ± 7.6	34.6 ± 7.0	33.8 ± 6.5	67.1 ± 5.4	65.5 ± 6.0	74.5 ± 3.5	75.3 ± 4.3
12	107	95	146.7 ± 8.7	146.7 ± 8.4	36.0 ± 7.3	36.4 ± 6.9	69.2 ± 5.4	69.5 ± 6.4	76.8 ± 4.8	77.8 ± 4.1
13	113	110	150.8 ± 7.1	150.6 ± 6.4	39.7 ± 7.0	40.5 ± 7.7	70.5 ± 7.9	71.7 ± 8.0	78.9 ± 3.7	79.9 ± 3.8
14	100	97	158.2 ± 8.7	153.5 ± 8.2	45.6 ± 8.8	44.6 ± 7.0	75.0 ± 6.6	76.5 ± 6.6	83.3 ± 4.8	81.7 ± 3.5
15	97	104	162.8 ± 7.1	155.4 ± 6.5	47.2 ± 5.9	47.2 ± 6.2	76.0 ± 4.4	79.1 ± 5.9	85.4 ± 3.4	82.8 ± 3.0
16	102	111	164.7 ± 6.6	155.0 ± 5.1	50.4 ± 6.7	48.3 ± 6.0	77.5 ± 7.5	79.7 ± 5.4	87.1 ± 3.1	82.8 ± 3.5
17	107	108	167.6 ± 4.7	156.1 ± 5.0	53.5 ± 7.7	50.8 ± 7.1	80.6 ± 6.0	81.0 ± 5.9	88.6 ± 3.0	84.0 ± 2.9
18	109	96	166.7 ± 5.1	157.1 ± 4.9	53.0 ± 5.8	49.3 ± 7.0	81.9 ± 3.6	80.1 ± 6.0	88.9 ± 3.2	84.1 ± 3.5
19	68	58	166.6 ± 5.8	157.6 ± 5.4	55.3 ± 7.4	50.6 ± 6.7	82.5 ± 5.6	79.4 ± 7.6	89.7 ± 3.3	84.1 ± 2.9
20	49	27	168.1 ± 5.2	158.3 ± 5.6	55.5 ± 7.0	52.4 ± 8.2	83.3 ± 4.5	80.5 ± 6.8	89.7 ± 2.0	85.0 ± 3.1
21	24	17	167.1 ± 6.2	156.7 ± 6.7	56.9 ± 9.6	52.0 ± 5.7	84.5 ± 6.7	83.0 ± 4.8	89.5 ± 3.6	85.1 ± 3.1

Tibetan males from Tibet and India was greater than those from Peru and Chile, relative sitting heights did not vary much between Tibetans from Tibet and other samples. Nevertheless, the relative sitting height of male Tibetans from India tended to be somewhat greater than those in the current study and from the Andes. Among females, Tibetans from India and Peruvians tended to have slightly larger relative sitting heights than Tibetans from Tibet and Chileans prior to puberty. Afterwards, all groups were alike.

DISCUSSION

It is well known that lung volumes of children and adolescents living at altitude increase rapidly with age (de Meer et al., 1995) due to both normal growth and development, and due to an adaptive enlargement for hypoxia. Sitting height and chest circumference may reflect these increases. Our results indicate that sitting height and chest circumference increased with age and relative to stature. Comparison of highland Asian populations with highland Andean populations revealed that the chest circumferences of high altitude Asian populations are smaller than their Quechua age mates in Peru (alt. 4000 m) (Pawson, 1976; Miklashevskaia et al., 1973; Malik et al., 1978; Frisancho et al., 1970), while sitting heights are somewhat larger although differences relative to stature are minimal.

The results in this study are similar to those reached by Beall (1982) who also showed that Asian highlanders do not have the large chest characteristic of Andean highlanders. The two populations have adapted differently to the same stress, manifesting different chest morphologies or different timing of expression of the same morphology (Beall, 1982). Many factors, including genetics, climate, hypoxia, nutrition, strong ultraviolet radiation, and socioeconomics may have contributed to the physical growth and development in high altitude populations (Baker, 1969). Among these environmental stresses, high-altitude hypoxia is the only condition which traditional technology is incapable of mediating (Peng et al., 2011). However, populations in Tibetan-Qinghai plateau and Andes plateau have sets of physiological and morphological traits that are distinct from each other. This has led to the conclusion that Andean and Himalayan populations have

TABLE 2 Stature, weight, chest circumference and sitting height of children and adolescents in other regions at altitude ($\bar{X} \pm SD$ cm/kg)

Age (years)	Tibetan					Andes			
	China Qinghai		India			Peru Nunoan		Chile Putre	
			Choglam Sar		Ladakh Males				
	Males	Females	Males	Females		Males	Females	Males	Females
6	113.2 ± 5.7	112.8 ± 5.3	113.2 ± 6.7	111.6 ± 6.8		107.8 ± 5.6	104.8 ± 4.8	117.8 ± 4.4	116.8 ± 3.8
	19.0 ± 2.5	18.6 ± 2.2	18.8 ± 2.1	17.7 ± 2.3		19.7 ± 2.0	18.8 ± 2.7	23.0 ± 3.1	21.2 ± 2.1
	56.1 ± 0.4**	54.0 ± 0.3**	—	—		62.2 ± 2.7	60.4 ± 3.4	—	—
	—	—	64.8 ± 2.6**	63.3 ± 3.2**		60.4 ± 3.5**	58.6 ± 8.1	63.4 ± 3.3*	62.9 ± 2.6*
7						112.8 ± 6.6	110.9 ± 3.9		
						21.3 ± 2.7	20.8 ± 1.7		
						63.9 ± 3.4	62.7 ± 2.8		
						63.0 ± 3.7**	62.7 ± 3.1**		
8	121.6 ± 5.8	122.9 ± 5.5	124.5 ± 7.3	122.8 ± 4.7		115.4 ± 5.1	116.5 ± 5.1	123.1 ± 7.2	123.1 ± 4.9
	22.6 ± 3.1	21.7 ± 2.5	23.6 ± 3.7	22.2 ± 3.1		22.8 ± 2.2	22.9 ± 2.1	25.0 ± 3.0	24.9 ± 3.6
	58.7 ± 0.3**	57.0 ± 0.4**	—	—		65.0 ± 2.9	64.9 ± 3.2	—	—
	—	—	70.2 ± 3.7	68.7 ± 3.4**		65.0 ± 3.6**	64.8 ± 3.1**	66.6 ± 4.0*	66.3 ± 2.6**

（续　表）

Age (years)	Tibetan					Andes			
	China Qinghai		India			Peru Nunoan		Chile Putre	
			Choglam Sar		Ladakh				
	Males	Females	Males	Females	Males	Males	Females	Males	Females
9						121.6 ± 5.9	120.1 ± 6.4		
						25.3 ± 2.7	24.8 ± 2.8		
						67.2 ± 3.1	66.1 ± 2.9		
						67.4 ± 3.1**	66.5 ± 3.1**		
10	131.1 ± 5.7	132.9 ± 8.4	131.8 ± 7.2	133.4 ± 9.5		125.4 ± 4.8	124.2 ± 5.8	135.1 ± 5.0	134.2 ± 5.2
	27.1 ± 3.6	26.7 ± 4.7	26.2 ± 3.8	28.4 ± 5.4		27.2 ± 2.8	27.0 ± 2.9	31.5 ± 2.7	32.5 ± 3.9
	61.6 ± 0.4**	60.7 ± 0.4**	—	—		68.4 ± 3.1	67.7 ± 3.5	—	—
	—	—	71.6 ± 3.5**	73.6 ± 4.8**		67.7 ± 6.4**	68.9 ± 3.4**	71.3 ± 2.9*	70.9 ± 3.0*
11					128.0 ± 7.4	129.0 ± 7.4	129.0 ± 7.4		
					24.5 ± 1.6	29.6 ± 4.3	30.3 ± 4.7		
					64.7 ± 2.7	70.8 ± 4.1	70.0 ± 4.2		
					—	70.1 ± 3.1**	70.4 ± 3.7**		

（续　表）

Age (years)	Tibetan					Andes			
	China Qinghai		India			Peru Nunoan		Chile Putre	
			Choglam Sar		Ladakh Males				
	Males	Females	Males	Females		Males	Females	Males	Females
12	142.3 ± 6.6	144.5 ± 7.6	139.5 ± 8.8	144.4 ± 5.7	133.7 ± 3.5	133.0 ± 5.9	137.9 ± 7.6	146.8 ± 8.5	147.2 ± 5.3
	33.2 ± 5.1	34.4 ± 6.2	32.4 ± 4.4	38.5 ± 8.3	26.9 ± 1.9	31.7 ± 4.3	35.7 ± 4.9	42.3 ± 5.8	43.0 ± 6.1
	66.5 ± 0.4**	66.1 ± 0.6**	—	—	67.0 ± 3.3	71.7 ± 2.6	73.6 ± 3.6	—	—
	—	—	75.2 ± 4.4**	77.6 ± 4.3**	—	72.6 ± 3.0**	75.0 ± 3.3**	77.2 ± 4.0	77.5 ± 3.1*
13					135.9 ± 5.7	138.5 ± 7.2	138.9 ± 5.9		
					27.8 ± 3.0	34.9 ± 5.2	37.6 ± 6.0		
					68.8 ± 3.5	73.9 ± 4.0	73.1 ± 4.4		
					—	73.7 ± 3.1**	75.7 ± 2.1**		
14	153.1 ± 8.7	150.7 ± 5.1	153.6 ± 10.2	151.2 ± 4.0	143.4 ± 5.3	143.2 ± 6.7	143.3 ± 6.9	156.7 ± 3.4	151.54 ± 2.3
	41.5 ± 6.9	41.6 ± 5.4	41.9 ± 8.6	45.1 ± 5.9	32.2 ± 4.0	38.8 ± 6.0	43.1 ± 6.3	50.1 ± 6.1	48.3 ± 5.0
	73.0 ± 0.4**	72.2 ± 0.6**	—	—	71.5 ± 4.0	77.4 ± 4.2	78.3 ± 2.8	—	—
	—	—	83.0 ± 5.7**	81.3 ± 1.9	—	75.6 ± 3.1**	77.4 ± 4.4**	83.5 ± 3.2*	80.6 ± 2.0*

（续　表）

Age (years)	Tibetan					Andes			
	China Qinghai		India			Peru Nunoan		Chile Putre	
			Choglam Sar		Ladakh Males				
	Males	Females	Males	Females		Males	Females	Males	Females
15					150.3 ± 7.7	149.4 ± 9.3	147.0 ± 5.7		
					37.7 ± 6.9	44.5 ± 8.1	46.5 ± 6.8		
					75.0 ± 5.0	80.1 ± 7.0	77.4 ± 1.9		
					—	78.5 ± 5.0**	78.1 ± 5.1**		
16	164.4 ± 6.2	154.7 ± 4.8	160.0 ± 6.2	153.1 ± 4.7	157.9 ± 5.9	155.5 ± 5.8	149.1 ± 3.6	164.5 ± 6.1	152.8 ± 5.5
	52.7 ± 5.6	49.2 ± 5.1	51.2 ± 7.1	50.3 ± 4.1	43.9 ± 5.8	50.9 ± 4.8	51.7 ± 5.5	59.3 ± 5.8	52.8 ± 6.0
	81.3 ± 0.5	78.7 ± 0.6*	—	—	79.9 ± 4.7	84.4 ± 4.8	—	—	—
	—	—	86.2 ± 4.5**	81.9 ± 4.8**	—	82.1 ± 3.0**	79.9 ± 2.3**	86.3 ± 3.5*	80.2 ± 2.9**
17					161.1 ± 5.9	157.1 ± 6.8	148.0 ± 4.5		
					46.8 ± 6.0	51.0 ± 5.5	51.6 ± 4.6		
					80.9 ± 4.7	85.2 ± 3.9	—		
					—	83.2 ± 3.5**	79.9 ± 2.7**		

（续　表）

Age (years)	Tibetan					Andes			
	China Qinghai		India			Peru Nunoan		Chile Putre	
			Choglam Sar		Ladakh Males				
	Males	Females	Males	Females		Males	Females	Males	Females
18	166.6 ± 5.0	155.7 ± 6.2	164.5 ± 8.5	154.1 ± 6.0	163.7 ± 3.9	160.0 ± 6.0	148.9 ± 4.8	160.3 ± 4.0	151.2 ± 4.7
	55.8 ± 4.7	52.1 ± 5.9	63.9 ± 15.1	53.2 ± 8.54	51.9 ± 4.6	57.2 ± 4.3	52.9 ± 4.0	60.3 ± 0.6	58.3 ± 5.6
	82.5 ± 0.5	80.1 ± 0.7	—	—	82.8 ± 4.4	89.3 ± 3.5	—	—	—
	—	—	89.4 ± 3.9	87.3 ± 3.8**	—	84.2 ± 3.1**	80.4 ± 3.3	85.1 ± 2.2*	79.9 ± 1.9
19					167.8 ± 2.8	159.6 ± 5.0	148.1 ± 4.0		
					57.5 ± 2.5	58.1 ± 6.3	51.3 ± 6.2		
					84.8 ± 3.5	89.9 ± 3.4	—		
					—	85.2 ± 3.3**	80.3 ± 4.1*		
20	167.8 ± 5.9	157.3 ± 4.1	164.5 ± 8.5	154.1 ± 6.01		158.2 ± 5.3	147.2 ± 3.5		
	57.7 ± 6.0	50.7 ± 5.7	63.9 ± 5.13	53.2 ± 8.54		58.5 ± 4.0	52.6 ± 6.2		
	86.9 ± 0.8	81.9 ± 0.7	—	—		91.0 ± 3.2	—		
	—	—	89.4 ± 3.9	87.3 ± 3.8		84.3 ± 3.4**	79.1 ± 3.3**		

（续　表）

Age (years)	Tibetan					Andes			
	China Qinghai		India			Peru Nunoan		Chile Putre	
			Choglam Sar		Ladakh Males				
	Males	Females	Males	Females		Males	Females	Males	Females
21						161.2 ± 7.0	147.4 ± 5.6		
						60.5 ± 7.2	48.8 ± 6.2		
						92.2 ± 3.2	—		
						86.7 ± 3.7	77.5 ± 3.4**		
Altitude(m)	3200–4300		3521		3514	3860		3530	
Author	Weitz et al.		Tripathy et al.		Malik et al.	Leatherrman et al.		Dittmar	
Year	2004		2007		1978	1995		1997	

Four sets of data in boxes from top to bottom are stature, weight, chest circumference and sitting height, respectively. Comparing to Tibetan in the same sex and age group $^{*}P<0.05$; $^{**}P<0.01$.

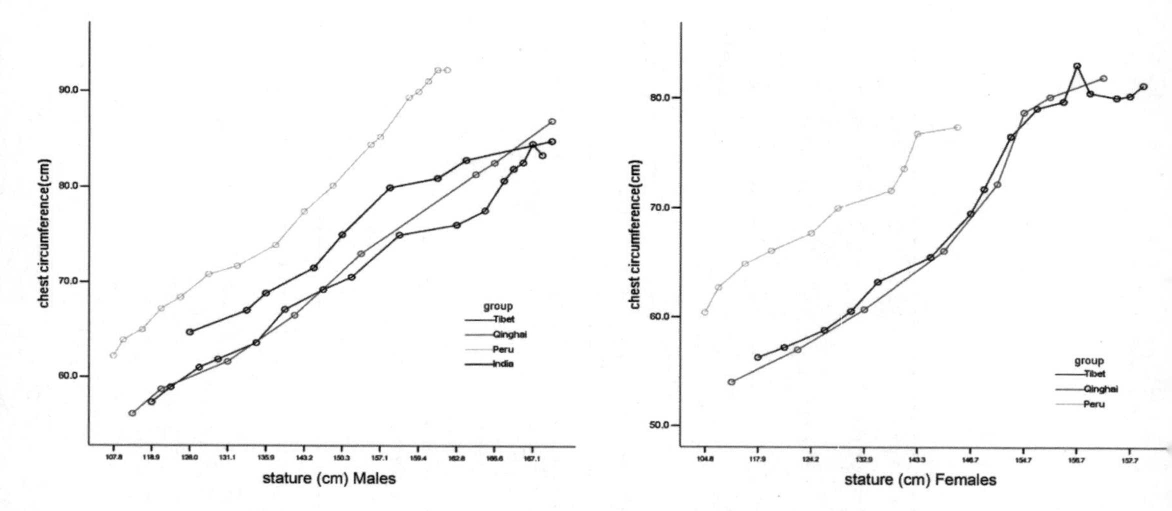

Fig. 1　Chest circumference with stature for children and adolescents at high altitude.

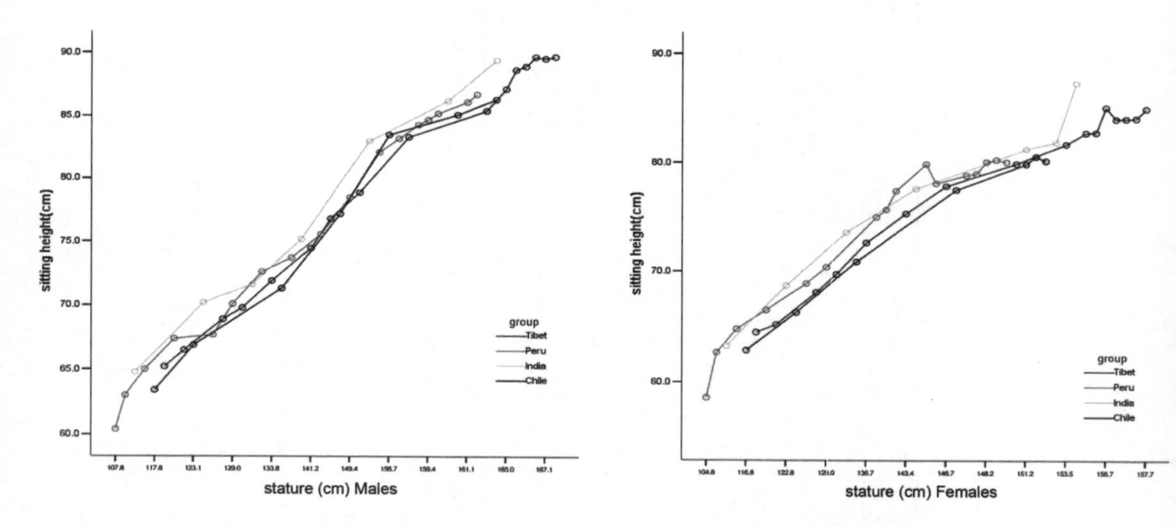

Fig. 2　Sitting height relative to stature for children and adolescents at high altitude.

undergone natural selection for different adaptive pathways in response to chronic exposure to high-altitude hypoxia over the course of millennia (Weinstein, 2005, 2007). Highland Tibetans have undergone selection for higher ventilation rates and lower hemoglobin concentration to enhance oxygen delivery (Beall, 2001). Highland Andean populations, in contrast, have responded to highaltitude hypoxia through selection for enlarged chests relative to stature (Beall, 1982, 2001; Beall et al., 1977; Brutsaert et al., 2004; Frisancho et al., 1997; Greksa et al., 1989). Morphological changes of the chest in response to high-altitude hypoxia were formed through a long evolutionary process. Studying Andean prehistory, Weinstein (2007) found that Atacama highlanders have larger sterna and clavicle proportions, and ribs with the largest area and least curvature in comparison to lowlanders. In addition, highlanders have an antero-posteriorly deep and medio-laterally wide thoracic skeleton, and lowlanders exhibit narrowed and shallower chests, which indicates that an environmental stressor shaped the biology of the highland Andean. Thus, an enlarged chest, as measured by chest diameters, is probably a genetically determined trait among Andeans in response to high-altitude hypoxia and cannot be acquired through processes of developmental adaptation or adult acclimatization (Frisancho et al., 1997; Greksa et al., 1989).

Recent genomic studies also have indicated that the largely distinct genetic adaptation patterns between Tibetan and Andeans reflect different patterns of adaption to altitude hypoxia (Bingham et al., 2010; Jay, 2010; van Patot, 2011). Molecular biology studies indicate that HIF−1a, EPAS−1 represent a key genetic adaptation to altitude in Tibetan populations (Simonson et al., 2010; Peng et al., 2011). The observed indicators of natural selection on EPAS1 and EGLN1 suggest that during the long-term occupation of high-altitude areas, the functional sequence variations for acquiring biological adaptation to high-altitude hypoxia have been enriched in Tibetan populations (Peng et al., 2011). These characteristics may be a consequence of earlier human occupation of the Tibetan plateau than the Andean Altiplano (Jay, 2010).

Our research indicated that relative sitting heights are similar among populations living at high altitude in the Andes or on the Qinghai-Tibetan Plateau. This may be related to the fact that children and adolescents living on the two plateaus are all faced with altitude hypoxia stress. Previous studies indicated that hypoxia retards growth and development of children and adolescents at high altitude (de Meer et al., 1995). A study conducted in 2010 indicated that stature, weight, sitting height and chest circumference of Tibetans were all smaller than the average values from nine low-altitude provinces in China (Zhang et al., 2010; Ji, 2007). Besides hypoxia, nutritional stress (Leonard, 1990), infectious diseases (Martorell, 1980) and poor socio-economic conditions (Frisancho et al., 1973) have been discussed as sources of growth retardation in highlanders. Analyzing families, ethnicity, living place and altitudes of four Andean countries during 1995–1998, Larrea (2002) found that the rate of growth retardation among children and adolescents was three times less in the best socioeconomic district than that in the worst socioeconomic district. The important contribution of poor socio-economic conditions to growth retardation at high altitude has been noted by studies in other populations (de Meer et al., 1995; Haas et al., 1982; Greksa 1986, 2006; Stinson, 1982; Tripathy et al., 2007; Weitz et al., 2004).

While Andean trunk length may be influenced by genetic ancestry (Stinson, 2009), it is becoming increasingly clear that body shapes are also influenced by nutritional effects on linear growth (Frisancho, 2007). Because relative sitting height is indicative of stress during growth, similar sitting-height-to-stature ratios among Andeans and Tibetans may reflect similar living conditions. Small differences in sitting height among children and adolescents in different regions of Tibet may be the result of differences in local socioeconomic status.

In summary, during growth and development at high altitude, the plasticity of morphology can be simultaneously affected by multiple factors, such as climate, ecology, nutrition, heredity, social economics, and culture, but hypoxia may have contributed both to the longer growth of the body as well as the shape of the chest as an adaption to low oxygen levels. Our study finding is consistent with previous publications on late and retarded development and growth. The finding of different patterns of chest enlargement in response to hypoxia in different populations suggests that different morphological adaptations occurred during human evolution in facing the hypoxia challenge.

ACKNOWLEDGMENTS

The authors are grateful to all volunteers, hospitals and schools in Tibet and their colleagues, post graduate students in Liaoning Medical University for help in this study. They thank Zhaohong and Dr Hurong for collecting some literatures.

LITERATURE CITED

Baker PT. 1969. Human adaptation to high altitude. Science 14: 1149–1156.

Beall CM. 2001. Adaptations to altitude: a current assessment. Annu Rev Anthropol 30: 423–456.

Beall CM. 1982. A comparison of chest morphology in high altitude Asian and Andean populations. Hum Biol 54: 145–163.

Beall CM, Baker PT, Baker TS, et al. 1977. The effects of high altitude on adolescent growth in Southern Peruvian Amerindians. Hum Biol 49: 109–124.

Bigham A, Marc B, Dalila P, et al. 2010. Identifying signatures of natural selection in Tibetan and Andean populations using dense genome scan data. Natural Selection of high altitude. PLoS Gene 6: 1–14. www.plosgenetics.org.

Brutsaert TD, Parra E, Shriver M, et al. 2004. Effects of birthplace and individual genetic admixture on lung volume and exercise phenotypes of Peruvian Quechua. Am J Phys Anthropol 123: 390–398.

de Meer K, Heymans HSA, Zijlstra. 1995. Physical adaptation of children to life at high altitude. Eur J Pediatri 154: 263–272.

Dittmar M. 1997. Linear growth in weight, stature, sitting height and leg length, and body proportions of Aymara school-children living in a hypoxic environment at high altitude in Chile. Z Morphol Anthropol 81: 333–344.

Frisancho AR. 1969. Human growth and pulmonary function of a high altitude Peruvian Quechua population. Hum Biol 41: 365–379.

Frisancho AR, Baker PT. 1970. Altitude and growth: a study of the patterns of physical growth of a high altitude Peruvian Quechua. Am J Phys Anthropol 32: 279–292.

Frisancho AR, Velasquez T, Sanchez J. 1973. Influence of developmental adaptation on lung function at high altitude. Hum Biol 45: 583–594.

Frisancho AR, Frisancho H, Albalak R, et al. 1997. Developmental, genetic, and environmental components of lung function at high altitude. Am J Hum Biol 9: 191–203.

Frisancho AR. 2007. Relative leg length as a biological marker to trace the developmental history of individuals and populations: growth delay and increased body fat. Am J Hum Biol 19: 703–710.

Greksa LP. 1986. Growth patterns of European and Amerindian high altitude natives. Curr Anthropol 27: 72–74.

Greksa LP, Beall CM. 1989. Development of chest size and lung function at high altitude. In: Little MA, Haas JD, editors. Human population biology: a transdisplinary science. New York: Oxford University Press. pp 222–238.

Greksa LP. 2006. Growth and development of Andean high altitude residents. High Alt Med Biol 7: 116–124.

Haas JD, Moreno-Black G, Frongillo EA Jr, et al. 1982. Altitude and infant growth in Bolivia: a longitudinal study. Am J Phys Anthropol 59: 251–262.

Jay FS. 2010. Genes for high altitudes. Science 329: 40.

Ji C. 2007. Health of children and adolescents. People Health Publisher: Beijing (in Chinese).

Larrea C, Freire W. 2002. Social inequality and child malnutrition in four Andean countries. Rev Panam Salud Publica 11: 356–364.

Leatherman TL, Carey JW, Thomas RB. 1995. Socioeconomic change and patterns of growth in the Andes. Am J Phys Anthropol 97: 307–321.

Lohman TG, Roche AF, Martorell R. 1988. Anthropometric standardization reference manual. Human Kinetics Books: Champaign, Illinois.

Leonard WR, Leatherman TL, Carey JW, et al. 1990. Contributions of nutrition versus hypoxia to growth in rural Andean populations. Am J Hum Biol 2: 613–626.

Malik SL, Singh IP. 1978. Growth trends among male Bods of Ladakh—a high altitude population. Am J Phys Anthropol 48: 171–175.

Martorell R. 1980. Interrelationships between diet, infections disease, and nutritional status. In: Greene LS, Johnston FE, editors. Social and biological predictors of nutritional status, physical growth and neurological development. Academic Press: New York. pp 81–101.

Mueller WH, Schull VN, Schull WJ, et al. 1978. A multinational Andean genetic and health program: growth and development in a hypoxic environment. Ann Hum Biol 5: 329–352.

Miklashevaskaia NN, Solovyeva VS. and Godina EZ. 1973. Growth and development in high altitude regions of Southern Kirghizia, U.S.S.R.Field Reseach Projects. Miami, Florida.

Palomino H, Mueller WH, Schull WJ. 1979. Altitude, heredity and body proportions in northern Chile. Am J Phys Anthropol 50: 39–50.

Pawson IG. 1976. Growth and development in high altitude populations: a review of Ethiopian, Peruvian, and Nepalese studies. Proc R Soc Lond B 194: 83–98.

Peng Y, Yang Z, Zhang H, et al. 2011. Genetic variations in Tibetan populations and high-altitude adaptation at the Himalayas. Mol Biol Evol 28: 1075–1081.

Simonson TS, Yang, Y, Huff CD, et al. 2010. Genetic evidence for high-altitude adaptation in Tibet. Science 329: 72–75.

Stinson S. 2009. Nutritional, developmental, and genetic influences on relative sitting height at high altitude. Am J Hum Biol 21: 606–613.

Stinson S. 1982. The effect of high altitude on the growth of children of high socioeconomic status in Bolivia. Am J Phys Anthropol 59: 61–71.

Stinson S, Frisancho AR. 1978. Body proportions of highland and lowland Peruvian Quechua children. Hum Biol 50: 57–68.

Tibet Autonomous Region Bureau of Statistics. 2007. Tibet Autonomous Region Statistics Yearbook. (in Chinese)

Tripathy V, Gupta R. 2007. Growth among Tibetans at high and low altitudes in India. Am J Hum Biol 19: 789–800.

van Patot MC, Gassmann M. 2011. Hypoxia: adapting to high altitude by mutating EPAS-1, the gene encoding HIF-2 α. High Alt Med Biol 12: 157–167.

Weinstein KJ. 2005. Body proportions in ancient Andeans from high and low altitudes. Am J Phys Anthropol 128: 569–585.

Weinstein KJ. 2007. Thoracic skeletal morphology and high-altitude hypoxia in Andean prehistory. Am J Phys Anthropol 134: 36–49.

Weitz CA, Garruto, RM, Chin C-T, et al. 2004. Morphological growth and thorax dimensions among Tibetan compared to Han children, adolescents and young adults born and raised at high altitude. Ann Hum Biol 31: 292–310.

Wu R, Xinzhi W, Zhenbiao Z. 1984. Anthropometric methods (Chinese). Science Publisher: Beijing.

Zhang Y, Hui L, Chengye J. 2010. Study on body proportions of 0–16 years old children and adolescents in China. China J Evid Based Pediatr 5: 349–354. (in Chinese)

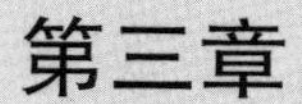

第三章

格沙致理万古器

2019年人类学终身成就奖获得者

——陈　淳

人类学生涯回顾

老三届的身世

我今天能够在人类学和史前考古学上做出一些成绩全属运气，在变幻无常的平凡生活中抓住了转瞬即逝的机会，尽了自己的力量。自己的成功轨迹并无目的和方向可言，似乎也不取决于天资禀赋，因为影响个人人生轨迹的因素和机会过于复杂。相较于现在年轻人发展的多种机会和选择，老三届一代人基本只能听天由命。作为“文革”十年被荒废的一代，自己今天有幸能在无数落魄知青中跻身教授之列，并在这里写一篇成就回顾纯属偶然。与生物进化一样，个人的命运也是物竞天择、社会选择的结果。

我于1948年5月3日出生在上海，祖籍浙江奉化。父亲很早就来上海读书，起先就读于教会学校圣方济学院，后来从东吴大学法学院毕业。他的英语非常好，一生从事英语教学工作。外祖父家境优裕，所以母亲算是小家碧玉，在家乡读到初中毕业，从未参加过工作。从我懂事开始，就住在虹口区四平路的新陆村。这是一片典型的日式砖木结构连体住宅小区，1940年代由日本人营造，名叫“高风寮”，供侵华日军佐级军官居住。抗战胜利后成为上海师范专科学校和新陆师范学校的教师宿舍，改称“新陆村”。新中国成立后，成为虹口中学的教师宿舍。因为当时父亲在虹口中学执教，我从小就住在这里，而且一住就是三十多年。不久前新陆村拆迁，难免令人感怀，有人便钩沉抉微，梳理出新中国成立前后在此住过的不少著名文人学者。比如中国现代象征派诗人戴望舒、莎士比亚专家复旦大学教授孙大雨、复旦大学中文系教授

童年照

吴文琦、华东师范大学中文系教授和作家施蛰存、华东师范大学教授钟钟山、翻译家周煦良、上海师范大学教授和历史学家程应镠、复旦大学历史学家邱汉生等。可惜，当这些文人住在新陆村时，我还蒙昧无知，甚至还没有出生。唯独对住在斜对面的孙大雨教授印象颇深，常见他在二楼窗口探出微胖的身躯向下张望，住在这里的人都知道他是翻译莎士比亚作品的大家。

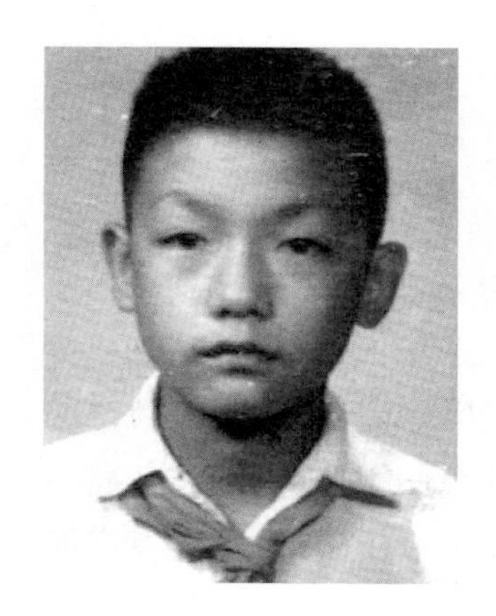

小学毕业照

我的小学和初中年代平淡无奇，乏善可陈。我是那种不很用功，成绩一般，偏好文科而数理化较弱的学生。记得五年级时，有一次语文老师把我叫到办公室，让我把自己的作文抄在一块小黑板上，他上课专门作为样板用来点评，但我当时实在不知道这篇作文好在哪里。我喜欢画画，父母和老师认为我画画有点天赋，所以买了《芥子园画谱》和齐白石的画来临摹，潜心学习山水画和花鸟画的传统技法。初中毕业，在美术老师的积极鼓动下，我报考上海美术学校，结果在考试中遭遇滑铁卢。因为当场拿到的题目是根据“上海的早晨”创作一幅画，这使得山水花鸟的训练毫无施展余地。这真可谓一题选人才，一考定终生，如果当时碰巧临场发挥正常，那么我就不会有写这篇文章的机会了。

后来我高中进了北虹中学，校址就是原来的圣方济学院，校门对面是座教堂。我属于1967届的高中生，高二那年夏天，学校组织大家下农村参加双抢（抢收、抢种）。虽然“文革”是1966年5月16日开始的，但要到8月18日红卫兵进京接受检阅才变成“群众运动”，我们从农村被叫回学校闹革命。上海也举行了大规模的火炬游行，记得游行的那天晚上下大雨，我们从学校列队拿着火炬前往人民广场集会。不能撑伞，浑身淋得湿透，一路上有人供应姜茶。在灯光暗淡的人民广场排好队后，见市长曹荻秋出来，说了几句话后就匆匆离去。没过几天，曹荻秋就被打倒。

可能上海的中学里高干子弟比较少，所以“文革”开始时不像北京那么血腥。基本上写写大字报就算投入了，没有听说学生揪斗和侮辱老师的情况，也没有见打砸抢。只听说其他班级有位激进的同学到社会上去造反，不幸从高楼上失足殒命。非工农和干部出身的学生也无缘参加什么造反派组织。我在"文革"期间基本上是所谓的"逍遥派",没有参加过红卫兵。所以,能在家里看点书,跟着父亲学习英语(学校里教的是俄语)。但是,"文革"中家家户户都免不了担惊受怕。记得家里有不少很珍贵的字画,虽然当时不是很懂,但是对一幅八大山人的墨迹印象很深,在破四旧中都偷偷销毁了,非常可惜。家里藏书不少(主要是古文的线装书和外文书),所以在无"书"可读的年代,还能够有自习的资源。

1968年10月,在"逍遥"了两年之后,我被分配到上海杨浦区长阳路上的一家铸造厂当工人。12月22日,毛主席发出"知识青年到农村去,接受贫下中农的再教育,很有必要"的指示。当时,还没有分配的学生以及后来毕业的学生全部都上山下乡,所以我比较幸运地留在了上海。铸造厂就是做翻砂工,又脏又累,矽肺是常见的职业病。厂领导根据知识越多越反动的标准,认为高中生比初中生更需要吃苦锻炼,所以除一名女生外,我和其他两名高中生全部被分到最辛苦的浇铸班。每天上夜班,两人搭档,一前一后,扛着一百多斤的铁水,在车间里来回穿梭,把白天翻砂师傅做好的模子全部浇完为止,要忙到深夜才下班。这种高强度的工作整整做了十年。

"文革"期间所有杂志停刊,出版业凋零,科学文化园地一片荒芜。1971年7月,郭沫若给周恩来总理写报告,请求《考古学报》、《考古》和《文物》杂志复刊,并得到同意批复。1972年,《化石》杂志试刊,1973年创刊,是全国唯一的一本科普杂志。《化石》是由中国科学院古脊椎动物与古人类研究所(以下简称古脊椎所)主办的半年刊。我从创刊号开始订阅,还订阅了两年《考古》与《文物》杂志。当时的出发点完全是消遣,不料这几本杂志居然影响到我后来的人生轨迹。由于当时文化生活贫乏,所以《化石》杂志广受欢迎,创刊号发行5 000份,到1975年增加到了51 150份,民间粉丝极多。我到古脊椎所后听到过一则趣闻,说某位同仁违反交规被警察拦住,询问是什么单位的,说是《化石》杂志的,警察反赔笑脸,马上放行。

1973年第1期《化石》刊登了周国兴先生"现代的猿能变成人吗?"一文。1975年第1期《化石》发表了武汉一位工人和一位中学生的一篇商榷文章——"我们对'现代的猿能变成人吗'一文的看法",批评周国兴先生有关环境决定论的偏颇。当时极左思想流行,这种文章无非就是找岔子,扣帽子,很难说有什么学术性。于是,我也写了篇文章"怎样理解人类起源中的'特定的环境条件'",并于1975年第2期刊出。这篇文章让我和周国兴先生建立起良好的关系，并一直得到他的大力帮助和扶携。通过周国兴先生，我也认识了上海自然博物馆徐永庆、黄象洪、陈翁良和林嘉旺等人类学专家。

《化石》杂志还有一段掌故。1975年7月23日毛主席做了白内障手术后可以阅读书刊了，作为消遣要求看《古脊椎动物学报》和《化石》。于是，《化石》编辑张锋专门为他编排大字繁体版的《化石》。由于办普及读物在科学院并不受重视，所以张锋借此机会给毛主席写了封诉苦信，希望在科学普及工作上得到中央的支持。9月16日，毛主席对张锋的信做了批示，并与张锋的信件一起印发给中央各部委负责人，最后在胡耀邦的安排下得到解决。当时，突然有《文汇报》编辑向我约稿，要求写一篇介绍《化石》杂志和人类起源方面知识的文章，这才侧面了解到有这样一个背景。写科普文章也逼我花很多时间来补习相关的专业知识。《化石》编辑部的两位编辑刘后一和张锋两位先生常到上海开会，也曾约我去宾馆见面，相谈甚欢。刘后一后来为我考研提供了非常有用的建议和指点。

1976年"文革"结束，1977年恢复高考，1978年开始恢复研究生招生。我便向刘后一先生询问考研的可行性，他建议我考贾兰坡先生的旧石器时代考古方向，就专业知识方面，相对于古人类和古脊椎动物较为容易。他也代我向贾先生转达了我准备考他研究生的意愿。在"文革"进行了四年之后的1970年，北大、清华等高校开始招收工农兵大学生，是根据政治条件和表现保送。所以，1977年恢复高考和考研时候，考生的学历很乱，所以只能以同等学力来对待。

考研的参考书是吴汝康等编撰、1978年1月刚出版的《人类的进化》一书和《十万个为什么》(人类卷)。两本都是科普读物，现在研究生听到当年用的考研参考书不免大为惊讶，好似太过容易。当时确实没有可供学习和考研的专业教材和参考书，甚至连一本考古学通论的教材都没有。记得20世纪80年代初，有人要报考复旦大学文博专业研究生班，向我请教相关的参考书，我只能提供北京大学50年代初油印的几本考古内部教材，内容也十分简单。

考研报名照

所以考研从这些参考书来掌握人类进化和旧石器考古的基础知识不算很难，主要的障碍是英语，因为无法速成，好在我一直坚持学习。那年有75人报考贾先生的研究生，最后只录取了我一个人。这么多考生中不少是专业基础比我好得多的工农兵学员，而我能够考上，可能占了英语成绩的便宜。后来听师母说，恢复研究生招生时，贾先生因“文革”中受苦，拒绝再带学生，后来经研究所领导反复动员才勉强答应。到北京参加面试，在贾先生办公室参加面试的好像有李炎贤、黄慰文和邱中郎等几位专家。面试聊到最后，贾先生似乎挺高兴，还特地问我，“你字写得不错，怎么学的？”我告诉他小时候临颜真卿的《多宝塔碑》，后来临文徵明的《滕王阁序》，他听了很高兴的样子。他告诉我，学习旧石器并不难，主要需掌握地质学和古脊椎动物等专业知识。

1979年与首届古脊椎所研究生

跟贾兰坡先生读研

到古脊椎所读研，第一年在中国科技大学研究生院（后来叫中国科学院研究生院，现在改为中国科学院大学）学习基础课程，当时的临时地址在五道口附近的北京林学院旧址，集中了在北京的中国科学院各研究所的研究生。第一届入学研究生的年龄差距很大，最大的已近四十岁，是“文革”前中断学业的研究生，最小的刚中学毕业，二十岁出头。当时研究生

在贾兰坡先生办公室

院的院长是严济慈先生，李佩女士则是我们的英语老师。我选的专业基础课程包括动力地质学、第四纪地质学、地貌学、沉积学、古脊椎动物学等，都由北大地质系的老师授课。

周末，我常去贾先生的家看望他。他和师母住在祁家豁子古脊椎所原址的旧宿舍楼里，面积20多平方米的一间斗室，堆满了东西，烧饭都在走廊里。在二里沟家属楼完工后，他才搬进比较宽敞的住房里安度晚年。

研究生院学习结束就回到研究所学习专业知识和准备论文。研究所的学习比较特殊，基本上就是师徒传授，并不开课，也没有人管。我常跟师兄卫奇先生跑野外，石器上有疑问则向黄慰文老师请教，他们是贾先生的两大弟子。有时自己也会到资料室借阅资料，自学旧石器考古的基本知识。为了看懂旧石器，我拿着一本本考古报告，一个人钻在标本室里对照报告中的分类描述观察具体标本。最后，把标本室里的标本全部看了一遍，基本领悟了旧石器的分类、描述和分析要点。标本室还有裴文中先生从法国带回的旧石器标本、1927—1932年中瑞西北考察团袁复礼教授在戈壁沙漠采集的石器标本，以及美国考古学家赠送的中美洲史前人类打制的柱状石核、石叶标本。所以师傅领进门，修行靠自身，此话不虚。

贾先生接待外国学者时常让我陪同当翻译，让我帮他翻译回复外国学者和友人的信件。所以，这也给了我了解和认识这门学科的国际的机会。贾先生书房里的专业书籍很多，大多是世界各国同行的赠书，需要时也向他借阅。他也保留着周口店的发掘记录，这些记录是当年日本人占领协和医院时，在日本兵岗哨眼皮下一点点偷运出来的。他有保持信件档案的好习惯，他给我看过20世纪30年代裴文中在法国留学时给他邮寄的明信片，还有国外学者和政要与他交流北京人化石下落信息的来往信件。

我还和贾先生登过一次山，那年他应该已经有72岁。当时北京的报纸报

与贾老、黄慰文和欧阳志山

道，在房山县周口店西面不远的上方山云水洞发现了动物化石，而且有人将洞里找到的化石拿到古脊椎所让他过目。由于离周口店很近，所以怀疑是否会有古人类活动的遗迹。于是，贾先生决定亲自前往勘探。云水洞的位置很高，要爬一段260多级的陡峭云梯。他撑着拐杖，慢慢地向山上攀登，路上游人见了年龄这么大的老人还来登山，无不惊叹。到了云水洞看了化石的堆积情况，大多是很小的动物化石，看似啮齿类的骨骼，甚至有的还未石化，因此没有什么价值。现在回想起来，从古人类适应的环境条件来看，周口店猿人洞山前海拔只有50米左右，而云水洞海拔有860米。古人类的生存流动性很大，其活动范围取决于水源和食物的可获性。高山是可食生物量很低的生态环境，他们是不大可能爬到这么高的地方来觅食或活动的。

跟随贾兰坡先生考察云水洞

1982年我从古脊椎所毕业，到复

观察苏州太湖三山岛旧石器

1985年在太湖三山岛发掘地留影

旦大学分校(后改为上海大学文学院)历史系任职，当时历史系的文物考古教研室有很强的师资力量。在此期间，我申请到了国家自然科学基金项目，与南京博物院的张祖方先生寻找苏南的古人类和旧石器遗址。当时在太湖三山岛刚好发现了哺乳动物化石，现场的张先生打电话让我马上赶去一起调查，看看是否有古人类活动遗迹。岛上的石灰岩裂隙出土了许多大型哺乳类化石，其中有猕猴、棕熊、黑熊、狗獾、猪獾、虎、豪猪、猞猁、野猪、水鹿、斑鹿和牛等。我根据“当地有较好燧石”这一线索，在小岛一隅的清风岭发现一处旧石器地点，并在1985年12月进行了发掘。小岛只有1.6平方千米，不可能维持这么一个动物群乃至人类的生存。打制石器和动物群的发现和相关地质背景的研究表明，太湖和小岛的形成是比较晚近的事件。

1984年，我收到何传坤先生来信，说美国考古学家路易斯·宾福德准备来华访问，会见贾兰坡先生商谈合作事宜。最后，宾福德的访问因与贾先生意见发生冲突以不欢而散而告终。一方面这与宾福德对中国的国情一无所知有关；另一方面这次访华的失败也许是一种必然，因为中西考古学在认识论和方法论上都有巨大的鸿沟。宾福德采用演绎法，想根据埋藏学和动物考古学的原理来审视和检验过去对北京人之家的解释，而贾先生觉得宾福德对北京人的观点全凭自己想象，并带有否定中国学者成果的不良企图。

我后来在学习和介绍国际考古学界的理论方法时，深深感受到这种隔阂的无奈。虽然中国考古学是在20世纪初从西方传入的，但是在经过几十年的封闭发展之后已被我国自身的传统所同化和固化，其最大的特点就是成了历史学证经补史的工具。结果，这种被同化和固化的考古学范式在80年代重新与西方考古学相遇时，却将对方视为异类。由于不了解国际考古学范式变化的历史背景，于是在相当长的一段时间里，我国不少学者视美国新考古学为旁门左道而予以抵制，坚持认为我们从事的才是真正的和最纯粹的考古学。

第二年，作为“文革”后中美考古学界之间的初步交流，我和几位考古学者作为宾福德和何传坤先生的客人参加了1985年4月在美国丹佛召开的美国考古学会第50届年会，并参观了加州大学伯克利分校的人类学博物馆和怀俄明大学人类学系，了解到美国考古学家的最新工作。正是通过这次访问，我了解到了美国的新考古学，知道了考古学方法除了类型学之外还有埋藏学、民族考古学和中程理论，深切感受到中外考古学研究的显著差距，也使我萌生了到北美留学的念头。

麦吉尔大学的留学生涯

在复旦大学分校工作期间，有一次接待的加拿大客人是麦吉尔大学的井川史子和蒙特利尔大学菲利普·史密斯夫妇。他们都是国际知名的考古学家，两位都从哈佛大学毕业。史密斯教授是加拿大皇家学会的会员，井川史子教授时任麦吉尔大学的副校长、东亚考古学学会主席，专攻东亚旧石器时代考古。他们访问后不久，我向井川先生提出去麦吉尔留学的意向，她表示

加拿大学者菲利普·史密斯访问复旦大学分校

麦吉尔大学留影

欢迎。在办妥了申请手续之后，我于1986年10月前往蒙特利尔，开始了自己的留学生涯。

适应国外的留学和生活环境是一种巨大的挑战，虽然自己有点英语读写的基础，但离运用自如的水平还相差很远，特别在听和说的方面捉襟见肘。因此，开始的一年基本是在适应语言和学习的环境。井川先生让我旁听特里格先生的"考古学思想史"一课，这是为本科生开设的一门基础课。由于对于课程的背景知识几乎是空白，所以听起来就非常吃力，于是买了录音机把上课内容录下来课后复习。其他的课程阅读量很大，所以除了上课以外就是看资料和文献。由于受训背景不同，老师讲课不会介绍基本的概念、定义和背景，所以有些学术思想和概念的来龙去脉只能在阅读和向同学询问当中慢慢领会。特别是北美考古学强烈的人类学取向，考古学理论受文化人类学理论的影响很大。当时，我对美国新考古学兴起的背景一无所知。为什么叫过程考古学？它和传统考古学有何不同？为何考古学文化已经不再流行？如何研究人类的行为？对于这些方面的知识，我都是从头开始学习。起步阶段读得十分艰苦，对于有些问题完全是一头雾水。在一段时间的积累和适应之后，才慢慢知道事情的来龙去脉，这才算渐入佳境。在对考古学理论和思想史的了解上，特里格教授是我的启蒙老师。他为人非常低调与和蔼，不喜欢参加各种会议，他是一位高产学者，在国际考古学界享有盛誉，对我后来的学术走向影响很大。

另外令人获益良多的是跟迈克尔·比森教授学习打制石器的研究和实验考古。比森教授是美国人，从加州大学圣巴巴拉分校博士毕业，专攻欧洲旧石器时代考古和非洲铁器时代考古。他曾经参加过以色列塔邦洞穴的考古发掘，这是一处旧石器时代中期著名的尼安德特人文化遗址。他讲授旧石器的类型学和技术分析方法，并在每周二晚上在系会议室向同学们指导打制石器。他从美国和加拿大安大略省运来许多黑曜石和燧石作为实验打制的原料，我跟他学习打制手斧、勒瓦娄哇石片和压制箭镞。他很耐心地指点打片的技巧，特别遇到石核或毛坯不顺手时，讲解如何处理的方法，比如选择棱脊走向和打击点、处理和加工台面、用力的方向和轻重等一些关键点。这种经验传授对于掌握打制的技巧十分重要，学习石器打制技术不是随便拿着一块石头胡乱敲砸就能学到的，这也是一门需要师承的精致手艺。对于史前人类来说肯定也是如此，有的技法如两面器和石叶技术应该是古人类上万年剥片经验言传身教的积累。

我的博士论文写的是更新世末细石叶技术从东亚向北美的扩散。在硕士论文的基础上，我把北美的细石叶材料也纳入比较和分析的范畴。为了观察和分析北美的考古材料，我参观了阿拉斯加大学、西蒙菲莎大学和俄勒冈大学的人类学系的考古藏品。我的博士论文还是比较传统，采用了所谓的类型-技术分析，比较华北细石核与北美细石核之间类型和技术特点的异同，并参考了西伯利亚和日本的研究成果。我将东北亚和西北面的生态环境因素也考虑在内，将这类技艺精湛的打制工业，看作是流动性极大的狩猎群体对冰后期极地苔原环境的适应。

麦吉尔的博士生奖学金只能申请四年，后来两年我到同学在安大略省的考古公司从事合同考古学工作，算是半工半读。这项工作主要为基建进行野外调查、勘探和抢救考古发掘。这段经历让我了解到北美的文物保护现状、政策以及合同考古学的国际发展趋势。

世界考古学有三大战略性基石，分别为人类起源、农业起源和文明起源研究。我对人类起源部分的旧石器时代考古有了一定程度的了解，但是对农业起源和文明起源的国际现状不甚了了。于是，我利用留学的这段时间大量阅读和积累有关农业起源和文明与国家起源的著作与文献，特别是西方考古学界对这两个研究领域在理论上的发展和突破。我在这方面补课和下功夫，是因为考虑到回国后不一定能从事旧石器考古。有了这些方面的知识储备，我可以在这两个战略性研究课题上施展拳脚。

复旦大学的工作经历

我是1996年回国的，之前先通过了复旦大学的访问学者联系了复旦大学文博系，并告知贾兰坡先生这个消息。他听说我要回国，非常高兴，并鼓励我去古脊椎所工作。如果要搞旧石器考古研究，古脊椎所是不二的选择，于是我回国后先去北京看望了恩师，并由他出面与所领导谈了我的情况，希望研究所能作为人才引进。但是所里意见是只能作为一般进人的条件考虑，于是我来到了复旦大学。回国后进行的一项重要研究是对河北阳原县泥河湾小长梁遗址的发掘。这是古脊椎所陈万勇和汤英俊两位先生的国家自然科学基金项目，他们两位主要从事地质学研究，我应邀作为旧石器研究人员参加。意识到石器微痕分析的重要性，我邀请了多伦多大学人类学系毕业、后就职

与卫奇和沈辰在泥河湾

于安大略皇家博物馆的沈辰一起参加，因为他在硕士生阶段跟随美国考古学微痕巨擘乔治·奥德尔研习微痕分析。小长梁遗址的发掘在1998年夏季正式进行，发掘成果分两篇报告在《人类学学报》上发表，这是我首次采用国外旧石器的范式来研究我国的旧石器材料。可惜小长梁的埋藏环境是河湖相的二次堆积，没有人类活动的遗迹，如果是陆相的湖滨堆积，可能会发现早期人类生存和活动的迹象。小长梁是我国的早更新世遗址，距今136万年。它的发现是我国史前史的一座里程碑，它的名字被镌刻在北京中华世纪坛青铜甬道的第一阶，代表了中华文明的开始和人类在东亚的最早证据。虽然后来又有更早考古遗存的发现，但是其意义在于超越了基于周口店中更新世认知的旧石器时代年代学的重大突破。

小长梁遗址的研究成果在2000年4月于费城召开的美国考古学会第65届年会上以“中国更新世考古的理论和实践”为主题分会上做了介绍，得到了与会学者的好评。分会主持人之一的加拿大资深考古学家舒特勒教授在

1998年发掘小长梁遗址

2006年重访小长梁遗址

老中青三代泥河湾旧石器学者合影

评述中指出，“中国旧石器考古研究已经进入了利用现代技术和理论解释文化遗存的新阶段，中国旧石器考古学家已成功走上了运用现代考古技术和实验方法的研究道路”。我感到中国旧石器时代考古学已趋成熟，其前进的步伐是迅速的。国际著名考古学家、哈佛大学教授巴尔·约瑟夫也在点评中指出，“中国旧石器考古研究的层位学、年代学、石器工业技术等，长期以来难以摆脱早年研究的窠臼，而最新的研究试图纠正这种状况，打破前人的桎梏，为我们提供由国外和国内学者联手奉献的最新成果”。

2002年，浙江萧山跨湖桥遗址被评为2001年度的中国十大考古新发现。2005年，萧山市文化局和萧山博物馆为筹建跨湖桥遗址展示馆制定了信息展示型的方针，就是要以考古发现所蕴含的信息来介绍和表现跨湖桥遗址的文化内涵及其历史意义。我应邀承担了博物馆陈列文本的起草工作，感到对人地关系的细节方面做一些深入分析，提炼更多有关环境变迁和体现遗址科学价值的信息是值得的。跨湖桥遗址位于钱塘江口，为了了解全新世初大河河口与海岸线变迁对新石器时代早期人类生存环境的影响，我邀请了陈中原教授来共同研究这个问题。陈教授承担着更新世末以来全球大河口海岸线变迁的国际性课题，非常适合这项任务。陈教授的合作者宗永强教授也一同前往跨湖桥采样。宗教授是硅藻方面的专家，而硅藻正是海水和淡水交替最敏感的指示物。研究的结果与宗永强和陈中原教授一起署名，在2007年9月27

日的《自然》杂志上发表。论文通过花粉与非孢粉微生物化石对跨湖桥遗址古环境进行了深入观察，以高精度的时间尺度分析人地关系和环境变迁，为我们重建人类活动和遗址废弃过程提供了新的信息。

此外，我还写了许多文章介绍农业起源的理论，认为竞争宴飨理论比较适合解释水稻驯化的动力机制。同时，我也将农业起源与社会复杂化问题结合起来讨论，将文明起源置于农业经济的发展之上来予以考虑，并在郑建明博士的论文指导中，设法从人地关系和水稻栽培来分析长江下游良渚文明的兴衰以及宁绍平原与太湖流域文明发展的不平衡。考古材料分析表明，由于长江三角洲野生食物资源非常丰富，在跨湖桥、河姆渡、马家浜文化时期，人们主要还是利用野生资源，对于水稻栽培的重要性认识并不明显。精耕细作的稻作农业要到崧泽文化晚期才开始，表现为深耕和耘田工具的大量出现和野生资源物种和数量的明显下降。良渚文明这种大型复杂社会对剩余产品的需求显然远远超过自给自足的部落社会，野生资源已无法满足社会运转的需求，而酒类也可能成为社会祭祀活动不可或缺的消费品，于是强化粮食生产自然成为经济发展的一个趋势。在良渚文明崩溃之后，马桥文化又再次倒退到以狩猎采集为主的简单社会。长江下游文明大约经过了1 500年到越国的兴起才再次崛起，但是越国不像良渚那样是一种原生文明，而是受了中原文明刺激和影响的次生国家。

我在2001年申请到了国家教委文科科研基金的资助，从事“中国国家探源的理论探索”的研究课题。其间发表了多篇介绍国际文明探源的理论文章，最后形成了《文明与早期国家探源——中外理论、方法与研究之比较》一

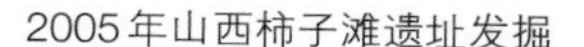

2005年山西柿子滩遗址发掘

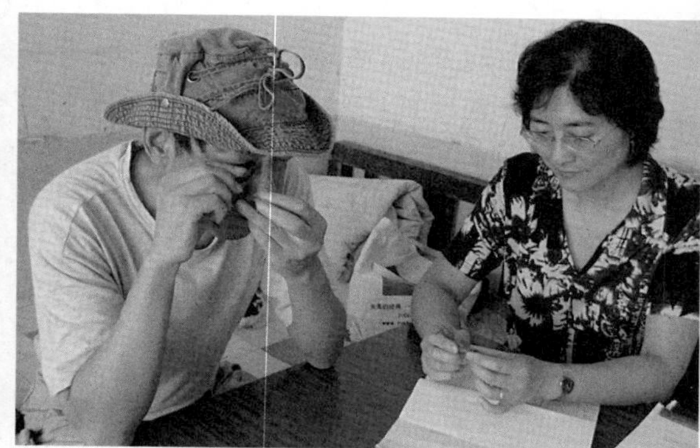

与安家瑗观察小南海石器

与陈星灿、刘莉和王益人在壶口

陪同加拿大学者克劳福德参观河姆渡遗址

与吴新智院士、李隆柱在萨拉乌苏

与体质人类学家刘武

书这一成果，并由上海书店出版社于2007年出版。我还获得上海哲学和社会科学学术著作出版基金资助，于2003年由学林出版社出版了论文集《考古学的理论与研究》，该书在学界产生了较大影响，荣获2004年上海市哲学社会科学优秀成果奖著作类二等奖，并在2005年中国文物报主办的十佳文博图书评选中被评为2003年最佳论著。后来，《中国文物报》刊登了该书的书评，题为“考古学的新视野”，给予了“高标独树、卓尔不群，在广大中青年学生中享有很高的声誉”的评价。2014年8月，该书在纪念上海市学术著作出版基金成立25周年之际入选《上海市学术著作出版基金25周年精选丛书》，由上海人民出版社出版。2003年，我因为尿毒症进行了肾移植手术。感谢现代医学科技之赐，给了我第二次生命，使得我的一些主要成果后来得以陆续完成。也是这一年，我被评为复旦本科教学名师。

从2006年开始，我开始和陈星灿一起主持《南方文物》杂志的"域外视野"栏目，每期提供两篇精选的翻译论文，介绍国外考古学的理论、方法与研究，为国内学界借鉴国外的研究成果提供了一个窗口。这些文章大多来自我的研究生课程，从学生阅读的一些国外学术经典中，挑选一些优秀论文让他们翻译成中文，在仔细校对之后再供杂志发表。我发现，这样的翻译锻炼对学生的阅读翻译能力以及学术水平的提高都有很大帮助。

2007年是英国考古学家戈登·柴尔德逝世五十周年。上海三联书店和我联系，希望我能够翻译出版他的系列著作。于是，我挑选了柴尔德五本有代表性的著作，包括《欧洲文明的曙光》《历史发生了什么》和《人类创造了自身》，于2008年出版。中国考古学目前采用的范式就是柴尔德在20世纪20年代初所建立的，但是大部分人对这个范式的背景、概念和操作并不十分清楚，因此介绍柴尔德的系列著作对于我们了解文化历史考古学的要义很有帮助。

2006年和2007年被评为优秀研究生导师。

2013年，我开始主持国家社科重大项目《外国考古研究译丛》。该项目出于引介国外考古学经典的目的，选取了欧美考古学界具有里程碑意义或国际影响力的五本著作，内容分别涉及：① 学科最新发展的整体介绍；② 考古学与族属的关系；③ 考古学田野工作与方法论的变革；④ 国家与文明起源研究；⑤ 农业起源研究。这五本著作的内容与我国考古学研究的发展密切相关，它们在国际学界的经典地位能够为我国考古学研究提供值得借鉴的他山之石。刘庆柱先生在为译丛写总序中说："译丛的五本书堪称在世界考古学范围内的相关学术领域的重要著作，其内容涉及考古学学科综述、聚落考古、农业起源、文明探源、民族身份考古五个方面，涵盖了当今世界考古最为重要、最为关注、最为前沿的学术问题"。五本著作已经由上海古籍出版社陆续出版，该项目也顺利结项。

此间，我的论文集《考古学研究前沿：理论与问题》入选国家哲学社会科学文库，2016年由北京师范大学出版社出版。这本书在2018年获第十四届上海市哲学社会科学优秀成果著作二等奖。

我于2018年5月正式退休，但是工作仍然照常进行。2020年我的一本书稿《从史前到文明》由中州古籍出版社出版，该书介绍了从人类起源到文明起源的史前史。我还整理和重译了何传坤先生的遗稿《柴尔德：考古学的革命》一书，已由中国人民大学出版社出版。商务印书馆也在2020年出版了我的《沙发考古随笔》，里面收集了我过去30多年来报刊上发表的随笔和短文，是我努力搭建中西学术沟通桥梁而尽的一点微薄之力。

叹人生之须臾，羡长江之无穷。回国以来的三十年里，深感国外考古学发展迅猛，自己知识日趋陈旧。始终想努力跟上时代潮流，多读些新的著作和文章，再做些不同的探索和研究。但作为时间的过客，虽有鸿鹄之志，已无鸿鹄之力了。

2016年首届中国考古大会颁发裴文中、贾兰坡奖

2016年上海考古大会访伦福儒

2018年参观石峁遗址，与发掘者邵晶合影

2018年参观宁夏考古所姚河源遗址

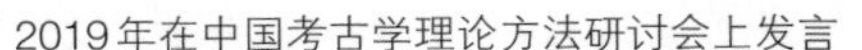

2019年在中国考古学理论方法研讨会上发言

2019年9月柳州白莲洞新馆开馆仪式与周国兴先生合影

代表性论著

(一)代表著作

[1] 贾兰坡,陈淳.中国猿人[M].上海：上海科技教育出版社,1998.
[2] 陈淳.考古学的理论与研究[M].上海：学林出版社,2003.
[3] 陈淳.考古学的理论与研究[M].上海：上海人民出版社,2014.
[4] 陈淳.当代考古学[M].上海：上海社会科学院出版社,2004.
[5] 陈淳.考古学理论[M].上海：复旦大学出版社,2004.
[6] 陈淳.考古学理论(修订版)[M].上海：复旦大学出版社,2015.
[7] 陈淳.文明与早期国家探源[M].上海：上海书店出版社,2007.
[8] 陈淳.考古学研究入门[M].北京：北京大学出版社,2009.
[9] 陈淳.远古人类：我们是猿人的后裔吗[M].上海：上海科学普及出版社,2015.
[10] 陈淳.考古学前沿研究：理论与问题[M].北京：北京师范大学出版社,2016.
[11] 陈淳.从史前到文明[M].河南：中州古籍出版社,2020.
[12] 陈淳.沙发考古随笔[M].北京：商务印书馆.2020.

(二)论文

[1] 陈淳,郭璐莎.文明探源的经典之作——塞维斯《国家与文明的起源》推介[N].中国文物报,2019-9-13.
[2] 陈淳.农业起源的考古学探索——从安诺遗址到圭拉那魁兹洞穴[J].文物春秋,2019,4：3-12.
[3] 陈淳.美国的文化资源管理与公共考古学[C].//曹兵武,赵夏,何流主编.他山之石：国际文物保护利用理论与实践.北京：文物出版社,2019：115-126.
[4] 陈淳.国家探源的他山之石[N].中国社会科学报,2019-8-5.
[5] 陈淳.科学探寻夏朝与最早中国[N].中国社会科学报,2019-6-10.

[6] 陈淳.考古学理论：回顾与期望[M]//中国考古学会编.中国考古学年鉴2016.北京：文物出版社,2017.
[7] 陈淳.戈登·威利《秘鲁维鲁河谷的史前聚落形态》述评[J].考古学报,2018,2：287-298.
[8] 陈淳,张萌.旧石器时代考古与栖居及生计形态分析[J].人类学学报,2018,37(2)：306-317.
[9] 陈淳,张萌.细石叶工业研究的回顾与再思考[J].人类学学报,2018,37(4)：577-589.
[10] 陈淳.酋邦概念与国家探源——埃尔曼·塞维斯《国家与文明的起源》导读[J].东南文化,2018,5：19-25+128.
[11] 陈淳.从考古学理论方法进展谈古史重建[J].历史研究,2018,6：4-20+188.
[12] 陈淳,张萌.西学中用与时俱进——陈淳教授访谈录[J].南方文物,2017,3：23-36.
[13] 陈淳.远古人类的世界——没有文字的考古学探索领域[N].文汇学人,2017-08-25.
[14] 陈淳.丁村遗址六十年与旧石器考古范式的变迁[C]//山西省考古研究所编.砥砺集——丁村遗址发现60周年纪念文集.山西：三晋出版社,2017：37-51.
[15] 陈淳.文化与族群[N].中国文物报,2016-01-26.
[16] 谢银玲,陈淳.考古学文化功能研究的战略性起点——戈登·威利的《秘鲁维鲁河谷的史前聚落形态》译介[J].东南文化,2015,4：6-14.
[17] 陈淳.谈考古研究的问题意识与阐释[J]//复旦大学文物与博物馆学系,复旦大学文化遗产研究中心编.文化遗产研究集刊.第7辑.上海：复旦大学出版社,2015.
[18] 陈淳.继往开来的鸿篇巨制[N].中国文物报,2014-10-17.
[19] 陈淳.谈考古学阐释[N].中国文物报,2014-9-12.
[20] 陈淳.考古学的昨天与今天[J].科学,2014,66(3)：35-39+4.
[21] 陈淳.谈考古学的一般性和特殊性研究——《考古学：理论、方法与实践》中文第二版译后记[J].南方文物,2014,2：5-9+14.
[22] 陈淳.考古学发展的历程及其代表人物[J].大众考古,2014,1：19-20.
[23] 陈淳.聚落考古与城市起源研究[J].杭州师范大学学报(社会科学版),2014,36(1)：47-57.
[24] 陈淳.考古学与人类起源[J].大众考古,2013,1：57-60.
[25] 陈淳.民族学对考古学阐释的贡献[J].考古学研究,2013：6-17.
[26] 陈淳,潘艳.农业起源概念和理论的变迁[J]//郑卓主编.岭南考古研究.第13辑.香港：中国评论学术出版社,2013.
[27] 陈淳.探究早期文明社会的世界观[J]//复旦大学文物与博物馆学系,复旦大学文化遗产研究中心编.文化遗产研究集刊.第6辑.上海：复旦大学出版社,2013.
[28] 陈淳,杨茜.陶瓷器研究的进展与思考[J]//复旦大学文物与博物馆学系,复旦大学文化遗产研究中心编.文化遗产研究集刊.第5辑.上海：复旦大学出版社,2012.
[29] 陈淳.早期文明的标准与解释[J].东南文化,2012,3：17-24.
[30] 陈淳.从"专业"到"通业"：当前文明探源的理论、方法与实践[J].历史研究,2012,3：172-188+192.

[31] 潘艳，陈淳.农业起源研究的实践与理论[J].江汉考古，2012，2：51-69.
[32] 潘艳，陈淳.农业起源与"广谱革命"理论的变迁[J].东南文化，2011，4：26-34.
[33] 陈淳.人类如何了解自己的来历[M]//韩昇，李辉.我们是谁.上海：复旦大学出版社，2011.
[34] 陈淳.美国性别考古的研究及启示[J].东南文化，2010，6：39-47.
[35] 陈淳.泥河湾旧石器早期工业与人类行为[M]//袁宝印，夏正楷，牛平山主编.泥河湾裂谷与古人类.北京：地质出版社，2011.
[36] 陈淳.马家浜文化与稻作起源研究[J].嘉兴学院学报，2010，22（5）：16-21.
[37] 陈淳.考古学的范例变更与概念重构[J].南方文物，2011，2：78-84+76.
[38] 陈淳.发扬传统，锐意创新——重译《时间与传统》记[J].南方文物，2011，1：47-53.
[39] 陈淳.文明探源需多学科协作[N].中国社会科学报，2011-08-16.
[40] 陈淳.考古学认识论的思考[J]//考古一生安志敏先生纪念文集编委会.考古一生——安志敏先生纪念论文集.北京：文物出版社，2011.
[41] 陈淳.谈考古科技与科学考古学[J].南方文物，2010，4：1-7.
[42] 陈淳.马家浜文化与稻作起源研究.嘉兴学院学报，2010，22（5）：16-21.
[43] 陈淳，陈洪波.科学思潮与早期国家探源[J]//山东大学东方考古研究中心编.东方考古.第3集.北京：科学出版社，2010.
[44] 陈淳，陈虹.西伯利亚发现未知古人类[J].江汉考古，2010，3：137-139.
[45] 陈淳，陈虹.线粒体DNA确定一未知古人类[J].人类学学报，2010，29（3）：302.
[46] 陈淳.北京猿人新传——读《龙骨山——冰河时代的直立人传奇》[J].江汉考古，2010，2：131-138.
[47] 陈淳.考古研究的经验主义与理性主义[J].南方文物，2010，1：13-18.
[48] 李琴，陈淳.公众考古学初探[J].江汉考古，2010，1：38-43.
[49] Chen C, An J, Chen H. Analysis of the Xiaonanhai lithic assemblage, excavated in 1978 [J]. Quaternary International, 2009, 211(1): 75-85.
[50] 陈淳.思想开放、积极开拓的一生——纪念旧石器考古学专家王建先生[N].中国文物报，2009-10-9.
[51] 陈淳.作为历史科学的考古学[J]//复旦大学文物与博物馆学系，复旦大学文化遗产研究中心编.文化遗产研究集刊.第4辑.上海：复旦大学出版社，2009.
[52] 陈淳.中国古人类及其文化研究的问题与思考[C].2002现代人类学国际研讨会论文集，2002.
[53] 潘艳，陈淳.生态学视野中的跨湖桥文化[C].//林华东，任关甫主编.跨湖桥文化论集.北京：人民出版社，2009.
[54] 陈淳.考古学史首先是思想观念的发展史——布鲁斯·特里格《考古学思想史》第二版读后感[J].南方文物，2009，1：37-45.
[55] 陈淳，潘艳，魏敏.再读跨湖桥[J].东方博物，2008，2：14-25+5.
[56] 陈淳.材料积累与历史重建[N].中国文物报，2008-11-14.
[57] 陈淳.里程碑式的人物考古学变革的杰作——柴尔德与宾福德译著推介[N].中国文物报，2008-10-22.

[58] 陈淳.能动性：当今考古研究的热点[N].中国文物报,2008-2-15.
[59] 陈淳.为史前考古学带来一场变革的杰出论著——柴尔德经典代表作补读[J].南方文物,2008,3：21-28.
[60] 陈淳.安阳小屯考古研究的回顾与反思——纪念殷墟发掘八十周年[J].文史哲,2008,3：5-23.
[61] 陈淳.新中国旧石器考古学回顾[C]//天道酬勤,桃李香主编.贾兰坡院士百年诞辰纪念文集,北京：科学出版社,2008：136-156.
[62] 陈淳,安家瑗,陈虹.小南海遗址1978年发掘石制品研究[J].考古学研究,2008,：149-166+602.
[63] Chen C. Techno-typological comparison of microblade cores from East Asia and North America[M]//Kuzmin YV, et al. eds. Origin and Spread of Microblade technology in Northern Asia and North America. Burnaby: Archaeological Press of Simon Fraser University, 2007: 7-38.
[64] 陈淳.跨湖桥环境变迁的新认识[N].中国文物报,2007-11-2.
[65] Zong Y, Chen Z, Innes J B, et al. Fire and flood management of coastal swamp enabled first rice paddy cultivation in east China[J]. Nature, 2007, 449(7161): 459-462.
[66] 陈淳.考古学与未来：工业文明的忧虑[J].复旦学报(社会科学版),2007,6：34-43.
[67] Chen C. Decipherment of Bronze Objects from Sanxingdui[J]. Fudan Journal of Humanities and Social Sciences, 2006, 3(2): 1-11.
[68] 陈淳.纪念布鲁斯·特里格教授[N].中国文物报,2007-3-2.
[69] 陈淳.夏商是奴隶社会吗？[N].中国文物报,2007-3-30.
[70] 陈淳.良渚文明崩溃探究——社会动力及与玛雅崩溃之比较研究[C]//浙江省文物考古研究所编.浙江文物考古研究所学刊.第8辑.北京：科学出版社,2006.
[71] 陈淳.疑古、考古与古史重建[J].文史哲,2006,6：16-27.
[72] 陈淳.社会进化模式与中国早期国家的社会性质[J].复旦学报(社会科学版),2006,6：125-132.
[73] 陈淳,孔德贞.性别考古与玉璜的社会学观察[J].考古与文物,2006,4：31-37.
[74] 陈淳.考古研究的哲学思考[N].中国文物报,2006-8-11.
[75] 陈淳.人类探源：谁是我们的祖先[J].科学,2006,58(4)：8-12.
[76] 陈淳.人类探源的新进展[N].中国文物报,2006-11-17.
[77] 陈淳.从东亚最早陶器谈跨湖桥和小黄山遗址年代[N].中国文物报,2006-3-6.
[78] 郑建明,陈淳.环太湖与宁绍平原史前社会复杂化比较研究[J].南方文物,2005,4：19-27.
[79] 陈淳,殷敏.三星堆青铜树象征性研究[J].四川文物,2005,6：38-44.
[80] 陈淳.旧石器考古学的新进展[J].自然科学与博物馆研究,2005,1：117-131.
[81] 陈淳.中国考古学的历程与成就[M]//姜义华、武克全主编,上海市社会科学界联合会编.二十世纪中国社会科学(历史学卷).上海：上海人民出版社,2005.
[82] 陈淳,殷敏.三星堆青铜树象征性研究[J].四川文物,2005,6：38-44.
[83] 陈淳.提升研究水准,更好保护文化遗产[J].中国文化遗产,2005,3：8.

[84] 郑建明，陈淳．马家浜文化研究的回顾与展望——纪念马家浜遗址发现45周年[J]．东南文化，2005，4：16-25.
[85] 陈淳，郑建明．环境、稻作农业与社会演变[J]．科学，2005，57（5）：34-37.
[86] Chen C, Gong X. Earlitou, Xia and the study of early state of China[J]. Fudan Journal of Humanities and Social Sciences, 2005, 1: 235-257.
[87] 陈淳．中国旧石器研究的进展和差距[N]．中国文物报，2005-4-15.
[88] 陈淳．考古学文化概念的变迁与思考[N]．中国文物报，2005-1-28.
[89] 郑建明，陈淳．环太湖与宁绍平原史前社会复杂化比较研究．南方文物，200，4：19-27.
[90] 陈淳，郑建明．稻作起源的考古学探索[J]．复旦学报（社会科学版），2005，4：126-131.
[91] 陈淳，龚辛．二里头、夏与中国早期国家研究[J]．复旦学报（社会科学版），2004，4：82-91.
[92] 陈淳，孔德贞．玉璜与性别考古学[N]．中国文物报，2004-7-9.
[93] 陈淳．古蜀金器“射鱼纹”之我见[N]．中国文物报，2004-8-27.
[94] 陈淳，潘艳．三峡古文化的生态学观察[N]．中国文物报，2004-11-26.
[95] 陈淳．考古发掘：谁来关注它背后的遗产保护？[N]．中国文物报，2004-11-12.
[96] 王轶华，陈淳．考古学的人文关怀[J]．文物世界，2004，1：11-13.
[97] 韩佳瑶，陈淳．三星堆青铜器巫觋因素解析[J]．文物世界，2004，3：19-25.
[98] 陈淳．张光直先生的遗产[C]//复旦大学文物与博物馆学系，复旦大学文化遗产研究中心编．文化遗产研究集刊．第2辑．上海：复旦大学出版社，2001.
[99] Chen C. Retrospect of fifty years of paleolithic archaeology in China[J]//Shen C, Keates S G eds. Current research in Chinese pleistocene archaeology. British Archaeological Reports, Series 1179.Oxford: Archaeopress, 2003.
[100] 陈淳，高书勤．西阴村发掘与中国考古学的历程[C]//复旦大学文化遗产研究中心编．文化遗产研究集刊．第3辑．上海：上海古籍出版社，2003.
[101] 陈淳，顾伊．文化遗产保护的国际视野[J]．复旦学报（社会科学版），2003，4：122-129.
[102] 陈淳．酋邦概念与中国早期国家探源[J]．中国学术，2003，2：214-233.
[103] 陈淳．墓葬：个体考古学的视野[J]．上海文博论丛，2003，2：32-35.
[104] 陈淳．中国考古学80年[J]．历史教学问题，2003，1：33-37+45.
[105] 陈淳．中国国家起源研究的思考[J]．史学月刊，2002，7：5-12.
[106] 陈淳．考古学应当“与时俱进”[J]．文物世界，2002，5：23-29.
[107] 陈淳．文明与国家起源研究的理论问题[J]．东南文化，2002，3：6-15.
[108] 陈淳，沈辰，陈万勇，等．小长梁石工业研究[J]．人类学学报，2002，1：23-40.
[109] 陈淳．中国文明与国家探源的思考[J]．复旦学报（社会科学版），2002，1：45-52+70.
[110] 陈淳．文明进程研究的思考[N]．中国文物报，2002-11-15.
[111] 陈淳．“西汉纸”的质疑[N]．中国文物报，2002-7-17.

[112] 陈淳.谈酋邦[N].中国文物报,2002-3-29.
[113] 陈淳.检视器物类型学[N].中国文物报,2002-1-2.
[114] 陈淳.考古学的定位、视野与研究[N].中国文物报,2001-12-14.
[115] 沈辰,陈淳.微痕研究(低倍法)的探索与实践——兼谈小长梁遗址石制品的微痕观察[J].考古,2001,7:62-73+103-104.
[116] 陈淳."夏娃理论"与中国旧石器时代考古[N].中国文物报,2001-11-2.
[117] 陈淳.21世纪中国文明起源研究的思索[N].中国文物报,2001-10-19.
[118] 陈淳.谈考古学的学术定位[J].文物世界,2001,6:10-15.
[119] 陈淳."操作链"与旧石器研究范例的变革[C]//第八届中国古脊椎动物学学术年会论文集.北京:海洋出版社,2001:235-244.
[120] 陈淳.重视社会史的通史陈列[N].中国文物报,2001-5-30.
[121] 陈淳.周口店与贾兰坡[N].中国文物报,2001-5-2.
[122] 陈淳.增强考古报告的科学性[N].中国文物报,2001-5-2.
[123] 陈淳.文物学、考古学与文化遗产保护[C]//复旦大学文物与博物馆学系,复旦大学文化遗产研究中心编.文化遗产研究集刊.第2辑.上海:复旦大学出版社,2001.
[124] 陈淳.岭南史前研究的思考[C]//张镇洪主编.岭南考古论文集.第6辑.广州:岭南美术出版社,2001.
[125] 高蒙河,陈淳.考古学次生堆积的研究与探索[J].华夏考古,2001,2:94-101.
[126] 顾伊,陈淳.美国"文化资源管理"的镜鉴[J].文物世界,2001,1:18-24.
[127] 陈淳.资源,神权与文明的兴衰[J].东南文化,2000,5:14-19.
[128] 陈淳.Settlement archaeology与聚落考古[N].中国文物报,2000-9-27.
[129] 陈淳.再谈繁昌人字洞人工制品的真伪[N].中国文物报,2000-7-12.
[130] 陈淳.谈谈考古学的科学性[J].文物世界,2000,4:26-31.
[131] 陈淳.考古学变革的理论建树——《时间与传统》导读[C]//复旦大学文物与博物馆学系,复旦大学文化遗产研究中心编.文化遗产研究集刊.第1辑.上海:复旦大学出版社,2000.
[132] 陈淳.考古学研究与信息提炼[C]//复旦大学文物与博物馆学系,复旦大学文化遗产研究中心编.文化遗产研究集刊.第1辑.上海:复旦大学出版社,2000.
[133] 陈淳.北京猿人的后裔是谁?[N].中国文物报,2000-4-26.
[134] 陈淳.繁昌人字洞旧石器真伪问题和建议[N].中国文物报,2000-2-16.
[135] 陈淳.中石器时代的研究与思考[J].农业考古,2000,1:11-20.
[136] 陈淳.尼安德特人——人类演化之谜[N].中国文物报,2000-2-29.
[137] 陈淳.旧石器类型学的理论与实践[C]//徐钦琦,谢飞,王建主编.史前考古学的新进展:庆贺贾兰坡院士九十华诞国际学术讨论会文集.北京:科学出版社,1999.
[138] 陈淳.居址考古学的探索与启示[J].文物世界,1999,4:33-36.
[139] 陈淳.中国文明起源探索的新视野[J].科学,1999,51(2):10-12+2.
[140] 陈淳.旧石器时代考古学的昨天与今天[J].第四纪研究,1999,2:148-154.
[141] 陈淳.谈谈考古学的新与旧[J].文物季刊,1999,1:18-23.
[142] 陈淳,沈辰,陈万勇,等.河北阳原小长梁遗址1998年发掘报告[J].人类学学报,

1999,3：225–239.
[143] 陈淳.早期国家之黎明——兼谈良渚文化社会政治演化水平[J].东南文化，1999，6：20–25.
[144] 陈淳."西汉纸"公案及其他[N].中国文物报，1999–5–19.
[145] 陈淳.周口店遗址又起争议[N].中国文物报，1999–4–7.
[146] 陈淳.文物考古忧思录[N].中国文物报，1999–3–17.
[147] 陈淳.夏娃理论与中国人的起源[N].中国文物报，1999–2–10.
[148] 陈淳.酋邦的考古学观察[J].文物，1998，7：46–52.
[149] 陈淳.城市起源之研究[J].文物季刊，1998，2：59–65.
[150] 陈淳.两极法与pièceesquillées[J].人类学学报，1998，1：74–81.
[151] 陈淳.聚落·居址与围墙·城址[J].文物，1997，8：43–47.
[152] 陈淳.考古学文化与文化生态[J].文物季刊，1997，4：88–94.
[153] 陈淳.谈旧石器精致加工[J].人类学学报，1997，4：59–65.
[154] 陈淳.考古学方法论的思考与借鉴[J].东南文化，1997，4：37–42.
[155] 陈淳.稻作、旱地农业与中华远古文明发展轨迹[J].农业考古，1997，3：55–57+90.
[156] 陈淳.古纸研究与考古学实践[J].中国造纸，1997，2：69–74.
[157] 陈淳.再谈旧石器类型学[J].人类学学报，1997，1：75–81.
[158] 陈淳.莫斯特文化新发现[J].化石，1997，1：6–7.
[159] 陈淳.旧石器研究：原料、技术及其他[J].人类学学报，1996，3：268–275.
[160] 陈淳.国家起源之研究[J].文物季刊，1996，2：83–90.
[161] 陈淳.谈史前学的方法论[J].文物季刊，1995，4：27–33.
[162] 卫奇，陈淳.泥河湾盆地旧石器考古展望[J].文物季刊，1995，1：73–80.
[163] 陈淳.谈中石器时代[J].人类学学报，1995，1：82–90.
[164] 陈淳.为未来保存过去——美国、加拿大的文化资源管理与合同考古学[J].东南文化，1994，5：60–66.
[165] 陈淳.完善古纸研究的考古学范例[J].中国造纸，1997，4：68–71.
[166] 陈淳.东亚与北美细石叶遗存的古环境[J].第四纪研究，1994，4：369–377.
[167] 陈淳.史前技术之演变[J].文物季刊，1997，3：50–55.
[168] 陈淳.最佳觅食模式与农业起源研究[J].农业考古，1994，3：31–38.
[169] 陈淳.考古学文化概念之演变[J].文物季刊，1994，4：18–27.
[170] 陈淳.谈旧石器类型学[J].人类学学报，1994，4：374–382.
[171] 陈淳.石器时代分野问题[J].考古，1994，3：239–246.
[172] 陈淳.考古动物群数量统计[J].化石，1994，3：11–13.
[173] 陈淳.谈旧石器打制实验[J].人类学学报，1993，4：398–403.
[174] 陈淳.类型、形制与类型学[N].中国文物报，1993–12–5.
[175] 陈淳.古人类骨骼分析与史前研究[J].化石，1993，3：2–3.
[176] 陈淳.几何形细石器和细石叶的打制及用途[J].文物季刊，1993，4：72–78.
[177] 陈淳.埋藏学与骨骼破损分析[J].化石，1993，2：2–3.
[178] 陈淳.考古学之演变[J].文物季刊，1993，2：33–43.

[179] 陈淳.废片分析和旧石器研究[J].文物季刊,1993,1: 10–15.
[180] 布鲁斯·特里格,陈淳.考古学理论,环境与诠释[J].东南文化,1992,6: 57–64.
[181] 陈淳.微磨损分析和旧石器用途[J].化石,1992,2: 2–3.
[182] 陈淳.考古学文化的功能观[N].中国文物报,1992–8–23.
[183] 布鲁斯·特里格,陈淳.美国新考古学述评[J].文物季刊,1992,3: 87–95.
[184] 陈淳."民族考古学"的译法、涵义及其他[J].东南文化,1992,(Z1): 40–41.
[185] 陈淳.埋藏学与遗址动态分析[N].中国文物报,1991–12–1.
[186] 陈淳.考古学"相关性"浅论[N].中国文物报,1990–8–9.
[187] 陈淳.谈谈民族考古学[N].中国文物报,1990–5–10.
[188] 陈淳.谈谈考古研究中的"类同"与趋同[N].中国文物报,1990–2–22.
[189] Chen C, Wang X Q. Upper Paleolithic industries in North China and their relationship with Northeast Asia and North America. Arctic Anthropology, 1989, 26(2): 127–156.
[190] 陈淳.文物考古界改革开放之我见[J].考古与文物,1988,4: 101–102.
[191] 陈淳.当代北美史前考古学介绍[N].中国文物报,1988–10–28.
[192] 陈淳.保护不可再生的地下文物资源[N].中国文物报,1988–10–1.
[193] 陈淳.旧石器考古的非标准件及其他[N].中国文物报,1988–8–19.
[194] 陈淳.太湖地区远古文化探源[J].上海大学学报(社会科学版),1987,3: 101–104.
[195] 陈淳,张祖方,王闽闽,等.三山文化:江苏吴县三山岛旧石器时代晚期遗址发掘报告[C]//南京博物院集刊(九).南京:江苏美术出版社,1987: 7–20.
[196] 陈淳,张祖方.磨盘墩石钻研究[J].东南文化,1986,1: 139–141.
[197] 陈淳.从考古学谈人类的早期文明[J].上海大学学报(社会科学版),1984,1: 78–84.
[198] Chen C. The microlithic in China[J]. Journal of Anthropological Archaeology, 1984, 3(2): 79–115.
[199] 陈淳.中国细石核类型和工艺初探——兼谈与东北亚、西北美的文化联系[J].人类学学报,1983,4: 331–341+407–408.

(三)译作

[1] 约翰.内皮尔/塔特尔.手[M].陈淳,译.上海:上海科技教育出版社,2001.
[2] 戈登·柴尔德.欧洲文明的曙光[M].陈淳,陈洪波,译.上海:上海三联书店,2008.
[3] 肯尼斯·费德.骗局、神话与奥秘:考古学中的科学与伪科学[M].陈淳,译.上海:复旦大学出版社,2010.
[4] 布鲁斯·特里格.考古学思想史.第二版[M].陈淳,译.北京:中国人民大学出版社,2010.
[5] 诺埃尔·博厄兹,拉塞尔·乔昆.龙骨山:冰河时代的直立人传奇[M].陈淳,陈虹,沈辛成,译.上海:上海辞书出版社,2011.
[6] 布鲁斯·特里格.时间与传统(重译本)[M].陈淳,译.北京:中国人民大学出版社,2011.

[7] 科林·伦福儒，保罗·巴恩.考古学：理论、方法与实践[M].陈淳，译.上海：上海古籍出版社，2015.
[8] 希安·琼斯.族属的考古：构建古今的身份[M].陈淳，沈辛成，译.上海：上海古籍出版社，2017.
[9] 肯特·弗兰纳利.圭拉那奎兹：墨西哥瓦哈卡古代期觅食与早期农业[M].陈淳，陈虹，等译.上海：上海古籍出版社，2019.
[10] 布鲁斯·特里格.柴尔德：考古学的革命[M].陈淳，何传坤，译.北京：中国人民大学出版社，2020.

（四）译文

[1] 迈克尔·希弗.关于遗址形成过程研究[J].陈淳，译.南方文物，2015，2：182–192，181.
[2] 布鲁斯·布拉德利.石器剥片顺序：术语及讨论[J].陈淳，译.江汉考古，2012，4：125–128.
[3] 德怀特·里德，理查德·沃森，帕蒂·齐·沃森.类型学讨论三题[J].陈淳，译.南方文物，2012，4：49.1.
[4] 威廉·亚当斯.考古学分类的理论与实践[J].陈淳，译.南方文物，2012，4：39–48.
[5] 波希丝·克拉克森.考古学之想象：图像的情景化[J].陈淳，译.南方文物，2012，3：185–188+195+184.
[6] 唐·克雷布特利.石器技术的潜力[J].陈淳，译.江汉考古，2012，3：127–130.
[7] 奥法·巴尔—约瑟夫，斯蒂夫·库恩.石叶的要义：薄片技术与人类进化[J].陈淳，译.江汉考古，2012，2：120–128.
[8] 加里·韦伯斯特.文化历史考古学述评[J].陈淳，译.南方文物，2012，2：53–61.
[9] 贾洛斯拉夫·马里纳，泽德奈克·瓦希塞克.考古学概念的考古[J].陈淳，译.南方文物，2011，2：93–99.
[10] 布鲁斯·特里格.兼具“伞兵”和“蘑菇采集者”特质的当代伟大的考古学家——布鲁斯·特里格自传[J].陈淳，译.南方文物，2011，1：145–155.
[11] 约翰·霍克斯，米尔福德·沃尔波夫.现代人起源六十年之争[J].陈淳，译.南方文物，2011，3：158–165+157.
[12] 迈克尔·加拉蒂.欧洲区域聚落形态研究[J].陈淳，译.南方文物，2010，2：113–123+145+112.
[13] 乔伊斯·马库斯（J. Marcus）.社会进化的考古学证据[J].陈淳，译.南方文物，2009，2：115–121+114.
[14] 罗泰.论中国考古学的编史倾向[J].陈淳，译.文物季刊，1995，2：83–89.
[15] 亚瑟·杰利内克.石器分析的式样、功能与形制[J].陈淳，译.文物季刊，1995，3：88–95.
[16] 布鲁斯·特里格.论文化的起源、传播与迁移[J].陈淳，译.文物季刊，1994，1：81–94.

[17] 布赖恩·海登.驯化的模式[J].陈淳,译.农业考古,1994,1: 25-30+40.
[18] 布赖恩·科特雷尔,约翰·坎明加.石片之形成[J].陈淳,等译.文物季刊,1993,3: 80-102.
[19] 路易斯·宾福德.论新考古学[J].陈淳,译.东南文化,1992,1: 41-46.
[20] 弗朗索瓦·博尔德.旧石器类型学和工艺技术[J].陈淳,译.文物季刊,1992,2: 83-103.
[21] 布鲁斯·特里格.世界考古学展望[J].陈淳,译.东南文化,1991,5: 239-241.
[22] 布鲁斯·特里格.考古学与未来[J].陈淳,译.东南文化,1990,4: 166-169.

获奖情况

[1]《考古学前沿研究：理论与问题》2018年获上海市第十四届哲学社会科学优秀成果著作二等奖。
[2]《考古学前沿研究：理论与问题》2016年入选国家哲学社会科学成果文库。
[3]《考古学：理论、方法与实践》2016年获第十四届上海图书奖二等奖，2016年获第十五届优秀引进版图书。
[4]《考古学的理论与研究》入选庆祝上海哲学社会科学出版基金成立25周年精丛。
[5]《龙骨山——冰河时代的直立人传奇》2011年被《中华读书报》评为该年度十佳图书。
[6]《考古学研究入门》2011年获上海市教委评选的上海普通高校优秀教材二等奖。
[7]《考古学思想史（第二版）》2012年获中国大学出版社协会第二届优秀教材评选一等奖。
[8]《骗局、神话与奥秘》2010获全国文化遗产十佳图书评选活动办公室（《中国文物报》）评选的年度全国文化遗产优秀图书。
[9]《考古学的理论与研究》2004年获上海市第七届哲学社会科学优秀成果著作二等奖；并被《中国文物报》评选为2004年最佳文博学术论著。
[10]《二里头与中国早期国家研究》一文获上海市第八届哲学社会科学（2004—2006年度）优秀论文三等奖。

代表作

从考古学理论方法进展谈古史重建*

陈　淳

（复旦大学）

摘要：王国维的“二重证据法”被认为是中国新史学的肇始和中国考古学的特色。但是将文献与考古材料简单比附印证的简单做法，已经造成了古史重建的重大弊端。本文从考古学理论方法的进展回顾了这门学科从材料的分类描述转向社会重建的过程，并讨论了对文献应持的评判性态度。文章最后从认识论角度指出，历史并不是一种客观和价值中立的构建，而是学者根据残留至今文字和物质材料，根据集体认同和当代需要而编织的产物。因此，有学者提出的夏代信史其实是意识形态的历史，而非科学的历史。文章认为，当代考古学独立研究所能获得的信息可以从各个方面补充文献研究的政治史，并为文献研究提供一种全新的视野。这种科学和脱胎换骨的“二重证据法”应该成为21世纪中国古史重建的必由之路。

关键词：二重证据法　考古学理论方法　文献批判　认识论　古史重建

中国考古学发轫于20世纪初的古史辨运动。疑古思潮对上古史传说成分的质疑，令中国学界意识到地下文物的重要性，开启了“上穷碧落下黄泉，动手动脚找东西”①的中国新史学先河，也确立了中国考古学从属于历史学的

* 由于本文的影响较大，故选用为本人的代表作。它刊载于2018年第6期《历史研究》。南京大学《新古史辨》公众号在转载本文时加了一段编者按，作为本文的评价转录在此：时值新春佳节，《历史研究》刊载了陈淳先生大作，经期刊网发布。之前，我们已推送过先生多年前《文史哲》上相关古史重建的大作。后出者精而新。这篇文章更为翔实深入系统。因为上古史研究受“二重证据法”影响太深，2018年“夏”为信史说烽烟再起，成为一大热点，这篇文章是年终的一个重量级回应。陈先生的主张也是我们新古史辨一直坚持的理念。我们相信，陈先生大作在《历史研究》的发表，标志中国上古史的一重要转向，一个重视科学求真，理论多元繁荣的上古史研究时代悄然来临，将由此融入世界史学的大潮。

学术定位。夏鼐指出，考古学和历史学"同是以恢复人类历史的本来面目为目标，是历史科学（广义历史学）的两个主要的组成部分，犹如车子的两轮，飞鸟的两翼，不可偏废"[②]。而中国考古学的使命感和学术定位又因王国维提出"幸于纸上之材料外，更得地下之新材料"的"二重证据法"而得到加强，甚至一直被认为是中国考古学的特色。

实际上，王国维所谓的地下之材料仅限于商周的甲骨文和金文，它们都是文字资料而非纯物质遗存。而且这种方法只强调"证实"而排斥证伪，具有片面性和学术思想的保守性。王国维本人的立场是站在古史辨运动的对立面，而这一方法因殷墟发掘令商代成为信史而得到普遍推崇并定于一尊。于是疑古的理性思维受到全面打压，地下之新材料被扩大到所有考古材料，并不加审视和批判地以穿凿附会来重建古史，埋下了很大的弊端和无穷的乱象[③]。

其实，文献资料结合考古发掘的研究方法并非中国首创和独有。在欧洲，文献记载与考古学有着非常密切的关系。从1870年起，谢里曼根据《荷马史诗》的线索来寻找和发掘特洛伊即是用田野考古验证文献记载的先声。谢里曼的发掘和研究使学界认识到古希腊之前的灿烂文明，从而揭开了欧洲古代史研究的新篇章。欧洲的古典考古学、中世纪考古学和以色列的圣经考古学都是文字记录与考古材料的互补和结合，或以考古发现来印证文字记载，或以文字记载来解释考古材料。

一个多世纪以来，考古学和历史学这两门学科无论在理论方法、基础材料、研究对象和范围还是学者的自我意识上都有了重大的进展。而考古学的这种进展完全超越了文献资料与考古发现简单互证的范畴，它早已不再是"历史学的侍女"，而是充分借鉴艺术史、民族志、语言学、口述传统乃至自然科学的多重证据来全方位研究过去人类的生活和社会变迁。

我们现在认识到，文字记录和物质文化是在不同的社会和历史背景中产生的，代表着独立的证据和线索，它们不能简单合并[④]。而且，历史文献和考古材料都是历史的产物而非历史本身，必须予以理性梳理才能利用。加上历史重建本身就是一种政治过程，是我们今天构建的一种产物。如果没有考古学理论指导下的严格评判和置信度考量，简单用"二重证据法"将这两种材料简单撮合，必然会遇到两重证据的不协调。这样的古史重建难免出现一片乱象。

文献资料在当代考古学研究中只不过是一种不同的证据或有价值的线索，而非一种从属和依附关系。本文尝试从考古学理论方法的进展和文献批判精神来反思“二重证据法”和质疑当下试图重建夏代信史的做法，进而从认识论的角度审视考古学在历史重建中的作用，以期为考古学和历史学的学科交叉与互补提供一种较为客观的视角和可行的努力方向。

一、考古学的进展

历史学家研究的大部分材料都是文字记录，这种材料是对人类思想和行为的直接说明，他们一般认为，没有这种文字就没有真正的历史。考古学家研究的材料主要是人工制品，常被叫做“物质文化”，他们主要是根据古代先民遗留至今而免于损毁，并有幸被我们发现的有限物质遗存来了解过去的。历史学和考古学能彼此互补，以充实人类发展的编年史。但是，它们在利用的证据类型以及重建过去的方法上差异很大，因此所能重建的人类历史所侧重的方面也必然不同。

考古学从其本质而言是研究古代匿名者的遗存，其大部分努力是在推断古代文化的产物而非文化本身。这些物质遗存更多体现了古代先民的生活方式、技术和经济，并追溯它们历时的发展。由于证据的物质性，考古学家只能对它们的含义做出非常一般的推断。加上这些证据常常是残缺不全和碎片化的，所以需要考古学理论方法在解释这些材料中发挥着更大的作用。考古学家掌握的理论方法越是粗糙和单一，处理的材料越有限，越容易对人类历史产生错误认识和不当解释[⑤]。

俄国考古学家利奥·克莱因指出考古学和历史学研究存在很大区别，考古材料是物质遗存和人工制品，并非是由语言锁定的概念。历史学家处理的是思想和语言领域的材料。文献资料处于历史学家相通的语言领域，但考古材料则不是，它们处在另一个领域。考古学家的任务就是要把考古材料（特别是它们的信息）转换到思想和语言的领域中来[⑥]。而这种转换的关键不但在于材料的积累，更在于理论方法的完善。

英国考古学家科林·伦福儒和保罗·巴恩讨论了考古学研究中理论、方法和材料这三个关键组成部分的相对重要性。他们把理论和观念置于最重要的地位，而考古材料只有在理论方法充分完善的情况下才能被了解[⑦]。英国考古学家马修·约翰逊指出，考古学家和没有头脑的废铜烂铁收集者之间的差别在于，我们要用一套法则将考古发现转化为对过去有意义的解释。这些法则从根本上说是理论的。事实固然重要，但是没有理论，它们不会吐露真言[8]。

在欧洲的古物学阶段，学者们利用艺术品和古建筑来增进对文献历史的了解，他们并不想研究不见经传的人群。那里的历史记录基本不超过罗马时期。在英国，对于文字记录之前的新石器时代和青铜时代遗迹和遗存，学者将这些物质文化与凯尔特人的德鲁伊特相联系是当时流行的做法。1803年，波拿巴·拿破仑建立了《凯尔特学院》，鼓励学者们从考古学、历史学、民俗学和语言学来构建高卢人与现代法国人之间直接传承的历史意识[9]。

考古学诞生后，这门学科的理论方法是不断借鉴其他学科而发展起来的，而且材料的解读和历史的重建与考古学家受训的专业背景密切相关。在19世纪和20世纪初的进化考古学阶段，许多考古学家都是地质学出身，而且在历史重建的理论上受进化论影响很大。他们把考古材料看作是古生物学中的“标准化石”以划分不同的时代，并以直线进化的观点解释文化的发展。比如法国考古学家莫尔蒂耶就认为，法国的旧石器文化序列可以解释世界上所有地区的旧石器文化发展。这一时期考古学的历史重建带有浓厚的地质学色彩，即人类历史的发展可以用单一的序列来标示，并能够从某洞穴的一个剖面上予以观察[10]。

从19世纪60年代开始，欧美考古学与民族学建立起密切的关系，认为民族学能够为考古学提供他们想要知道的所有事情。英国博物学家约翰·卢伯克用现代原始人群的生活方式和习俗来解释古代遗存，并提出了一种文化发展的社会达尔文主义解释，把世界各地不同民族文化的差异看作是自然选择的结果。卢伯克的综述对20世纪上半叶的殖民主义和种族主义考古学阐释产生了很大的影响[11]。

在考古学的草创阶段，其基本工作仅限于构建物质文化的年代序列和分期，对物质文化的解读大体参照民族志材料的类比。由于考古学发轫较早的国家如丹麦、瑞典、英国和法国都缺乏悠久的编年史，使得这门学科更加依赖独立的方法进行研究。而以古希腊和罗马为对象的古典考古学则保持着以文献、艺术史和族属为导向的特点，与考古学理论方法的关系比较疏远。

20世纪上半叶，英国考古学柴尔德在他《欧洲文明的曙光》一书中创建了文化-历史考古学的范式。这就是用类型学构建“考古学文化”的一种分析单位，用相似即相近的原理把一批相似考古学文化的遗址归组，以对应史前与历史时期的族属。柴尔德说："文化是一种遗产，它对应于享有共同传统、共同社会机构以及共同生活方式的一个社群。这群人可以顺理成章地被称为某人群（people）……。于是考古学家能够将一种文化对应于该人群。如果以族群来形容这群人，那么史前考古学就有望建立起一部欧洲的民族史[12]。"在此之前，英国还没有一种系统方法来研究考古学材料。通过这本书，考古学文化成为所有欧洲考古学家的一个研究工具，并被格林·丹尼尔誉为"史前考古学的一个新起点"[13]。

20世纪初随着西学东渐引入中国的就是这种文化-历史考古学范式，这种范式在我国本土进而发展成为用考古学文化区系类型来将编年史向史前期延伸。但是，由于上古史的文字资料非常有限，所以常会以考古学文化来指称夏商周三代的物质遗存，或以夏商周三代来指称全国范围大体相同时期的不同文化。比如，夏鼐在一次有关夏文化讨论会上所做的总结就体现了这种范式的无助。首先，他预设夏王朝是存在的。第二，夏文化应该有其特点，是夏王朝时期夏民族的文化。第三，夏时期的其他文化不能被视为夏文化。第四，现有的考古材料不足以分辨夏文化[14]。在这里，夏鼐把夏王朝的存在等同于夏文化和夏民族，但是坦陈从考古材料上难以分辨。

布鲁斯·特里格在讨论商代的政治结构时指出，现在学界采用的多种术语如商文明、商时期、商民族、商代、商国和商文化是范畴不同的概念，它们彼此之间不能互换。商代国家要比考古学定义的商文化和商文明小得多。研究商代国家好比盲人摸象，有的分析十分有用，而有时却相互矛盾[15]。同理，我国考古学家习称的"夏文化"至少也包含夏代、夏朝和夏族三义。楚文化的"楚"也含有地域、国家、族属和文化四个概念[16]，它们彼此之间也不能互换。

现在我们已经明白，考古学文化并不能简单与族群划等号，更无法与早期国家相对应。从世界各地民族志证据来看，许多不同民族会共享许多物质文化，只有部分材料才会体现族属的身份。民族志证据表明，没有一个原始国家可以从单一族群或酋邦独立发展的基础上产生，早期国家的形成普遍见证了不同酋邦和族群的征服与融合[17]。而苏美尔和玛雅文明林立的城邦国家，从考古学看只是一种文化。而且，族属认同大体是一种自我定义系统，许多自认为同族的人群在习俗和物质文化并不完全相同。考古学文化并非铁板一块和界线分明的实体，因此考古学家不应认为，物质文化的异同可以提供一种族群关系和身份的直接证据[18]。

由于文化–历史考古学除了类型学和传播论外没有太好的系统方法，主要凭借学者自己的灵感做"想当然耳"的解释[19]。"二重证据法"倡导纸上与地下材料结合而成为古代史研究的不二法门。殷墟发掘确立了考古学证经补史的作用，将地下新材料的范围从文字扩充到非文字，正式奠定了历史学与考古学结合的典范[20]。然而，将这种方法在文献记载对应考古材料之间不加限制和不加批判的误用和乱用，其结果并不能提供一种可信的古代史，而只会产生更多的争议和制造更大的混乱。

20世纪60年代美国兴起的新考古学是范式的革命，新考古学的学术定位是人类学，因此与文化–历史考古学的旨趣大不相同。新考古学又称过程考古学，主要采用文化生态学、系统论和文化功能主义等理论研究文化的适应。它强调实证论的科学方法，将探索社会文化发展规律置于比历史重建更重要的地位。新考古学对文献历史的贡献甚少，但是它所秉持的文化功能观则对文化–历史考古学的范式大有裨益，功能观视文化是人类对环境的一种适应手段。同一批人群在不同季节或不同地点开发不同资源可能会采用不同的工具，因此物质文化的差异未必能指认不同的人群。

生态位（ecological niche）概念为考古学将文化与族群的对应提供了一个新的观察维度。生态位概念认为，文化区的划分基本上是生态学的，因此详细的生态学而非地理学的考虑应该作为研究的起点。生态位以一种镶嵌的视角来看待社群与文化的分布，即许多不同文化的族群以各种亲密的共生关系在一个地区比邻而居。各族群仅开拓整个环境中的某一部分，而将其他部分让给其他族群开发，比如农牧、农耕与渔猎社群的共生[21]。

美国考古学家戈登·威利的聚落形态研究为考古学透物见人和研究社会文化变迁提供了一种可行的方法。威利从四种聚落形态：生活居址、仪式建筑、防御建筑和墓地质量和数量分布的历时变化来了解秘鲁维鲁河谷的史前社会变迁，标志着考古学文化功能性阐释的战略性起点[22]。威利的聚落考古又被称为"社会考古学"[23]，代表了考古学自丹麦考古学家汤姆森建立三期论以来最重要的方法论突破[24]。

20世纪七八十年代在英国兴起的后过程考古学也是一场范式的革命。后过程考古学并不认为物质文化是人们对环境被动适应的产物，它也具有意识形态上的象征意义，并重视个人在导致文化变迁中发挥的主观能动性。英国考古学家伊恩·霍德根据民族考古学研究成果指出，物质文化并非是对社会结构的一种被动反映，它还是用来掩饰、颠倒和扭曲社会关系的一种积极因素。霍德的观点被认为是对考古学阐释的一个重要贡献，并可与汤姆森的三期论和威利的聚落形态研究的贡献相媲美[25]。

霍德还提出了背景考古学（contextual archaeology）的概念，认为物质文化与其含义之间的关系在根本上是主观的，但是其含义至少部分源自其背景和先前的背景。这种背景是指文化及历史的背景。于是，背景考古学再次强调文化历史的重要性以及考古学与历史学的紧密关系。考古材料不仅是推断证据的基础，而且是需要读懂的一种"文本"[26]。后过程考古学还强调一门学科操作的自我批判意识，批评考古学（critical archaeology）意在揭示考古学家在解释中因自身立场所产生的偏见。由于考古学在不同国家的意识形态中是为政治服务的，因此它特别要求考古学家关注在当下的政治讨论中如何看待过去。

从考古学理论方法的发展可知，这门学科研究对象的物质性和不完整性使得它不可能提供与文献资料相同的信息。因此，"二重证据法"试图用地下之材来印证文献中的族群和国家是无望的。因为这不是一个技术问题，比如材料不够充分，而是物质文化的分布未必与社群或政治组织相一致[27]。于是，考古学开始利用自己材料的特点和理论方法的创新来探索这门学科最擅长的领域，并在以下方面取得了长足的进展。测年技术的发展使得考古学能够更加仔细地观察文化变迁；环境考古学能够详细了解生态变迁和人地关系，人类生计方式和经济形态的特点与演变，并了解复杂社会兴衰的经济基础；聚落考古能够从家户、村落到区域的不同尺度了解社会结构的历时变迁与国家及城市的形成过程；人工制品的研究能够了解各种器物的生产方式，劳力投入与分工，专业化程度，实用品与奢侈品的制造、使用与分配；葬俗的研究可以了解社会等级和分层，管窥古人的世界观和灵魂观；显赫物品的象征性研究能够了解社会地位、联盟、竞争和贵族的权力结构；体质人类学分析能够了解人群的食谱、病理、劳动强度和营养状况；分子人类学能够提供基因的流动、人群的血缘关系和社会结构等。

当今考古学提炼的这些信息已经基本上能够独立重建史前期和原史时期社会文化发展的历史过程，如果加上文献证据，就能够为这种历史重建提供更加清晰的图像。以玛雅考古为例，象形文字的破译为考古学提供了更加具体的历史信息：了解了物质文化相同的玛雅文明并非一个统一的帝国，而是许多相互竞争的城邦国家；了解了许多城邦国家和国王的名字、头衔以及王朝序列；了解了玛雅的天文学、历法和宇宙观；了解了各城市国家之间发生的战争事件和一些国家兴替；了解了一些国家的王位继承事件与时间[28]。

由此可见，文字资料与考古材料的历史重建应该是各自领域所获信息的互补，而非两门学科之间对具体材料的穿凿印证。目前，文化-历史考古学仍然是我国考古学采用的主导范式，虽然近年引入了国外过程考古学的一些技术，但是在信息提炼和整合上还有待提高。

二、文献批判精神

任何文字记录都是在一定社会背景中，为了某些人的某种目的而产生的，历史事件的记载也会在流传过程中发生改变。我们自己经历的历史有时也真相难辨，一二十年后就已说法不同，几千年前的历史文献更须进行梳理和判别才能谨慎使用。这就是历史文献的批判精神。

王国维的“二重证据法”将殷商的甲骨文和西周的金文看作是历史研究的地下之材，无非是证实了《史记》中提到的三代并非虚构。从书写的历史来看，早期文字的功能极其狭窄，因此很难提供后世想要了解的史实。即使是这些文字记载本身，也不能从它们说什么内容，而要从它们创造和使用的背景来了解。

柴尔德说过，文字的发明是人类进步的一个新纪元。它对我们现代人看来之所以意义重大，是因为它提供了一个机会来透视我们祖先的思想，而不必从他们的物质表现来推断。但是，文字的意义不应被夸大，它并非为公共交流而发明，而是为了实际管理的需要。早期文字作为表达思想的工具相当笨拙，难免成为一种高难度的特殊技艺，需要一个漫长的师承才能掌握。阅读则是一种秘境，只有经过长期学习才能入其堂奥[29]。

世界各古代文明的早期记录系统至少有三种不同的功能。一是纪念国王和他们的事迹，如古埃及和玛雅的碑铭。二是行政管理，如美索不达米亚庙宇群里的账目、契约和收据。三是宗教目的的早期记录系统，如商代与占卜相关的甲骨文和美索不达米亚早王朝时期的祭祀祷文。在商代，文字除了神谕之外的用途并不清楚。因为缺乏占卜和展示以外文字其他用途的直接证据。西周早期青铜器上的历史或仪式性文字看来也并非准确体现当时的口语，铭文极其简约。在早期文明中，刻意保持复杂的字体以确保它们由一小撮统治者或上层阶级所垄断。埃及的圣书体被称为“神的语言”，学会用这种方式书写乃至书写也被看作是一种不能分享的神圣艺术[30]。

大概在公元前一千年中叶，中国的文字记录开始从祭祀文本向历史事件的记录发展。两部这类年鉴分别是公元前一千年鲁国的《春秋》和魏国的《竹书纪年》。《尚书》是一本古代统治者言论的汇编，被视为道德规范和政治智慧的源泉。自秦汉开始，正史编纂成为一项政府职能。司马迁在《史记》中创建的历史叙述模式在作了一些改动后，成为正式朝代史志的范式。传统的中国编年史学关注统治者的想法，宣扬朝廷官方对历史事件的意见，并带有一种有力的政治宣导和道德说教[31]。

基于这种性质的文献记载，结合其他材料来重建历史现在已经受到了批评。这是因为人们认识到，这类历史记载并不能为过去的社会性质提供客观和完整的陈述。这不仅是由于这种记载体现的是有关过去局部和碎片化的视角，而且由于它们代表了社会特定阶层、主要是统治集团的观点。文献资料应该根据它们产生的社会与政治背景、作者和读者的地位和利益，还有文献在构建和调节文化认同中所发挥的作用来加以分析，而非按其表面说法而无条件接受。

这便产生了许多新的看法和立场，比如有的考古学家对利用文献表示疑虑，认为考古学是一种更加客观的证据来源，建议史料在考古学阐释中的作用应该降低。他们设法排除文献中所有的主观性（利益、政治、意识形态因素等）而不是认可它，并建立一种观察过去主观表现的评判性视角。实际上，这种途径导致了在历史学和考古学证据之间一种决定论关系的逆转，使得考古学证据变得优于历史学[32]。

有些考古学家认为，需要了解文献产生及它们在社会中发挥作用的不同的形式，而断文识字只限于社会的某些社会阶层，在历史上它只限于贵族和专业群体。一般认为，文献记录是贵族为了安排他们自己的社会生活和记忆方式而创造和使用的。而这种记录普遍漠视平民和农村生活。为了克服这一偏颇，考古学家必须将平民纳入学术研究，重新考虑文献记载与物质文化的关系。于是，文献被用作一种手段来构建一批预判，然后结合其他证据来源来进行探究[33]。

还有考古学家指出，许多“文献”实际上是被构建的档案，它们本身需要做出解释，而不是单从字面来“读懂”，这需要我们了解文献在其原始背景中产生、使用和废弃的生命史。这些学者甚至认为考古材料也是一种被构建的档案，考古材料从其埋藏的背景中提取后，经初步分类放到库房或实验室里就被“档案化”了。然后通过类型学研究和比较，各种器物会被赋予不同的价值和重要性。这种专业性很强的操作，又会给予研究者以价值和声誉的回馈。文献和器物这两种证据在与人类各种背景的互动中有着迥异的生命轨迹，并发挥着不同的作用，在对器物与文献两类证据之间进行整合时会有很大的不协调。由于文献和考古材料都是在独一无二的背景中被发现，并经历了档案化过程，于是它们就能被限制接触和操纵，进而有可能转变为一种重大成果而膜拜，并成为特定知识网络中的强行通道（obligatory points of passage）[34]。

在文献与考古材料相结合的历史重建中，当代学者无论对待文献还是对待自己都应该有一种批判和自我反省精神。比如对待夏的问题，我国有学者认为，既然《史记》中商代的记载被殷墟发现所证实，那么夏代应该也是可信的。还有学者认为，我国最早的文献《尚书》和《诗经》中多处提到夏，说明周人不仅知道夏的存在，还知道夏人政绩败坏，商人革了夏的命。而魏国史书《竹书纪年》中记录了夏代诸王所在的都邑、某些国王在位的年数以及夏的积年，因此，夏的存在应基本可信[35]。

也有学者持不同意见，认为西周对商代历史的叙述有所依据，但对夏代史实则大不相同，一是距西周为时过久，二是文献无凭。因此，杨宽、陈梦家和顾颉刚等认为西周统治者将夏夸大为一个统摄万邦的朝代，借以对“顽民”宣传灭商之合理，同时又作为“殷鉴”来教导臣民是十分可能的[36]。

对于西方学术背景的学者，就各种历史记载而言，首先要考虑其背景和档案化过程，而不是就其表面含义而认可其真实性。正如吉德玮在研究甲骨文时指出，问题不在于这些文字说了什么，而是要了解这样说的用意何在，文吏在他这样做时是怎么想的[37]。他还指出，东周的年谱作者和先哲对商代所知甚微。比如，孔子就哀叹从商的后人宋国那里得到的记录无法重建商代的仪式实践（“文献不足故也”，《论语》，《十三经注疏》）。汉代史家提供的详细信息也无非是诸王的次序和五次迁都[38]。

根据文献线索，徐旭生找到了二里头遗址。由于二里头遗址与夏的纪年有所重叠，于是得到了学界特别的惠顾并被赋予重要的价值，甚至被尊为“夏墟”。但是，从考古学证据来看，其物质文化所反映的显赫程度显然不及良渚、红山和凌家滩等新石器时代晚期遗址。罗泰指出，对中原地区早期朝代的定型看法已对历史造成了一种扭曲的图像。在与“三代”相当的时期里，许多独立的史前文化在中国各地发展。十分可能的是，夏朝的重要性，如果它确实存在过的话，已因为它在史籍中的幸存而被强调得过头。如果独立于史籍以外来进行探究，田野考古可能为之提供一种真正的新见解[39]。

李峰指出，即使第一手资料如铭文也带有偏见和主观性，而晚出史料的缺陷更加明显。因为在历史记忆的传递中，大量重要信息已经丢失，有关的最初记录也会被增饰和修改，甚至植入后世的观点和言论。于是，目前的文献研究不再将文献看作一个整体，而是把它当作不同时期的层位累积。文献的重要性在于它们在一个与其他类型证据的共同历史背景中能彼此联系，并揭示一个有关历史事件潜在和一贯的记叙。这种“文献批判”态度要求将后出的文献资料放到一个同时受到考古证据支持的历史背景中与铭刻和早期典籍一起使用。因此，古代史的重建并非是将历史学家和考古学家各自得出的结论简单叠加，而是需要重新对各自领域内存在的问题做根本的研究[40]。如果采取这种文献批判的态度，目前将有关夏的记载与二里头彼此对应的做法，无论从文字背景还是考古材料上仍找不到一个能够将彼此衔接到一起的坚实支撑点。

西山尚志讨论了运用出土文献的科学性问题，指出了王国维“二重证据法”在历史重建中存在的五条内在缺陷。(1)没有设想出土文献有“伪”。(2)出土和传世文献内容不一致。这有三种可能：a. 出土文献的内容是“真”，而传世文献是“伪”。b. 出土文献的内容是“伪”，而传世文献是“真”。c. 出土与传世文献的内容不一致，而且都是“伪”。“二重证据法”只考虑了情况a，而完全没有考虑b和c的可能。(3)即使出土与传世文献内容一致也未必是实录。(4)“二重证据法”并不证伪，而只有可证伪的陈述才是科学的。(5)“二重证据法”是典型的证实主义，认为没有得到证明的传世文献内容不能加以否定。西山进而认为，某些学者坚称只有“二重证据法”才能重建古史，把文献批判看作只做破坏，不能重建古史的见解是不对的。他指出，“批判”与“重建”并非对立的概念。相反，“批判”是“重建”的唯一方法。无法证伪的“二重证据法”不能带来任何的进步[41]。

对于文献与考古学的关系，罗泰提出的观点值得我们深思。他说，考古学必须从文献历史学的束缚中解放出来。只有在不受文献史学外在干扰的情况下，考古学才能提供一种认识论上的独立认识。解放了的考古学将会大幅度拓宽历史学的研究范围[42]。

三、讨论与结语

经过一个多世纪的发展，无论考古学还是历史学都得到了长足的发展。考古学已经从文化-历史学的分类描述转向科学实证的社会文化史重建，而历史学也突破了编年史和政治史的窠臼而开始关注整体史。不幸的是，当下在公众对考古发现兴趣日增的情况下，将考古发现与史料直接比附的“二重证据法”仍长盛不衰，甚至愈演愈烈。学者和媒体在公众场合普遍强调伴有文字和有案可稽的考古发现价值非凡、意义重大，不见经传的重大发现则迷雾重重、荒谬莫测。这有意或无意向公众误导了这样的看法：历史只能从文字来了解。许多学者也偏好甚至倾心用文献价值来拔高考古发现的重要性，比如建议将我国前国家的酋邦社会称为“五帝时代”，把陶寺遗址指称为“尧都”。而近来将二里头对应于夏墟的争论硝烟再起[43]。

有学者指出，用实物和文献相互参证来重建古史是不够的，还需要参照民族志蓝图来了解已逝的社会。即使有实物，研究者也未必能提供正确的解释。而文字记载植入了记述者的思维，未必反映真相。由于这种主客观因素的交集，而情况又复杂多样，即使不弄虚作假、蓄意歪曲、捏造篡改之类，正打正着、歪打正着、正打歪着、打而不着、空打空着的情况肯定不少[44]。

也有学者批评指出，各种古籍对上古“先王”的描述本就五花八门、自相矛盾，距今的年代和活动的区域都不能确认。考古发现的年代学也只是一个大概的时间范围，出土文物更复杂多样，本身就可以做不同的解释。在扩大的“二重证据法”指导下，对考古学新发现的一些资深解说家，不仅没有接受顾颉刚等对上古史的扫除工作，就连顾颉刚预言地下之材料“寻不出”的三皇五帝，也披上考古学的外衣，从他们的口中卷土重来。用“二重证据法”令考古来印证古籍，不仅穿凿附会，而且挑挑拣拣、各取所需，还常常会想不周到，顾此失彼。这样搞出的古史新证，鲜不成为秽史[45]。

更有甚者，在当今考古研究和文献证据的契合尚未取得显著进展的情况下，有学者著书立说，把传说当作史实，明确赋予夏以“信史”的地位[46]。这种信古依据是“疑罪从无”的推定，即如果文献记载中的夏无法证明其伪，就应该相信其真[47]。如果根据这种逻辑，司马迁《史记》里有不少明显属于传说和虚构且无法验证的内容，我们是否也都应该信其为真？我们是否也可以换个角度推定，对文献记载的“夏”因有疑而从无呢？因此，这种立足于并不证伪的“二重证据法”之上的夏代“信史”重建，并非科学论断，而只是作者主观的一家之言而已。虽然这种观点代表了中国考古学界的主流，但是科学界的共识并不能代表真理。

许宏认为，二里头文化主体与夏纪年的契合仍有很大的不确定性。传世文献中的夏是历史文本而非史实本身。虽然二里头遗址具有王气，但是没有文字仍然无法确定该遗址的属性。夏商世系并非史学意义上的编年史，而夏目前还是个既不能证实也无法证伪的问题。权且存疑，也不失为科学的态度。今天考古学早已跨过了证经补史的阶段，确定文献中的王朝未必是这门学科最重要的任务。中国考古学家应该去做自己擅长的对社会长期演变的研究，而不必纠结于纯属自己短板的对确切时间和事件的裁定[48]。

意大利历史学家克罗齐说，“一切真历史都是当代史”[49]。美国哲学家查尔斯·弗兰克尔说，“历史学解释根本不同于其他种类的解释，一种无可救药的主观‘阐释’因素会渗入到所有历史解释之中”[50]。傅斯年说，“一件事经过三个人的口传便成谣言。我们现在看报纸的记载，竟那么靠不住。则时经百千年，辗转经若干人手的记载，假定中间人并无成见，并无恶意，已可使这材料全变一翻面目；何况人人免不了他自己时代的精神，即免不了他不自觉而实在深远的改动[51]”。

我们研究历史的学者对我们所做历史判断的可靠性过于天真，但问题是我们不了解我们构建的历史其实是我们今天关注和想要的东西。这个真相就是，历史是一种构建，是一种我们根据自己学科编织而得到的幻觉。尝试回答有关你自己对过去的记忆，并尽量详细。你昨天这个时候在干什么？一周前？一年前？五年前？二十年前？当你越往前追溯，就会发现细节越发模糊，越发约略。如果你对自己过去的了解都不那么详细，不那么可靠，怎能希望别人详细提供你的历史？而且，笔者通过本人当时情绪、欲望和压力的筛子写下的日记，对于看日记的另一个人来说，因立场和体会不同，记录的事件肯定会有不同的意义和解释。

纳德尔（S.R. Nadel）对意识形态历史和科学历史做了区分。前者以过去人们造就的无数事件为特点，赋予这些事件以各种意义的程度取决于当下编撰历史事件的认识。这一编排过程是以当下的兴趣以及历史图像在社会中所发挥的意识形态作用为导向的。“真相”（reality）和“看法”（ideas）之间肯定存在差别，而且总是受到厚此（后者）而薄彼（前者）的对待。因此，不可能存在人们意识体验之外的历史过程，换言之，没有一种历史过程不是建立在集体记忆之上[52]。

由此视之，上面提及的夏代信史重建实际上是一种意识形态的历史，而非科学的历史。而且，这种上古史中的夏代并非真相，而只是看法而已。马克思和恩格斯说：“人们的观念、观点和概念，一句话人们的意识，随着人们的生活条件、人们的社会关系、人们的社会存在的改变而改变[53]。”这意味着，我们历史重建本身也是历史过程的产物，并受制于这一过程。

于是，历史重建遇到了两个未解的问题，第一，学者的主观性研究如何达到客观的效果，力求“返璞归真”？第二，我们如何获得这种客观知识，它既非相对主义的，也非虚幻的？

第一个办法是完全否定客观性的存在，客观条件如文献和考古材料只有当被研究者选择来从事研究时才是真实的，历史事实和学术重建都由我们大脑的分类过程所构建。于是，客观的文献和考古材料与科学研究的主观活动，都可以被看作是认知构建的产物。第二个办法以阿尔都塞马克思主义（Althusser Marxism）为代表。阿尔都塞将真相（reality）与认识（thought）严格区分对待。其中任何一方都无法被还原成另一方。第三个办法以波普尔的三个世界为代表。第一世界是物质和客体的世界，第二世界是意识和精神状态的世界，第三世界是科学思想、诗和艺术作品的世界。第三世界既不是客观的物质世界，也不是主观的精神世界，而是脱离了思想主体而存在的人类创造性思维活动的世界[54]。因此，文献与考古材料是人类思想和行为的创造，并非客体的世界，一旦产生就独立于真实历史之外，并塑造着世代传承的行为[55]。结果，历史作为民族认同的一种强有力支撑不是一成不变的东西，它不断在被重新评价，并反复进行着重构。

虽然历史学和考古学研究无法完全还原真实的历史，但是在提出某种看法和断言时应该独立评估证据的置信度和研究对象的不确定性，这就是理性主义的批判性思维。这种文献批判意识在我国学界最早体现在顾颉刚身上。他的古史辨就是用批判性思维考察古史记载的不确定性和历史记叙的前后逻辑关系。顾颉刚首先从尧舜禹的发生次序和排列系统之反背，提出古史形成的层累说，力图将传说与信史分离开来。他进而从古史形成过程来审视某种论述和某个事件的说法在二三千年里是怎么变化的，这样就能够去伪存真，将后世添加的层累部分剥离出来。如果我们认可“历史构建是当下回顾性意志行为强加于过去的形式”[56]，那么历代和现代史家不仅以其主观意志构建了他们所处时代的历史，而且会对过去的历史进行重构，以便使得这种记叙能够符合其所处时代的价值取向和要求。

今天“二重证据法”的历史研究，不应再从王国维纸上之材和地下考古之材相互印证的方法论来思考，认为文献史学占据着历史学的中心地位，而考古研究是为历史学提供佐证，是为后者服务的。我们可以将纸上之材看作是一种社会集体层累记忆的政治史，而地下考古之材大多是古人行为和日常生活的物质遗存，虽然它们无法像文字那样不证自明，但是能够通过学科交叉和科技手段的整合研究来提炼各种信息，重建生态环境、生计活动、经济贸易、手工艺技术、人口数量和变迁、社群规模与结构、食谱、营养与病理、种群分布与迁徙、社会结构、意识形态等方面的发展过程。考古材料与文献资料相比，较少受到人类观念和意志的选择性影响，能够提供比较客观、较少扭曲的历史图相，但缺点是残缺不全。因此，从考古材料获得的大量信息可以补充文献历史，并为文献研究提供一种全新的视野[57]。这种科学和脱胎换骨的“二重证据法”应该成为21世纪中国古史重建的必由之路，因为根据这两重证据的历史重建比较接近年鉴学派所预期的那种“整体史”。

从认识论来说，历史的重建无人能够做到真正的客观。即使学者力求尊重材料，秉笔直书，但这种叙述仍是个人或集体观点和立场的表达，总会受到价值观和流行偏好的左右。所以，在力求做到客观的同时，我们需牢记柏拉图的名言“认识自己，万事有度”[58]，对自身的局限和不足有清楚的自知之明。

注释

① 傅斯年.历史语言研究所工作之旨趣[M].湖南：湖南教育出版社，2003：11.

② 夏鼐.什么是考古学[J].考古，1984，(10)：931-935+948.

③ 乔治忠.王国维“二重证据法”蕴义与影响的再审视[J].南开学报(哲学社会科学版)，2010，4：131-140.

④ Patricia G. Material Culture and Text: Exploring the Spaces Within and Between[C]// Historical Archaeology. Oxford: Blackwell Publishing, 2006: 43.

⑤ Trigger B G. Beyond History: The Methods of Prehistory[M]. New York: Holt McDougal, 1968: 1–5.
⑥ Klejn L S. To separate a centaur: on the relationship of archaeology and history in Soviet tradition[C]. Antiquity. 1993, 67(255): 339–348.
⑦ 科林·伦福儒，保罗·巴恩.考古学：理论、方法与实践.第二版[M].陈淳，译.上海：上海古籍出版社，2015：3.
⑧ Johnson M. Archeological Theory: An Introduction[M]. Oxford: Blackwell Publishers, 2010: 7.
⑨ 布鲁斯·特里格.考古学思想史.第二版[M].陈淳，译.北京：中国人民大学出版社，2010：65+165.
⑩ Glyn D. A Hundred Years of Archaeology[M]. London: Duckworth, 1950: 244.
⑪ 布鲁斯·特里格.考古学思想史.第二版[M].陈淳，译.北京：中国人民大学出版社，2010：135–139.
⑫ Childe V G. Changing Methods and Aims in Prehistory: Presidential Address for 1935[J] //Proceedings of the Prehistoric Society. 1935, 1: 1–15.
⑬ Glyn D. A Hundred Years of Archaeology[M]. London: Duckworth, 1950: 247.
⑭ 夏鼐.夏鼐文集（中册）[M].北京：社会科学文献出版社，2000：4.
⑮ Trigger B G. Shang political organization, a comparative approach[J]. Journal of East Asian Archaeology, 1999(1): 43–52.
⑯ 杜正胜.考古学与中国古代史研究——一个方法学的探讨[J].考古，1992，4：335–346.
⑰ Elman R. Service, Origins of the State and Civilization[M]. New York: Norton Company, 1975: 104–164.
⑱ 希安·琼斯.族属的考古——构建古今的认同[M].陈淳，沈辛成，译.上海：上海古籍出版社，2017.
⑲ 张光直.取长补短、百家争鸣[N].中国文物报，1994–5–8.
⑳ 杜正胜.考古学与中国古代史研究——一个方法学的探讨[J].考古，1992，4：335–346.
㉑ Fredrik B. Ecological relationships of ethnic groups in Swat, North Pakistan[J]. American Anthropologist, 1956, 45(6): 1079.
㉒ Willey G R. Prehistoric Settlement Patterns in the Veru Velley, Peru. Washington DC, Bureau of American Ethnology[J]. Bulletin N, 1953: 155.
㉓ Trigger B G. Settlement Archaeology—Its Goals and Promise[J]. American Antiquity, 1967, 32(2): 149–160.
㉔ 布鲁斯·特里格.考古学思想史.第二版[M].陈淳，译.北京：中国人民大学出版社，2010：287.
㉕ 布鲁斯·特里格.考古学思想史.第二版[M].陈淳，译.北京：中国人民大学出版社，2010：342–344.
㉖ Ian H. Archaeology of Contextual Meanings[M]. Cambridge: Cambridge University Press, 1987.
㉗ 布鲁斯·特里格.时间与传统（重译本）[M].陈淳，译.北京：中国人民大学出版社，

2011：108.
㉘ 林恩·福斯特.古代玛雅社会生活[M].王春侠,译.北京：商务印书馆,2016.
㉙ 戈登·柴尔德.人类创造了自身[M].安家瑗,译.上海：三联书店,2008：140.
㉚ 布鲁斯·特里格.文字与早期文明[J].王哲昱,谢银玲,等译.南方文物,2014,4：184-189.
㉛ 陈淳.论中国考古学的编史倾向[J].文物季刊,1995,2：83-89.
㉜ Siân J. Historical categories and the praxis of identity: the interpretation of ethnicity in historical archaeology[M]//Historical Archaeology: Back from the Edge London. London: Routledge, 1999: 222-224.
㉝ Pedro P A, Funari S J, Martin H. Introduction: archaeology in history[M]//Pedro P A, Funari M H, Siân J, eds. Historical Archaeology: Back from the Edge London. London: Routledge, 1999: 1-10.
㉞ Patricia G. Material Culture and Text: Exploring the Spaces Within and Between[C]// Historical Archaeology. Oxford: Blackwell Publishing, 2006: 42-64.
㉟ 朱凤瀚.论中国考古学与历史学的关系[J].历史研究,2003,1：13-22+189.
㊱ 乔治忠.王国维“二重证据法”蕴义与影响的再审视[J].南开学报(哲学社会科学版),2010,4：131-140.
㊲ David N K. Marks and labels: early writings in Neolithic and Shang China[M]//Stark M T. Archaeology of Asia. Malden: Blackwell, 2006: 177.
㊳ Keightley D, Loewe M, Shaughnessy EL. The Shang: China's First Historical Dynasty[M]//The Cambridge History of Ancient China: From the Origins of Civilization to 221 BC. Cambridge University Press; 1999: 232-291.
㊴ 陈淳.论中国考古学的编史倾向[J].文物季刊,1995,2：83-89.
㊵ 李峰.西周的灭亡[M].上海：上海古籍出版社,2007：11-23.
㊶ 西山尚志.我们应该如何运用出土文献？——王国维“二重证据法”的不可证伪性[J].文史哲,2016,4：45-52.
㊷ 罗泰.宗子维城[M].吴长青,张莉,彭鹏,译.上海：上海古籍出版社,2017：12-13.
㊸ 刘周岩.寻找夏朝,中国从哪里开始[J].三联生活周刊,2018(23)：30-54.
㊹ 宁可.从“二重证据法”说开去——漫谈历史研究与实物、文献、调查和实验的结合[J].文史哲,2011,6：68-76.
㊺ 乔治忠.王国维“二重证据法”蕴义与影响的再审视[J].南开学报(哲学社会科学版),2010,4：131-140.
㊻ 孙庆伟.鼏宅禹迹——夏代信史的考古学重建[M].北京：三联书店,2018.
㊼ 刘周岩.孙庆伟:“信”比“疑”更难[J].三联生活周刊,2018(23)：56-59.
㊽ 刘周岩.许宏：无“疑”则无当代之学问[J].三联生活周刊,2018(23)：50-54.
㊾ 贝奈戴托·克罗齐.历史学的理论和实际[M].傅任敢,译.北京：商务印书馆,2005：2.
㊿ Charles F. Explanation and Interpretation in History[J]. Philosophy of Science, 1957, 23(3): 137.
51 傅斯年.史料略伦[M]//史学方法导论.北京：中国人民大学出版社,2004：5.

52 Nadel S R. A Black Byzantium[M]. Oxford: Oxford University Press, 1942: 72.
53 Marx K, Engels F. Collective Works. Vol. 1[M]. London: Lawrence and Wishart, 1959: 49.
54 卡尔·波普尔.客观知识——一种进化论的研究[M].上海：上海译文出版社，1987：114.
55 Rowlands M. Objectivity and subjectivity in archaeology[M]//Kristiansen K, Rowlands M. Social Transformations in Archaeology: Global and Local Perspectives. London: Routledge, 1998: 27–29.
56 Rowlands M. Objectivity and subjectivity in archaeology[M]//Kristiansen K, Rowlands M. Social Transformations in Archaeology: Global and Local Perspectives. London: Routledge, 1998: 26.
57 Lipe W. Value and meaning in cultural resources[M]//Clear H. Approaches to the Archaeological Heritage. Cambridge: Cambridge University Press, 1984: 6.
58 这是刻在古希腊阿波罗太阳神殿上的柏拉图名言：Know thyself, Nothing in excess。

第四章

▼

称象分石千里邦

2020年人类学终身成就奖获得者

——郑连斌

人类学生涯回顾

和大哥郑连康摄于北京外交学院(1958年)

我1948年11月出生于江苏淮阴五里庄,1954年离开故乡去了北京,所以对故乡的印象不深。记得小时候脑子里总出现奶奶的面容,还有麦田以及门前水塘(我们叫大柴汪)。我的父亲讳德麟,原来在南京大学上学,由于不喜欢所学的专业,当华北大学(中国人民大学的前身)俄文系来南京招生时便报名入学,1950年初去了北京,毕业后留校工作。1954年,母亲携哥哥和我北上京城去寻父亲,母亲没有文化,几千里的路程,还要带着我们兄弟二人汽车转火车,火车再转火车,一路上实在不容易。哥哥叫郑连康,长我四岁,在徐州火车站候车室等待转车时,由于旅客很多,秩序较乱,哥哥去上厕所,出来后差点找不着我们,要是把哥哥丢了,那可是要命的事情。记得刚到北京,我们住在西郊,那时西郊的中国人民大学只有几座红楼。没过多久,母亲又带我和哥哥回老家办理了户口迁移手续。之后我们又重新来到北京,住在了东四六条胡同的74号。1955年,我在东四五条胡同的一所私立学校普育小学上了一年级。1956年,父亲调到外交学院编译室工作,所以我们又搬到了阜成门外百万庄的外交学院校园中居住,我就成了北京展览馆路第二小学的学生。

1961年,北京各个单位纷纷下放干部,父亲支援边疆到了内蒙古呼和浩特,几经折腾在内蒙古师范学院外语系当了老师,可一时不知什么原因,暂

内蒙古巴盟师范学校同学合影（1974年）

时先被安排在师院附中教书。我1961年先考入呼市二中，又转到师院附中上学，这样我父亲就成了我的俄语老师。1963年，父亲离开附中，回到师院外语系。1964年我升入附中高中。1968年，我到内蒙古河套地区的磴口县上山下乡。1973年，我参加了高考，考得很好，我记得四门课考了418分，人们风传我是“状元”，可惜张铁生一封信，所有高考成绩作废，我的大学梦破灭了。所幸，河套地区急需教师来补充中小学师资，于是巴盟师范学校招收了我们这样一大批老三届的知青。我在数学班学习，班主任是王立洎老师。1974年我毕业后又回到磴口县，先后教授初中、高中的数学，这样我在河套地区一呆就是九年半。1977年底终于传来好消息，要全面恢复高考招生了。1978年春天，我进入内蒙古师范学院生物系学习，圆了自己的大学梦。1978年对于我们家是个吉祥的年份，春天我弟弟郑连勇考入中国矿业大学，夏天我妹妹考入内蒙古大学生物系。说实在，当时我对生物学并不熟悉，也就不喜欢，入学后一直想转到数学系，因此我向学校提出转系申请。一年后，学校教务处通知我，可以转到数学系了，此时，通过一年的学习，我对生物学已经有了初步的认识，所以我就继续留在生物系。1982年，我毕业留校，生物系安排我讲人体及动物生理学课程和实验。1997年，我调入天津师范大学生物系工作。如果从1981年作本科毕业论文开始算起，我从事中国

1979年夏天烟台野外实习

民族体质人类学研究已经近40年，40年间，科研已经成为我生命的一部分。

机缘巧合，初次接触人类学（1981）

大学四年级的时候，学校通知必须做毕业论文才能毕业。我们的“人体及动物生理学”教授杨国荣先生给了我和郃晓春、赛娜、苏荣四个学生的本科毕业论文题目是“蒙族、达斡尔族ABO血型的研究”。接到这个题目时，我们心中没有底，对如何开展科研工作一窍不通。老师给出题目，如何完成就全靠自己的努力了。在完成本科毕业论文过程中我初学了如何查阅文献（当时没有网络，只能泡在学校阅览室里查阅各种杂志），如何进行调查（经常在晚上敲门入户进行测量），如何进行数据处理（我的数学功底有了用武之地），如何撰写论文（我的文史地理知识派上用场）。经过不断地摸索，我们终于完成了论文。我们的毕业论文得到导师杨国荣教授的高度评价，并发表在1984年的《内蒙古师范大学学报》（自然科学版）上。这是我发表的第一篇学术论文，也是我科研生涯的序曲。大学毕业后，我留校师承杨国荣教授，讲授“人体及动物生理学”。在我国，医学院校是研究生理学的主力，一些著名的综合性大学多多少少也在开展人体生理学研究，但地方师范院校人体生理学研究力量很薄弱，几乎没有科研条件。教学和科研是高校教师的两条腿，中国的高校负有教学和科研双重任务，高校教师只从事教学，而不开展科学研究，是不合适的。可我未来研究方向是什么？我的科研之路如何走？这些摆在面前很现实的问题让我很茫然。

起步艰难，探索中学习（1982—1996）

我的科研起步用了14年，可见起步之艰难。万事开头难，当时高校科研气氛不浓，很多老师没有科研的意识，也没有科研方向。我的导师杨国荣先生是我国著名消化生理学家、细胞生物学家王堃仁先生的弟子，他的研究方向是消化道生理学。但当时内蒙古师范大学没有开展生理学研究的条件，所以我的科研方向都要靠自己摸索。留校任教第一步是过教学关。从生理学实验开始，逐步掌握每一个实验的技巧；理论课的教学是从听导师的课，逐步过渡到给专科班讲课，再到给本科生讲课。1986年，在留校4年之后，我终于被允许登上本科生课堂的讲台了。我的教学逐步走上正轨，科研成为摆在自己面前下一个现实的问题。

14年间在自己的努力下，从没有明确的科研方向，没有经费，没有技术，没有团队到有了明确的科研方向，初步掌握了研究方法、统计方法，在内蒙古师范大学建立了人类生物学研究团队。在这段时期，获得了第一笔科研基金资助；得到了朱钦、杜若甫二位先生的指导；参加了中国遗传学会首届人类及医学遗传学会议；参加了卫生部、国家民委、国家统计局联合开展的项目“全国少数民族人口健康素质抽样调查”。这一段是我科研的萌芽阶段。

坚定我在体质人类学研究道路上前行的，是一个偶然的机遇。我有一个爱好，喜欢泡图书馆，这是在内蒙古师范学院附属中学时期就养成的习惯。在中学阶段，我就在学校图书馆帮助借阅图书，这个工作没有任何经济报酬，但可以获得进入书库的机会，走进学校的书库，我发现图书馆真是知识的海洋。大学毕业留校后，泡图书馆的习惯保持下来。内蒙古师范大学图书馆馆藏图书还是很丰富的。一天我在图书馆里发现了一本云南人民出版社1982年出版的《中国八个民族体质调查报告》，书不厚，是上海自然博物馆和复旦大学一些从事体质人类学研究的专家写的。书中收录了对云南省境内的基诺族、布朗族、哈尼族、彝族、傣族、白族、藏族、瑶族等八个民族的体质形态所作的初步调查数据和分析，还介绍了体质人类学的研究方法。刘咸教授为这本书写了序，序中提到要建立中国人的体质数据库。我仔细看了这本书，做了详细的摘抄，隐隐感觉这些内容就是我未来希望做的。内蒙古自治区是多民族共同生活的地区，除了汉族、蒙古族外，还有一些特有民族，如达斡尔族、鄂温克族、鄂伦春族、回族、满族、朝鲜族、俄罗斯族等等。结合地区优势，在内蒙古开展中国民族体质研究应该是很好的科研方向。但是当时的我对体质人类学一窍不通，连书中提到的“眼耳平面（法兰克福平面）”是什么都不清楚。万事艰难在起步，应该说我是在摸索中开始走上了体质人类学研究的漫长道路。有了想法，可路怎么走呢？先从查阅论文开始，在发表的论文中寻找研究方法。后来听说内蒙古医学院解剖教研室的朱钦先生就是搞体质人类学的专家，我便去拜访朱先生，他热情地接待了我。朱先生是南通人，北京大学医学院毕业，高高的个子，很帅气，是位文雅的书生，讲话时略带有南方口音。因为是初次见面，所以我和朱先生只进行了简短的交谈。后来，我在新华书店看到了邵象清先生编的《人体测量手册》，如获至宝，买下仔细研读，遇到不懂的地方就向朱先生请教。我们大学虽然是师范院校，不是医学院校，但也很正规地开设过“人体组织胚胎解剖学”课程，我认真学了这门

1985年内蒙古师范大学师生合影

课。书中提到的一些名词，我还能看懂。遇到不清楚的解剖学名词，就翻开教科书，重新学习。

从1985年开始，我们主要在呼和浩特市开展科研工作，开始没有团队，就带领一批本科生工作。我利用指导本科生毕业论文的机会在高校和中学进行人类学指标的调查、测量工作。呼和浩特是一个多民族杂居的城市，回族居住相对集中在回民区，其他民族居住比较分散。不过，呼和浩特的大专院校和中学有不少少数民族学生，这样一来，寻找调查对象不是件太困难的事情。师院附中每年都专门招收内蒙古各个地区的优秀的蒙古族学生，这些学生就是我们的研究对象。我们又去了呼和浩特回民中学开展了回族的调查。此外，我们还委托郑明霞老师在内蒙古医学院调查，委托张心志同学和任茂盛同学在呼和浩特的中学调查，其间也委托在通辽地区的毕业生调查了科尔沁蒙古族的资料。这样，我开始了中国民族体质研究的漫漫之路。

1985年至1996年，我主要在内蒙古自治区开展工作。由于缺少研究经费，我在1981年开展了蒙古族、达斡尔族ABO血型研究后，就把大学生、中学生作为研究对象，开展一些简单的经典遗传指标的调查工作。这些指标调查不需要花费很多经费。发旋、手指毛、上眼睑皱褶、蒙古褶和肤纹都是体质人类学重要的研究指标。当时，蒙古族这方面的调查还没有见到相关研究资料的报道。苯硫脲是一种化学试剂，有些人是在浓度很低的时候，就能尝出苦味儿；有些人是在浓度较高时才能尝出苦味儿，还有些人则尝不出苦味。尝味能力的差异是由遗传决定的，存在个体之间、族群之间的差异。我最后在北京终于买到了苯硫脲试剂，开始测量蒙古族、汉族的苯硫脲尝味能力。工作几天终于完成了内蒙古蒙古族的苯硫脲尝味能力测量工作。现在回想起来，当时的工作虽然只是初步调查，也不深入，但确是我们科研中一个很重要的必经阶段。我们在困难中摸索、前行，学会一些科研的方法、数据统计的方法，也形成了科研团队的雏形。

1989年，我获批了第一个科研项目。有次我在生物系遇到了余诞年老师。余诞年老师是上海人，多年从事番茄育种工作，给我们讲授遗传学课程，学问很好。他要参加内蒙古教育厅科研项目指南的编写工作，问我有没有好的建议，我提出希望能够将内蒙古少数民族体质研究纳入指南中。不久我以“内蒙古蒙古族、汉族、朝鲜族和回族四个民族体质人类学与人类遗传学研究”为课题申报了内蒙古教育厅的科研项目并成功立项，获得资助经费八千元。这个项目要完成四个民族的青少年生长发育、营养状况、优势眼、眼部特征、耳部特征、苯硫脲味觉、指毛、肤纹学、月经初潮、听觉阈限等一系列指标的测量或调查。于是1989年我们正式组成研究团队，开始了比较规范、系统的体质人类学研究。1991年我们团队第一次离开呼和浩特，去兴安盟的乌兰浩特市开展东北汉族、科尔沁蒙古族、朝鲜族的经典遗传学指标的调查。这也是团队第一次野外工作，团队成员在兴奋的同时也有些迷茫、忐忑。所幸那次工作还是比较顺利，最终，我们用八千元科研经费顺利完成了课题。

1992年，在听说了中国遗传学会首届人类及医学遗传学会议将在南京师范大学召开后，我们将已经调查的数据进行汇总、统计，写成了十几篇论文摘要投给会议，内容涉及朝鲜族、蒙古族、回族、汉族的眼部特征、耳廓、月经初潮、耳垂、发旋、肤纹等内容。会议的主持人、中国著名的医学与人类群体遗传学家杜若甫先生很希望我们参加会议，愿意资助我们会务费。我久仰杜先生的大名，也非常感激杜先生对我们的支持。这是我第一次出席全国性的学术会议。我参加会议的目的主要是两个：一个是介绍我们研究组的科研成果，另一个是向同行请教、学习。我对会议充满着好奇，因为在会议上会认识很多知名的人类遗传学家，也想知道大家在这个领域都做了哪些工作，哪些工作我们也可以做。到了南京师范大学，首先惊讶的是校园之美。这个会议上我见到了强伯勤院士，强院士和我住在同一间房，这也让我更好地感受到强院士的朴素、低调；强院士介绍完美国的人类基因组测序计划就离会了。我见到了尊敬的杜老师，以及他的两位学生金锋和陈良忠，金锋是蒙古族，陈良忠曾经是内蒙古医学院的老师，所以我们虽是初次见面但还是有亲切感的；我还见到了艾琼华老师，艾琼华老师是四川人，在伊宁卫校工作，科研搞得很好；我也见到了昆明医学院的李明老师。会议上，我向各位专家求教，如我们那样的学校，人类遗传学科研工作将如何继续开展下去。通过问题的求教也让我感觉国内的知名专家待人都很诚恳、平和，很愿意奖掖后进。后来的科研工作中，我也秉持这样与同行交往的态度。会上，杜老师还让我给大家演示优势眼的调查方法，大家按照我的讲解方法操作，都感到很神奇。

会议中我还得到了一个重要的信息：卫生部、国家民委、国家统计局正在联合开展“全国少数民族人口健康素质抽样调查”，这个项目的具体工作由北京医科大学张致祥先生负责。这个重要的信息，影响了我一生的科研工作。

回到呼和浩特，我立即和北京医科大学科研处的韦老师联系，希望能够参加这个项目。韦老师把我的信转给了张致祥老师，张老师很快给我回信，邀请我到北京完成项目中关于成人体质测量部分的课题设计。后来我才知道原来项目中没有成人体质测量这部分内容，是采纳杜若甫先生建议后加上的。说实在，我在人体测量学领域也是个新手但是考虑到这是个机会，不能推辞，就应允了。我带上吴汝康等编著的《人体测量方法》以及邵象清编著的《人体测量手册》到了北京。在北京我拜访了杜若甫先生，杜先生又向我推荐了张振标先生。张振标先生详细为我分析人体形态学指标中，哪些指标是最重要的、反映种族差异的。我基本采纳了张老师的意见，最终完成了课题设计的任务。

后来，我和项目组成员一起到各地对全国参调人员进行培训。第一个培训点是兰州，在那里我见到了戴玉景先生。第二个培训点是南宁，在那里我见到了朱芳武先生。戴玉景先生和朱芳武先生都向我提供了培训工作所需要的测量仪器。第三个培训地点是延边，在那里我感受到了朝鲜族的能歌善舞。在培训中，我为全国各个地区的老师讲解各个指标的测点、测量技巧和注意事项。培训结束后我先是应内蒙古卫生厅邀请，到哲里木盟科左后旗参加了内蒙古卫生厅组织的科尔沁蒙古族体质测量工作。科左后旗是著名的晚清科尔沁亲王僧格林沁的故乡，也是著名的蒙古正骨医学之乡。后来又应宁夏卫生厅邀请，和朱钦先生结伴去银川参加回族体质测量组工作。1996 年我又参加了朱钦先生主持的国家自然科学基金项目，去了呼伦贝尔的莫力达瓦旗和鄂伦春旗，测量达斡尔族和鄂伦春族的体质数据。这几次野外测量工作，我分别负责体部和头面部的测量，也向朱钦先生学习了很多野外工作的知识，为我以后独立开展体质测量工作打下了比较坚实的基础。

在探索研究中，我们还对手肤纹、学生月经初潮、蒙古斑（骶部色素斑）、Heath-Carter体型法等进行了研究。

说起我们的手肤纹研究，还要特别感谢张海国老师。我们是通过张海国老师发表的论文开始接触肤纹调查并掌握测量方法的。1989—1991年期间，我们采用印油捺印了蒙古族、汉族、回族、朝鲜族4个民族的中小学学生手肤纹图样共计3 237例。我们报道了内蒙古地区汉族、蒙古族、回族、朝鲜族的肤纹特征与指纹白线特征，并对中国人群肤纹进行了主成分分析，研究证实，根据肤纹特征，可将中国人分为6个组，除南、北族群组外，还存在西藏组、新疆组、南北混合组与南北过渡组。我们提出，由于汉族与少数民族间、南北方民族间存在肤纹基因的交流，将中华民族分为两大类群的观点可能过于简单化了，据此，我们发表了《中国人群肤纹的主成分分析》一文。马慰国、杨汉民编著的《实用医学皮纹学》用了两页的篇幅全面地介绍了我们的论文："郑连斌（2002）的皮纹主成分分析结果提示，既往通过人群遗传、生化、体质特征的研究将中国人群简单分为南、北两大类群的观点可能过于简化了，值得商榷"。

1987年、1990年、1993年我们分3次在呼和浩特市区和郊区调查了女生月经初潮有关资料，用回顾法与现状法两种方法进行了分析。在《呼和浩特地区蒙、汉族女生月经初潮年龄变化》一文中，我们提出农村女性MMA（平均初潮年龄）提前速度略快于城市，蒙古族MMA早于汉族，城市女性MMA提前速度已经减慢，但农村仍较快，市区蒙古族、汉族MMA已降至我国经济发达地区大城市的水平，已接近遗传因素所控制的MMA值

1994年考察阳关遗址时狂风大作

下限。我们调查了8个群体月经初潮月分布状况，发现月分布曲线均呈双峰特征：1个位于12—2月，1个位于6—8月，认为双峰产生受生理、环境、心理、遗传等因素的影响。朱光亚、周光召主编的《中国科学技术文库：普通卷（生物学，医药卫生）》详细引用了我们这篇论文，并对我们论文中调查对象和方法、结果、分析三个部分进行了大量的介绍。

蒙古斑（骶部色素斑）是蒙古人种的形态特征之一。我详细向杜若甫先生询问了蒙古斑的测量、统计方法。我们和医院、幼儿园合作，对蒙古族、汉族新生儿及儿童的蒙古斑进行了近两年的调查，对其颜色、数量、分布、面积进行了详细的统计，发表了我国非常珍贵的蒙古族、汉族蒙古斑资料。由于蒙古斑研究难度大，所以至今尚未见其他学者新的资料发表。

Heath-Carter体型法是国际学术界常用的一种研究体型的方法。季成叶、赵凌霞最先将Heath-Carter体型法引入中国，并且开展了汉族的体型研究。我专程去了古脊椎动物所，找到英文原著，复印下来。回到呼和浩特，我向朱钦先生介绍了体型工作并提出可以在内蒙古开展。朱钦先生带领我们率先开展了少数民族和汉族的Heath-Carter体型法研究。此后我国民族的Heath-Carter体型法研究一度成为体质人类学研究的学术热点。

草创阶段，韧性的坚持（1996—2009）

20世纪90年代中期开始的十几年间是中国体质人类学发展的艰难阶段。由于从80年代开始活跃在中国学术界的老一批体质人类学家已经陆续隐退，再加上微观研究、细胞生物学技术、基因分析技术的发展，宏观的体质人类学研究的空间被大大压缩，课题难拿，没有经费，一些年轻人的科研开始转向，一些有很好研究基础的体质人类学团队解散了。传统的体质人类学研究出现了冷落、萧条的状态。在这种不利的大环境下，我们团队坚持了下来。在这一段时期，我先后主持了“内蒙古8个民族18项指标的人类群体遗传学研究”“中国11个少数民族体质特征的人类学研究”“中国僜人、克木人等6个人群的体质人类学研究”三项国家自然科学基金项目。我们的科研都是围绕这三个基金项目开展的。主持国家自然科学基金项目是一个很大的荣誉，也是很大的责任，它并不完全表明你有很高的科研水平，但它反映了你的科研方向得到了专家的认可。自从认识了杜若甫先生，参加了国家三部委主持的重大项目，又参加了朱钦先生的国家自然科学基金项目，我就琢磨，我可不可以也申报一个国家级项目呢?

1994年我已经46岁了,虽是科研新兵,但已经是年届中年了。那一年,我第一次申报国家自然科学基金项目,但是由于课题设计有问题,学术知名度也不高,工作积累不足,第一次申报失败了。1995年,我向杜若甫先生请教,杜先生很热情,帮助我设计项目内容,我进行了第二次申报,还是申报不果。我没有灰心,1996年春,杜若甫先生重新帮助我设计并增加了研究内容,紧接着,我进行了第三次申报。终于,在1996年9月,我的第一个国家自然科学基金项目"内蒙古8个民族18项遗传指标的调查"获得批准!我也成了国家自然科学基金项目的主持人。

1995年,我的女儿郑琪考入天津师范大学生物系,我也于1997年调入天津师范大学工作。1998年我就任天津师范大学生物学系主持工作的副主任一职,不久学校成立化学与生命科学学院,我又被任命为副院长。我一直觉得自己就是一位学者,有了职务,诸事缠身,再加上自己不适应,也不喜欢搞行政工作,于是任职两年后向学校提出了辞职。由于我辞职的决心很大,辞职最终获批,我又可以一心一意做自己喜欢的事情了。随即我在天津师范大学申报了遗传学硕士学位点,学位点获批后,我开始招收研究生,组建了天津师范大学人类生物学团队。天津师范大学和内蒙古师范大学两个人类生物学团队都是我组建的,这两个团队紧密合作,积极申请国家自然科学基金,研究对象也从内蒙古自治区诸民族逐渐扩展到全国范围的族群。

1997年,调入天津师范大学后,我开始实施1996年我在内蒙古师范大学申请获批的第一个国家自然科学基金项目。这个项目要求我们完成内蒙古8个民族18项遗传指标的调查工作,我们实际工作中将指标数量扩展到25项(5项舌运动类型指标、7项不对称行为特征、9项头面部指标、4项与手足有关的指标),族群数量也增加到18个:阿拉善蒙古族等9个蒙古族族群、伊盟汉族等4个汉族族群、鄂温克族、鄂伦春族、朝鲜族、达斡尔族、回族。为了完成这项工作,我和团队老师去了呼伦贝尔盟、兴安盟、锡林郭勒盟、伊克昭盟、巴彦淖尔盟、阿拉善盟、呼和浩特市,取得了大量的第一手资料。第一个国家基金项目也取得了丰富的研究成果。

我们最先开展了中国北方族群的舌运动类型研究,发现男女间各舌运动类型出现率不存在性别间差异。卷舌与其他4种舌型分别存在互相作用关系。卷舌基因对叠舌基因是隐性上位基因,对翻舌、尖舌、三叶舌是修饰基因。以后的多数群体的研究亦表明，各舌运动类型出现率不存在性别间差异，5种舌运动类型之间存在着较为密切的相关关系，不同族群5种舌运动类型出现率往往存在着差异。我们对内蒙古18个族群多元分析显示舌运动类型呈现地域特点、民族特点，民族内部的多元性及民族间出现一定程度的融合。尖舌性状是我们研究组最先发现并进行研究的。

与此同时，我们还开展了18个族群的扣手、利手、叠臂、叠腿、起步类型、利足、优势眼7项不对称行为特征研究。以往除零散利手资料外，我国学术界未见这方面的研究报道。起步类型是我们团队最先开始研究的性状，我们发现绝大多数不对称行为特征与性别无关，多数特征右型率>50%，多数特征的出现率男女间差异不显著。7项特征间彼此相关程度较高，利手与扣手、利手与叠臂之间存在着明显的互作关系，而叠臂与扣手之间则无相关。相关项目中右型-右型组合或左型-左型组合是属于亲和特征。其他族群资料也支持上述观点，左型-左型组合率低于右型-右型率。对内蒙古18个族群资料多元分析显示，不对称行为特征的出现率也表现出民族性、地域性和民族间基因交流的特点。国外学者对优势眼的研究很少，我国过去没有开展过研究，我们是最先开展中国人优势眼研究的团队。我们发现中国人右优势眼率为70%~80%，无性别间差异，优势眼与利手存在一定关系。

我们还对中国18个族群蒙古褶、上眼睑皱褶、铲形门齿、鼻梁侧面观、鼻孔形状、下颌类型、耳垂、额头发际、头发类型9项头面部和拇指类型、环示指长、指甲类型、足趾长4项手足遗传学指标特征进行了报道。指甲类型是我们团队在国际学术界最先开展研究的性状。我们发现拇指类型率无性别间差异，环示指长率性别间存在显著性差异，汉族、回族、蒙古族环示指长率高。我们对内蒙古18个族群的头面部和4项手足遗传学指标的多元分析显示，同一民族表现出较强的一致性。蒙古族诸族群窄鼻孔率、有耳垂率相对较高，内蒙古西部区汉族过伸拇指率、扁形指甲率、窄鼻孔率、有耳垂率均较低。

在报道各民族经典指标出现率时，我们对已知遗传方式的性状计算了基因频率。我们对兴安盟3个民族10对性状基因频率进行了计算。结果显示，汉族与朝鲜族间差异较大，蒙古族与朝鲜族间差异次之，汉族与蒙古族间差异最小。我们报道了鄂温克族、鄂伦春族、达斡尔族、蒙古族、汉族12对遗传性状的基因频率。研究显示：5个民族间，蒙古褶基因频率差异较大，叠舌次之，继之为上眼睑皱褶、鼻孔形状和耳垂类型，利手与鼻梁侧面观性状的基因频率民族间差异最小，男性环示指长性状民族间无显著性差异。应该说，我们第一个国家自然科学基金项目取得了丰硕的研究成果。

2002年，学术界已经报道了中国44个少数民族的体质数据，还有11个少数民族没有开展研究。所以在完成了第一项国家基金项目后，我们又申请到了国家自然科学基金项目“中国11个少数民族体质特征的人类学研究”，这11个民族一般都是位于边疆的山区，或者交通条件困难的地区。申请这个项目时，我还邀请了广西医科大学的魏博源教授写了推荐书。我和魏教授素昧平生，魏先生慨然应允，让我非常感动，后来听说魏先生因病去世，我也恻然良久，觉得这是我们体质人类学界的重大损失。为了使得中国55个少数民族都具有至少一份当代人的体质数据，我们开展了布依族、京族、仫佬族、佤族、乌孜别克族、俄罗斯族、塔塔尔族、怒族、独龙族、门巴族、珞巴族的研究。这个国家项目获得了16万元的项目经费，我们团队要去西藏、新疆、贵州、广西、内蒙古的呼伦贝尔大草原、云南，单靠这些经费是不够的，所以开展起来非常困难。后来我们又申请到了第三项国家基金项目，就把第二和第三个项目合在一起开展。团队克服了重重困难，最终完成了任务（塔塔尔族工作是崔静教授完成的）。

在测量数据的基础上，我们撰写了论文《我国23个群体体质的聚类分析与主成分分析》。经过聚类分析与主成分分析，我们认为中国人的体质特征除南北两大类型外，还有第三个类型——操藏缅语族语言的藏彝走廊类型，其体质特征介于南北两大人群之间，族源为古代氐、羌人。其实这个第三种类型是黎彦才老师最先提出的，我们予以证实。李法军在专著《生物人类学》（中山大学出版社，2007）书中引用了我们的第三种类型的说法，并认为藏彝走廊类型“在命名上或可商榷，但该地区存在一个南北类型的过度类型应当是一个较为客观的认识”。杜若甫老师在专著《中国人群体遗传学》用了近两页篇幅引用了我们本篇论文的聚类分析方面的主要研究成果，并引用了我们的男女聚类图。关于中国人体质类型问题，杜先生也有自己的看法，他从遗传学角度发现并不存在东亚类型体质，认为所谓的东亚类型人群或可归入北亚类型，或可归入南亚类型。我们的观点虽然有别于杜先生，但杜先生在专著中还是介绍了我们的研究成果。金力、褚嘉祐先生在专著《中华民族遗传多样性研究》中介绍我们选取了9项头面部测量指标和4项体部测量指标，对已发表的我国23个少数民族群体的数据资料进行了聚类和主成分分析，结果支持有第三种类型存在的观点。在南方组和北方组之间还存在一个中间类型组，即第三种类型组，也就是所谓的“藏彝走廊类型”。此外，我们的研究成果在《中华民族遗传多样性研究》、《实用医学皮纹学》等专著、论文中被肯定和大篇幅引用。金力、褚嘉祐先生指出“以上研究从民族体质人类学角度充分地证明了一点，即现代各地区的中国人（汉族和少数民族）并非各自独立发展，而是在密切的交往和不断的分化、融合演变中形成的”，表明作者对我们的探索的认可。

2007年带领科研团队在拉萨大昭寺

除了55个少数民族外，中国还有一些人，当你问他们的民族时，他们的回答可能出乎你的意料，“我是穿青人”“我是革家人”“我是僜人”“我是蔡家人”。这些称为“人”，而不称为“民族”的族群在学术上被称为“中国未识别民族”。未识别民族是指有些族群民族身份不明，至今尚未归属于中国56个民族中的某一个民族，其民族身份未被中国政府所承认的少数民族族群。还有些族群虽然已经被划入某一个民族之中，但在学术界对这些族群的民族身份始终存在明显的不同意见，至今这种争议并没有得到明晰的解决。他们对自己的民族身份提出质疑，不认可自己目前的民族身份。我国著名的体质人类学家席焕久先生建议用“未识别民族”一词来称号这一类特殊的族群。未识别民族一般具有强烈的民族愿望，几十年来一直要求国家承认自己的民族身份，为此而奔走、呼吁。未识别民族也是中华民族的一部分，我们在研究中国人体质时，不能忘记这些特殊族群。于是在完成第二项国家自然科学基金项目后我们申报了第三项国家自然科学基金项目“中国僜人、木人等6个人群的体质人类学研究”。获批后，我们开展了莽人、克木人、僜人、布里亚特人、图瓦人、云南蒙古族的特征测量工作。这项工作也取得了较多成果：我们认为僜人体质区别于周边民族，应该是一个独立的族群。图瓦人体质特征更接近哈萨克族等突厥语族民族。布里亚特人体质特征接近于蒙古族其他部

2007 年在呼伦贝尔草原调查布里亚特人

2007 年在西藏察隅调查僜人

2008 年在新疆喀纳斯调查图瓦人

2009 年在四川邛崃测量汉族头面部指标

2012年在贵州安顺调查屯堡人妇女

2010年研究团队抵达海南文昌

落，布里亚特人从体质特征角度分析应该是蒙古族的一个支系。云南蒙古族在中国南方边疆生活了700多年，其体质特征已经接近于南方少数民族，云南蒙古族总体上属于蒙古人种南亚类型体质，但在南方族群中，云南蒙古族属于体格比较强壮的类型，是南亚类型中体质相对接近于北亚类型的一个族群。云南蒙古族体质与阿昌、白族、彝族最为接近。我们认为云南蒙古族体质形成与其族源及与周边族群的基因交流有关。

汉族体质人类学研究（2009—2013）

真正反映研究团队科研实力的是看团队是否主持过国家自然科学基金重点项目、重大项目，看科研团队是否主持过国家科技部973项目、863项目或基础性工作专项。对于我们这样的地方院校，申报重大项目、973项目、863项目是不可能获批的，重点项目也是很难有机会轮上我们。1996年，中国解剖学会的学术年会（南宁会议）召开期间，我在朱钦先生鼓励下，提出开展中国汉族的体质研究想法。当时席焕久老师专门召开人类学专业委员会会议进行讨论，对此方向认可，并提出了很好的建议。会后的两年我都致力于申报关于汉族体质研究的面上项目，均未果，后来只好把研究对象又锁定为中国少数民族，但研究中国汉族体质的想法一直未泯。2007年我拜访朱钦老师时，又提起这件事，朱老师嘱我，一个面上项目难以完成汉族体质研究，不如干脆申报国家自然科学基金重点项目，来完成汉族体质的测量工作。受朱老师的启发，我路过北京时，专门去了国家基金委，见到了基金委的江虎军老师，谈了我的想法。江老师建议我先提交一份材料，回到天津，我就把材料很快写好，发给江老师。后来我又抽时间去北京商量此事，江老师给我讲解了国家自然科学基金重点项目申请的相关事宜。原来，当时重点项目申请方法有两个途径：多数是领域申请，领域申请是按照基金委公布的重点基金支持领域来申请，中标可能性大；少数是自由申请，自由申请项目的研究方向比较杂，中标率低。因为我很迟才向基金委提出汉族体质研究方向，错过了领域申请的申报时机，只能是申报自由申请了。汉族体质研究课题很大，单凭天津师范大学一个学校难以完成。开始我准备联合国内多所高校一起申报，后来知道合作单位只能有两个，因此我决定和内蒙古师范大学，以及辽宁医学院同仁合作来申报项目。撰写申请书的过程是理清思路的过程，也是完善研究方案的过程，写作起来比较困难，主要是要站在一定的高度上来评价这个研究意义，来规划这个项目的实施。重点要写明为什么要进行中国汉族体质研究？这个项目要解决哪些科学问题？有什么科学意义？研究的创新性在哪里？项目的申请书经过多次反复修改，最后定稿。在申报过程中，学校领导、学校科技处领导给予了很大的帮助，国内的一些专家对汉族体质研究也予以支持。2008年7月初，我申报的项目已经通过初审，将在7月下旬吉林大学会议上进行答辩，得此消息，我们很高兴，因为只要让答辩，就有可能立项。事后我才知道，我们项目虽然通过初审，但评分不是很高，而在答辩会上有相当一批申报项目会刷掉，首先刷掉的是一些初评时评分不很高的项目。我把通过初审的消息报告了席老师，席老师说，应该能够通过，我知道，他是在鼓励我。科技处吴晓荣处长决定和我一起去吉林大学参加答辩，这给了我很大的支持。我抓紧时间做好答辩前的各种准备工作，制作答辩用的幻灯片，准备回答专家可能提出的各种问题。我和科技处吴处长在规定的时间到了长春市，到指定的宾馆报到时，我看到细胞遗传和发育生物学方向准备答辩项目的单位名单，不由倒吸一口冷气，四十几个入围答辩项目中，基本上都是中国科学院系统、北京大学、清华大学、复旦大学和国内其他知名高校的学者申报的项目，天津师范大学能够名列其中已经是万幸。在宾馆里，我们仔细修改了答辩时所用的幻灯片内容。答辩那天，答辩委员会专家就有近三十位，而且有三分之一是从国外请来的专家。轮到我答辩时，我已经不太紧张了，我比较快速地介绍完项目，没想到专家对这个项目反应非常热烈，纷纷举手提问。有的专家问我取样地点和方法，有的专家问我是否到国外取样，总之，我们申报的项目引起了专家的重视。答辩完，我走出答辩室，心里感觉轻松。

后来得到消息，我们申报的国家重点项目获批，并得到150万元资助。"汉族体质人类学研究"（项目号：30830062）由天津师范大学、锦州医科大学（原辽宁医学院）、内蒙古师范大学人类生物学学者组成研究团队，将于4年内完成。后来，我听说这是国家自然科学基金委员会生命科学部解剖学方向上获批的第一个国家自然科学基金重点项目。

立项后研究工作随即展开，项目于2012年底结项。结项后，还剩有一些经费，我们又陆续前往黑龙江、吉林、河北开展了一年的汉族测量工作。研究组按照汉语方言对汉族北方话族群（东北方言族群、华北方言族群、西北方言族群、西南方言族群、江淮方言族群）、吴语族群、赣语族群、客家人、闽语族群、粤语族群和湘语族群进行了测量。5年间，团队在21个省、1个自治区的31个城市和36个乡村的汉族族群开展了指标较齐全、大样本量的人体测量学研究，共测量26 945例（城市男性5 048例，城市女性5 403例，乡村男性8 176例，乡村女性8 318例）。此外，还开展了一万多例汉族人的25项遗传学经典指标的研究。原来课题规划是在17个省、自治旗开展工作，为了更全面反映汉族的体质全貌，项目实施过程中我们又增加了黑龙江省、吉林省、湖北省、贵州省、河北省保定市清苑县汉族5个族群的测量工作。

汉族体质研究工作取得了丰硕的研究成果，共发表了约190篇学术论文，在国家自然科学基金重点项目完成的基础上，2017年我和李咏兰、席焕久编著的《中国汉族体质人类学研究》由科学出版社出版。这是第一部专门研究汉族体质的专著，为后人留下了21世纪之初中国汉族人比较详细的人体数据。

研究再次转向未识别民族和南方少数民族（2013—2016）

2010年我被聘为国家自然科学基金委员会第十三届生命科学部专家评审组成员。我想，这也是对我多年从事科研工作的肯定吧。不过，由于我那几年都连续参加课题申请，所以并没有参加每年基金委召开的申请书会评工作。

完成了汉族的体质人类学研究工作后，我把研究的对象又转向中国未识别民族和一些少数民族。应该说中国未识别民族的人类学特征的研究一直是我们团队的研究特色。2013年我们又开展了国家自然科学基金项目"中国革家人、摩梭人等9个族群的体质人类学研究”的相关工作，这个项目要对贵州革家人、西双版纳州傣族、川滇边界泸沽湖畔的摩梭人、新疆维吾尔族、西藏喜马拉雅山区的夏尔巴人、海南临高人、五指山的黎族、甘川边界大山中的白马人、四川阿坝州的羌族进行体质人类学研究，是一项很有意义的研究。2015年张兴华老师又主持了国家自然科学基金项目“中国蔡家人、木雅人等5个族群的体质人类学研究”，对四川雅安木雅人、尔苏人、云南西双版纳州空格人、八甲人、贵州蔡家人进行了测量、研究。总之，我们团队掌握了较为丰富的中国未识别民族的体质数据，最近已经完成《中国未识别民族体质人类学研究》一书的书稿，正在联系出版社准备出版。

科技部项目：中国民族表型特征调查（2015—2019）

2015年我们研究组参加了复旦大学金力先生主持的科技部基础性工作专项“中国各民族体质人类学表型特征调查”。我主持了课题四“藏缅语族等少数民族体质人类学基础表型特征调查”。在这个课题中，要完成哈尼族、基诺族、白族、拉祜族、彝族（四川彝族、云南彝族、贵州彝族）、土家族（湖南土家族、湖北土家族、贵州土家族）、羌族、珞巴族、门巴族、独龙族、佤族、仡佬族、德昂族、景颇族、阿昌族、怒族、傈僳族、独龙族、普米族、纳西族20个少数民族，共24个采样点约14 400例少数民族个体的体质人类学基础表型的调查和样本采集，其中我们天津师范大学完成16个采样点的任务，大连医科大学的徐飞教授团队完成8个点的采样任务。除了体质测量工作外，还有语音采集、人像采集、生化指标采集、遗传样本采集，最后还要负责编写并提交以上人群的体质表型调查报告，总之任务很重。

可以说这个科技部项目是我们科研工作开展以来遇到的最困难的项目，差旅费不足，测量的指标多，测量的族群多，样本量大。我们为了节约经费，一次野外工作要完成好几个族群的测量，设计好整个工作的路线和行程。以前我们野外工作时团队有六七个人就够了，因为科技部项目测量指标多，所以要求团队的人员就多，往往需要12个人左右。人多了，虽然有利于工作，但花销就大，团队转移、安排住宿都很麻烦，不过，这些困难最后都克服了。工作难度大，还与被测量者的积极性不高有关。尽管我们买了一些礼品送给被测量者，也给他们发放一些误工补贴，但农时不等人，村民都非常忙，经常无暇顾及“测量身体”这些他们看来的小事。还有一点，现在乡村年轻人很少，

2013 年在西昌民族中学调查彝族学生的扣手

2015 年在云南勐海县勐阿镇测量八甲人

2015 年在四川平武测量白马人

2016 年在西藏定结县陈塘镇研究夏尔巴人服装接近藏族

2016 年和黑水藏族一家相遇在拉萨罗布林卡

2017 年在贵州务川县和仡佬族小朋友聊天

2016年在西藏雅鲁藏布大峡谷突遇塌方

2017年在新疆喀什和维吾尔族妇女干部合影

2018年在贵州三都县九阡镇测量水族

2018年在贵州黔东南州丹寨县测量苗族

2017年在新疆那拉提哈萨克人的毡包前

2019年在云南景洪基诺山工作之余做回“老司机”

留守乡村的多是孩子和老人，我们人类学测量各个年龄组都有样本量的要求，这就是一个矛盾。我们在族群工作后期，往往要想方设法专门补测年轻样本，这样也加长了工作日程。

为此，几年间我们团队转战于云南、四川、贵州、西藏的大山里，辛苦甚于往常。我几乎参加了所有的野外工作，在宇克莉教授、张兴华老师、包金萍教授的协助下，在全体学生的努力下，比较圆满地完成了任务。

蓦然回首，甚感欣慰

近40年来我和团队成员走了很多地方，测量了大量中国族群的人体数据，也为历史留下了比较完整的中国人人体测量学资料，这份资料非常宝贵。

2002年以前我们对内蒙古地区的蒙古族、汉族、达斡尔族、鄂温克族、鄂伦春族、朝鲜族、回族开展了大样本的群体遗传学研究。2002年以后我们研究对象转向全国少数民族。我们和内蒙古师范大学团队一起开展了中国119个族群人体测量工作，被测量的总人数为63 452例(其中男性为29 419例，女性为34 033例)。总样本量中藏缅语族男性为6 706例，女性为8 329例；壮侗语族男性为4 516例，女性为5 844例；苗瑶语族男性为856例，女性为1 010例；南亚语系男性为1 621例，女性为2 118例；阿尔泰语系男性为2 499例，女性为3 025例；汉族男性为13 221例，女性为13 707例。我们研究团队拥有全国最全的人体测量学数据库。以下按照年代顺序，看看我们近40年来的工作历程。

1981年调查达斡尔族、蒙古族ABO血型。

1984—1990年调查呼和浩特蒙古族、汉族、回族遗传指标，调查通辽蒙古族和阿拉善蒙古族遗传指标。

1991年调查内蒙古乌兰浩特朝鲜族、蒙古族、汉族遗传指标。

1994年赴内蒙古科左后旗测量科尔沁蒙古族。

1995年赴银川市测量宁夏回族。

1996年赴内蒙古莫力达瓦旗测量达斡尔族，赴鄂伦春旗测量鄂伦春族，调查呼和浩特回族、汉族、通辽蒙古族遗传指标。

1997年赴内蒙古呼伦贝尔盟鄂温克旗和陈巴尔虎旗调查鄂温克族、达斡尔族遗传指标，赴鄂伦春旗调查鄂伦春族遗传指标。

1998年赴内蒙古兴安盟乌兰浩特市调查朝鲜族、东北汉族、科尔沁蒙古族遗传指标，赴巴彦淖尔盟调查乌拉特蒙古族遗传指标。

1999年赴阿拉善盟调查阿拉善盟蒙古族、阿拉善盟汉族遗传指标，赴巴彦淖尔盟调查巴彦淖尔盟汉族遗传指标。

2001年调查天津汉族，赴内蒙古锡林浩特市调查锡林郭勒蒙古族、察哈尔蒙古族遗传指标，赴伊克昭盟调查鄂尔多斯蒙古族、伊克昭盟汉族遗传指标。

2002年赴新疆伊犁测量乌孜别克族。

2003年赴内蒙古呼伦贝尔盟测量俄罗斯族、贵州布依族。

2004年赴广西东兴市和罗城县测量仫佬族、广西京族。

2005年赴云南西双版纳州勐腊县测量克木人，赴临沧市耿马县测量佤族、赴玉溪市通海县测量云南蒙古族。

2006年赴云南怒江州测量怒族、独龙族，赴红河州测量莽人。

2007年赴西藏林芝察隅县测量僜人，赴米林县测量珞巴族，赴山南地区错那县测量门巴族，赴内蒙古呼伦贝尔盟测量布里亚特人。

2008年赴新疆阿勒泰地区测量图瓦人。

2009年赴山东省寿光县测量山东汉族，赴内蒙古兴安盟测量东北汉族，赴邛崃、简阳测量四川汉族。

2010年赴海南文昌县、琼海县、万宁县测量闽语方言汉族，赴湖南宁乡县、双峰县测量湘语方言族群，赴安徽滁州和江苏淮阴测量江淮方言汉族。

2011年赴广东梅州和江西赣州测量客家人，赴福建漳州测量闽南语汉族，赴福州测量闽东语汉族。

2012年赴湖北荆门市、荆州市测量江汉平原汉族，赴云南呈贡测量云南汉族，赴贵州安顺测量屯堡人，赴贵州黄平县测量革家人。

2013年赴吉林榆树市测量东北汉族，赴黑龙江哈尔滨呼兰区测量东北汉族，赴河北清苑测量华北平原汉族，赴云南西双版纳州勐海县测量傣族，赴丽江宁蒗县测量摩梭人，赴四川西昌昭觉县测量大凉山彝族，赴内蒙古呼伦贝尔市测量巴尔虎人。

2014年赴内蒙古测量鄂尔多斯蒙古族，赴海南临高县测量临高人，赴五指山市测量黎族。

2015年赴四川平武测量白马人，赴茂县测量羌族，赴雅安市石棉县测量木雅人、尔苏人，赴云南西双版纳州测量空格人、八甲人、基诺族、布朗族，赴内蒙古额济纳旗土尔扈特蒙古族。

2016年赴湖北省、湖南省、贵州省测量土家族，赴西藏墨脱县和米林县测量门巴族、珞巴族，赴日喀则市定结县陈塘镇测量夏尔巴人，赴青海省海西州测量蒙古族，赴黑龙江杜尔伯特蒙古族自治县测量蒙古族，赴吉林前郭尔罗斯蒙古族自治县测量蒙古族，赴辽宁阜新县测量蒙古族，赴辽宁喀左县测量蒙古族，赴云南大理州鹤庆县测量白族，赴云南楚雄州测量彝族，赴四川大凉山昭觉测量彝族，赴阿坝州茂县测量羌族。

2017年赴云南西盟县测量佤族，赴澜沧县测量拉祜族，赴勐海县测量傣族，赴墨江县测量哈尼族，赴贵州毕节市威宁县测量彝族，赴贵州遵义市务川县测量仡佬族，赴新疆博尔塔拉蒙古自治州温泉县测量新疆蒙古族察哈尔部、精河县测量蒙古族土尔扈特部，赴喀什测量维吾尔族，赴内蒙古阿拉善左旗测量蒙古族和硕特部。

2018年赴怒江州泸水市、福贡县、贡山县测量傈僳族、怒族、独龙族，赴丽江州、迪庆测量白族、彝族、藏族，赴贵州三都测量水族和布依族，赴丹寨、雷山测量苗族，赴榕江测量侗族。

2019年赴云南西双版纳州普洱市测量基诺族、傣族、布朗族、哈尼族、拉祜族，赴西藏林芝市测量藏族、门巴族、珞巴族，赴内蒙古自治区正镶白旗测量察哈尔蒙古族。

可以说为了描绘中国人的体质地图，我们团队40年来几乎走遍了神州大地。

雏凤清于老凤声

我于2012年8月退休，后又两次返聘，继续申请并主持国家自然科研基金项目，参加科技部基础性工作专项的课题，一直处于工作状态，同时更关注团队后备力量的培养。近些年来，天津师范大学人类生物学研究团队在宇克莉教授、张兴华老师、包金萍教授、程智博士的带领下，体质人类学研究在发展中前进。张兴华老师、宇克莉老师先后申报了关于未识别民族体质测量和藏缅语族民族体成分研究的国家自然科学基金项目，并且都获得批准，进一步提高了团队的研究深度。

内蒙古师范大学人类生物学研究团队是我创立的科研团队。几十年来，我对这支团队给予了极大的支持，和他们一起探索研究方向，一起开展野外工作，早年，我和陆舜华、栗淑媛老师合作很多。2010年以后，陆舜华、栗淑媛老师陆续退休，我和李咏兰老师合作很多。李咏兰老师陆续主持了蒙古族体质研究、壮侗语族族群和苗瑶语族族群体成分研究、西北民族走廊民族体成分研究共三项国家自然科学基金项目，主持了科技部基础性工作专项的课题，参加了复旦大学李辉教授主持的科技部项目。李咏兰科研工作发展势头很好，她们开展的13个蒙古族族群、布依族、水族、苗族、侗族的体质测量工作都邀请我参加。同样，我们在云南、海南、西藏、新疆、贵州的一些族群测量工作也邀请内蒙古师范大学团队参加。

多年来我的成长过程离不开我们科研前辈的无私帮助，我也以他们为榜样，在科研上尽量帮助同行。不光我这样要求自己，我也是这样希望我们团队的其他人这样做。当徐飞教授团队因为教学任务很重，难以完成傈僳族、怒族、独龙族工作，向项目组求援时，我们团队接受了这个任务，第二次深入怒江大峡谷，克服重重困难，比较圆满完成了这三个民族的体质测量任务。

人种学问题与维吾尔族体质问题

关于是否存在人种这个问题，目前学术界有两种截然不同的看法。2016年第34期《科学通报》发表了吴新智和崔娅铭的文章，对不同人种的地理分布、其特征如何形成、对人种概念的质疑、现实生活中的人种、反种族主义斗争和如何看待人种等问题进行简要的介绍。美国体质人类学学术年会在1996年发表了一份宣言，称人种这一概念在生物学上没有合法地位。在这些论点的轮番攻击下，一些进化科学家完全摒弃了“种族”概念，转而使用群体、族群、民族等社会学名词。我们认为，目前人种这个名词在社会生活中仍然得到广泛使用。是否有人种之分，是一个学术问题，是可以讨论的。我们通过对蒙古族、北方汉族、傣族的头面部形态学指标以及身高的测量和观察，探讨亚美人种内部的北亚类型族群、东亚类型族群、南亚类型族群的体质差异。研究发现，北亚类型、东亚类型、南亚类型的头面部特征存在明显的差异。首先我们所说的人种并不是物种的概念，而是地理种差别。我们认为目前不同人种之间的性状差异还是很明显的，这种差异是不同地区现代人各自发生不同的遗传变异的结果，这些性状很多都是遗传性状。尽管人种之间存在基因的融合，几个大人种之间存在过渡类型，但主要人种之间的差异还是很明显的。尽管每一个人种内部也存在性状的多样性，但人种的主要特征并没有消失，目前人种差异还是会长期存在的。我们的论文《人种的客观存在》

一文发表在2020年的《科学通报》上。

杜若甫老师曾经和我说过，他准备写一篇文章，专门探讨维吾尔族的人种学问题。杜老师说，一些人认为维吾尔族是高加索人种，这是不对的，应该是亚美人种的成分是主要的。后来，我一直没有见到杜老师的论文。这件事情我一直记在心里，所以，在测量完喀什维吾尔族体质数据后，我便着手写《维吾尔族体质类型：来自喀什的资料》这篇论文。论文认为维吾尔族体质类型既具有亚洲东方人群特点，也具有欧亚人群特点。在亚洲东方各个类型中，维吾尔族体质更接近东亚类型族群，也具有一定的南亚类型、北亚类型族群的特点。维吾尔族体质特征反映出其多族群融合的历史进程。该论文已经发表在《中国科学：生命科学》上。

感恩

在我从事中国民族体质特征研究的过程中，得到了很多同仁、学者、朋友的多方面帮助。对于这些宝贵的支持和帮助，我永铭在心，不能忘怀！

我们队伍中有一批老师如宇克莉、陆舜华、李咏兰、张兴华、包金萍、栗淑媛、胡莹、旺庆、王双喜、程智等，他们多年来和我一起下村寨、爬高山、入森林、越河流，每次外出，少则二十天，多则一个多月，我们在一起筹划过、劳累过、苦恼过、高兴过。我和他们一起讨论申报课题，一起规划每一次外出工作的细节，一起分担野外工作的忧愁和喜悦。我从1996年招第一个研究生开始，培养了很多研究生。科研事业就是要有继承、要有发展，这些研究生与老师一起走遍了全国各地，为中国族群的体质调查，做出了卓越的贡献。我曾经和内蒙古师范大学的陆舜华教授、李咏兰教授、栗淑媛教授、王双喜教授、李玉玲教授一起工作过。李咏兰教授已经成为人类学方向的博士生导师，主持了多项体质人类学方向的国家课题。我和天津师范大学的宇克莉教授、包金萍教授、张兴华老师多年来合作，互相帮助，成为了极为良好的科研挚友。宇克莉教授、张兴华老师都主持了国家自然科学基金项目。他们为体质人类学科研事业做出了重要贡献，看到他们的成长，我由衷高兴。我们一起依次完成了一个个族群的体质测量，完成一个个国家科研项目我们一起走过了中国很多地方，汗水洒在了那一片片热土上。我们培养的一届届研究生，跟随我们耕耘在体质人类学沃土上。我从心里感谢天津师范大学和内蒙古师范大学研究团队的各位老师、学生，是我们一起风雨同舟，一起走内蒙古、赴宁夏、入云南，去四川，下广西、上西藏、驰青海、奔海南，进新疆，到贵州，一步一个脚印一起走到今天。我们在做汉族工作时，我们3个高校组成的研究团队在中国22个省取样，近四十年来，可以说我们几乎走遍祖国各地。在野外年复一年、一次又一次奔波中，我们忍受着工作的辛苦，接触到乡土的气息，领略少数民族风情，也欣赏山河的壮丽，这些经历是我们一生难以忘怀的。

我们的少数民族测量工作也得到了当地各级政府的大力支持，其中云南民族宗教事务委员会、西藏民族宗教事务委员会支持尤大。开展工作所去的各个县民族宗教局给我们的各种帮助是我们工作顺利开展的保证。

一生遇人无数，历事较多，有恩于我者，永远感恩于心！说实在，应该感谢的学者很多。在科研工作中，首先应该感谢的是中国著名的体质人类学家、内蒙古民族体质人类学研究开创者朱钦先生。我的人体测量学基础知识是跟朱钦教授学的，最初也是跟随朱钦教授去做野外工作，逐渐积累经验的。要感谢的第二位是中国著名的人类与医学群体遗传学专家杜若甫先生。杜先生在我科研工作遇到困难的时候，热情帮助我，把我视为“入室弟子”，为我指出研究方向，给我科研资料。还要感谢中国人类学泰斗、领军人物席焕久教授。席教授热情助人，关心、支持我和我们团队的研究，严格审阅我们的论文。特别应该感谢的是中国著名分子人类学家、中国人类表型组首席科学家金力院士。金力院士多年来一直积极支持我们研究团队的工作，邀请我们参加他主持的科技部项目。在我成长的道路上得到这些先生的提携真是倍感荣幸。我忘不了徐玖瑾、魏源博、花兆合、丁士海、朱芳武、戴玉景诸位先生多年的帮助。当然还要感谢学术同仁，如刘武、赵凌霞、陈昭、徐飞、余跃生、金锋、梁明康、黄秀峰、任甫、何玉秀、王传超、温有锋、杨亚军、谭婧泽、汪思佳、徐林、周丽宁、邓琼英、康龙丽、徐国昌、海向军、何烨教授等，我们为了同一个目标，互相扶持，这也是我们体质人类学界的传统。我觉得我们科研事业的发展，同样不能忘记《人类学学报》、《科学通报》、《中国科学》、《解剖学报》、《解剖学杂志》、科学出版社的巨大贡献，林玉芬、林圣龙、冯兴无、张玄、安瑞、许家军、李迪老师对我们帮助、支持极为重要！

几多喟叹

很长时间，一有闲暇，我就回忆起自己走过的科研历程，回忆起我们曾经撒过汗水的地方，几十年的往事有些真切，有些已经变得模糊，更多的则随风飘散，无影无痕了。

云南是民族体质人类学研究的沃土，同时也是风景优美的地方，群山逶迤，河流纵横，在高山、坡地、平坝、森林生活着很多少数民族同胞，他们的简朴、好客给我们留下很深的印象。云南各级政府部门（特别是云南民族事务委员会的杜处长）对我们工作给予了很大支持，保证了我们工作能够顺利完成。我们研究团队去了云南12次：第一次是2005年测量克木人、佤族、云南蒙古族；第二次是2006年测量怒族、独龙族、莽人；第三次是2012年测量云南汉族；第四次是2013年测量傣族、摩梭人；第五次是2015年测量空格人、基诺族、布朗族、八甲人；第六次是2016年测量白族、云南彝族；第七次是2017年春天测量佤族、拉祜族、傣族、哈尼族；第八次是2017年夏天宇克莉带学生与大连医科大学徐飞教授的团队一起测量景颇族、德昂族、阿昌族；第九次是2017年秋天宇克莉老师带队测量纳西族和普米族；第十次是2018年春天测量傈僳族、怒族、独龙族；第十一次是2018年秋天宇克莉老师带队测量白族、彝族、藏族；第十二次是2019年春天测量基诺族、傣族、布朗族、哈尼族、拉祜族。在短短的14年间，赴云南12次，几乎年年都要下云南，通过12次的工作，我们基本掌握了云南各个民族的体质数据。我们团队主持了藏缅语族17个民族、南亚语系3个民族、仡佬族的体质研究工作。这些民族主要分布在云南，特别是分布在云南边境地区，也有分布在边境地区的山区。几乎所有的民族都属于跨境分布的族群，这些对我们既有神秘感，又增加了工作的难度。我们在西双版纳度过了疯狂的傣族泼水节，度过了记载历史传说的拉祜族葫芦节。我们和云南各个民族结下了友缘，和云南山山水水结下了情缘。我们忘不了云南的崇山峻岭和坐落在山间的座座村寨，忘不了云南勤劳、淳朴的各个少数民族，忘不了那里浓郁、独特的民族风情。

呼伦贝尔像一颗明珠，镶嵌在祖国的东北，熠熠生辉。那是一片神奇的土地，辽阔无垠的草原，茫茫的兴安岭，达赉湖、贝尔湖镶嵌在绿色的草原上，浩浩的额尔古纳河水向北流去。呼伦贝尔是很多游牧民族诞生、成长的摇篮，有多少历史故事在这里演绎。因为工作的关系，从1983年开始，我到过呼伦贝尔六次。第一次是因为高校招生之事，与科研无关，但呼伦贝尔给我留下非常美好的印象，也是我后来陆续五次到呼伦贝尔大草原的动力之一。第二次随朱钦老师开展达斡尔族、鄂伦春族体质测量。第三次是1997年我们团队与朱钦老师团队一起在鄂温克自治旗完成了布里亚特人、巴尔虎人、达斡尔人、鄂温克人、鄂伦春人遗传指标的调查。第四次我们去额尔古纳河河的室韦镇、恩和镇测量俄罗斯族。第五次是在鄂伦春旗西苏木测量布里亚特人。第六次是在新巴尔虎左旗测量巴尔虎部蒙古人。说到六次呼伦贝尔之行，不能不提到沈向阳、谢宾夫妇，他们二位在呼伦贝尔学院工作，是我在内蒙古师范大学时的学生，每次到呼伦贝尔，他们都给与了最热情、最真挚的帮助，为我们接风洗尘，为我们招呼校友，为我们送行，陪我们下草原，参加我们的调查、测量工作，每回我们都是带着依依不舍的心情和他们告别。

我们团队五次进川，和复旦大学杨亚军团队在川北大山里的白马村寨里测量历史悠久的氐人后裔白马人，四月下旬的白马山寨寒风凛冽，我们围坐在火炉旁烤火，夜晚在篝火旁看白马人跳舞。我们在阿坝州茂县测量了乐观活泼的至今仍然保留“羌”族名的羌族，感受到了凤仪城的山风呼啸猛烈。在四川平原我们到了卓文君的故乡邛崃，在简阳的贾家镇的绵绵细雨中测量四川汉族的体质数据。在安顺场大渡河畔的高山上测量了尔苏人、木雅人。在大凉山昭觉的大街小巷、广场测量了彝族同胞。

贵州是我国民族众多的省份，山水秀美，民众质朴勤劳，给我们留下非常深的印象。我去了贵州五次：2002年我们第一次入黔，在三都县测量布依族，忘不了黔南地区喀斯特地貌的小巧玲珑的山峰。第二次是去安顺市七眼桥镇和天龙镇测量服饰独特的屯堡人，在细雨绵绵中度过了难忘的一个星期。随后我们在黄平县重安镇和重兴乡测量革家人。革家人生活还比较贫困，革家老妈妈用我们听不懂的革家话和我们聊天，我们只能“嗯，啊”地应付。第三次是在完成湘西土家族工作后转移到贵州铜仁地区的沿河县测量土家族，我们住的宾馆在乌江边，夜夜看着两岸的灯火。第四次是在毕节地区的威宁县测量乌蒙山彝族，去遵义地区的务川县测量仡佬族。忘不了乌蒙山彝族弹着悦耳的月琴，唱着动听的彝歌，忘不了洪渡河边的漫步。第五次我是应内蒙古师范大学人类生物学团队之邀请，在黔南州、黔东南州开展水族、布依族、苗族、侗族的体质测量工作，这次工作难度很大，我们一路循着都柳河去了三都县九阡镇、周覃镇、丹寨县、雷山县、榕江县。

神奇的西藏是众人仰慕的圣地，那是一片净土。我们团队三赴西藏，看过五彩的然乌湖、浩瀚的错那湖、平静的佩枯湖；深入僜人、门巴族、珞巴族、夏尔巴人、藏族的村寨；游走在雅鲁藏布大峡谷；领略过藏南的魅力；瞻仰过布达拉宫、大昭寺、罗布林卡、扎什伦布寺。飘香的酥油茶、青稞酒，虔诚磕长头的喇嘛教信众，都在展示神奇的藏地文化。南伊沟里的珞巴族同胞、云遮雾罩的南加巴瓦峰、松涛阵阵的鲁朗林海、八一镇香港路上身着藏袍的老阿妈、气势磅礴的雅鲁藏布大峡谷，都永存在研究组每一个队员的记忆中。

天津师范大学和内蒙古师范大学人类生物学研究团队为了测量乌孜别克族、图瓦人、维吾尔族、蒙古族察哈尔部、土尔扈特部的体质数据，曾经三次出塞入疆，长驱新疆各地，感受到新疆各个民族风情，饱览了天山及新疆南北大山大川之壮丽，体验到西域地区之辽阔。没有到过新疆你不知道什么是辽阔、空旷，你感受不到天下之大。我们到过神秘的阿尔泰山，拜访过喀纳斯湖畔古老的当地族群——图瓦人，宁静的边疆古城伊宁的小巷里留下过我们的足迹。我们听过伊犁河谷巩乃斯河水的喧嚣，吸吮过天上草原的草香，疾驰在独库公路上瞻仰过天山的雄姿。在充满西域风情的喀什，我们体会着西域古城昔日的光彩。在吐鲁番的葡萄沟、坎儿井我们贪婪地感受着凉意。昔日荒凉的可克达拉戈壁滩已经白杨行行、瓜果飘香。高山圣湖——赛里木湖湖水还是那样透蓝吗？湖边的草地还是那样青青吗？当年安静的霍尔果斯口岸是否开始变得热闹起来？

古人喟叹，往事如梦了无痕。历史已经记录我们这批中国体质人类学工作者测量的历程，让后辈学者了解前辈体质人类学工作的艰辛，并留下一份厚厚的记录。记录下在二十世纪与二十一世纪交替时期，有一批立志于体质人类学研究的人，是如何完成民族体质测量的。因为我们的工作，很多民族有了第一份完整的人体数据。我们这批人类学工作者记录下来了的第一份完整的、详细的当代中国人的体质数据，将成为千百年后宝贵的参考文献，我们工作的意义就在此吧。

每一次野外工作，有旅途的劳顿和对山川美景的惊喜；有和各级地方官员进行沟通时的腼腆、无奈以及得到支持后的喜悦；有重复无数次的机械式的观察和测量；有没有人来测量的尴尬和焦虑；也有被测者排起长龙时的喜悦和疲惫。最高兴的时候，就是宣布工作任务圆满结束的时候，一下子就心里放松了，浑身轻松了。几十年都是这样过来的。我不通诗词格律，有时又喜欢胡诌几句，我曾为此写了一首诗“近日无端忆旧时，如烟似梦总难知。苦吟正是吾家短，新韵方能学作诗”。这不，又有些忍不住了，想写几句，用新韵，算古风，也算打油诗，抒发一下感慨。有诗曰：

数卷诗词握在手，常羡李杜游九州。五岳四水留踪影，云霞万里眼底收。
又羡霞客辞故里，一部游记传后俦。掩卷愿踏古人迹，思绪已上岳阳楼。
上天怜我青云志，族群如星探体质。追随朱师下乡闾，晓昏从此无闲时。
求教奔波沙滩北，摸索喟叹科研迟。图书馆里查文献，遗传所中谒杜师。
兴安岭上起松涛，敕勒川前忆往事。组建团队走天下，高山草原亦是家。
年年奔波在远路，岁岁驱驰赴天涯。村寨仰头三春月，漫途满眼四季花。
韶华难禁到秋暮，人生不期看晚霞。蓦然回首白发现，卅载往事在眼前。
天南地北入村寨，五湖四海结友缘。愁累苦烦次第过，更兼几度命途蹇。
不愿人生平如水，忍将今生付流年。

与研究团队合影（天津，2019 年 12 月）
前排左起：包金萍、宇克莉、郑连斌、张淑丽、孙岳枫、陈媛媛
后排左起：李聃、程智、向小雪、张兴华、宋晴阳、丁博、张洪明、李珊

代表性论著

[1] 宇克莉，李咏兰，张兴华，等．维吾尔族族体质类型：来自喀什的资料[J]．中国科学：生命科学，2020，50(9)：983-995.
[2] 宇克莉，李咏兰，张兴华，等．人种的客观存在[J]．科学通报，2020，65(9)：825-833.
[3] 张兴华，宇克莉，李咏兰，等．18 ～ 97 岁中国人的超重与肥胖：来自 2002—2019 年

63449 例人体数据的分析[J].中国科学：生命科学，2020，50(6)：661–674.
[4] 张兴华，宇克莉，杨亚军，等.中国白马人的体质特征[J].人类学学报，2020，39(1)：143–151.
[5] 包金萍，郑连斌.中国汉族人的身体围度[J].人类学学报，2020，39(1)：152–158.
[6] 包金萍，郑连斌，席焕久，等.中国汉族人的皮下脂肪发育[J].人类学学报，2019，38(2)：285–291.
[7] 李咏兰，郑连斌.中国蒙古族的身体密度[J].解剖学报，2019，50(2)：259–263.
[8] 张洪明，魏榆，宇克莉，等.四川、云南、贵州彝族身体成分特征及其差异[J].解剖学报，2019，50(5)：651–655.
[9] 李咏兰，郑连斌，金丹.黎族的体成分与体质特征[J].人类学学报，2019，38(1)：77–87.
[10] 宇克莉，任佳易，李咏兰，等.临高人的体质特征[J].人类学学报，2019，38(2)：276–284.
[11] 梁玉，何玉秀，郑连斌，等.青海撒拉族乡村成人围度特征及其增龄性变化[J].解剖学报，2019，50(3)：374–382.
[12] 宇克莉，魏榆，张兴华，等.基于土家族成人头面部测量指标的身高估测分析(英文)[J].解剖学报，2018，49(4)：518–523.
[13] 宇克莉，贾亚兰，郑连斌.布朗族成人的身体成分分析[J].人类学学报，2020，39(2)：261–269.
[14] 李咏兰，郑连斌.城市汉族的瘦体质量和脂肪质量[J].人类学学报，2018，37(1)：121–130.
[15] 宇克莉，杜慧敏，贾亚兰，等.布朗族的体质特征[J].解剖学杂志，2017，40(5)：574–579+602.
[16] Zhang X H, Wei Y, Zheng L B, et al. Estimation of stature by using the dimensions of the right hand and right foot in Han Chinese adults[J]. Science China (Life Sciences), 2017, 60(1): 81–90.
[17] 张兴华，杨亚军，王子善，等.中国八甲人与空格人的体质特征[J].人类学学报，2017，36(2)：268–279.
[18] 宇克莉，郑连斌，李咏兰，等.海南临高人身体成分分析[J].人类学学报，2017，36(1)：101–109.
[19] 宇克莉，张兴华，李咏兰，等.中国乡村汉族脂肪质量指数与瘦体质量指数的地理性分布[J].人类学学报，2017，36(3)：388–394.
[20] 李咏兰，郑连斌.中国华北地区、西南地区汉族人与日本人的头面部形态学差异[J].解剖学杂志，2017，40(1)：60–62+81.
[21] 李咏兰，郑连斌，宇克莉.从体质特征看中国南方汉族的人种归属[J].人类学学报，2017，36(2)：248–259.
[22] 董文静，宇克莉，包金萍，等.海南临高人 Heath-Carter 法体型[J].解剖学杂志，2017，40(1)：68–71.
[23] 魏榆，宇克莉，郑连斌，等.基诺族成人围度与握力关系的探讨[J].解剖学杂志，

2017,40(2): 201–204,241.

[24] 魏榆,张兴华,宇克莉,等.中国基诺族、木雅人、尔苏人、八甲人与空格人5个族群的体型特征[J].解剖学报,2017,48(5): 605–609.

[25] 李咏兰,郑连斌,席焕久.中国南北方汉族人身高、身体比例差异的对比分析[J].解剖学杂志,2016,39(6): 727–729.

[26] 宇克莉,郑连斌,赵大鹏.用腹围、腹臀比评价超重、肥胖标准[J].解剖学杂志,2016,39(6): 730–733.

[27] 廉伟,李咏兰,郑连斌,等.蒙古族四个族群皮褶厚度的比较[J].解剖学杂志,2016,39(1): 108–112.

[28] 李咏兰,郑连斌.中国巴尔虎人的体质特征[J].人类学学报,2016,35(3): 431–444.

[29] 宇克莉,郑连斌,李咏兰,等.中国北方、南方汉族头面部形态学特征的差异[J].解剖学报,2016,47(3): 404–408.

[30] 包金萍,郑连斌,宇克莉,等.屯堡人7项不对称行为特征[J].人类学学报,2016,35(4): 608–616.

[31] 任佳易,张兴华,魏榆,等.从不对称行为特征探讨白马人的族源[J].解剖学杂志,2016,39(6): 709–711.

[32] 王文佳,张兴华,宇克莉,等.木雅人的手长宽、足长宽与身高的关系[J].解剖学报,2016,47(6): 829–832.

[33] Li Y L, Zheng L B, Xi H J, et al. Stature of Han Chinese dialect groups: a most recent survey[J]. Science Bulletin, 2015, 60(5): 565–569.

[34] Zheng L B, Li Y L, Xi H J, et al. Statures in Han populations of China[J]. Science China. Life sciences, 2015, 58(2): 215–217.

[35] 宇克莉,郑连斌,胡莹,等.汉族闽东语族群乡村成人体型[J].解剖学杂志,2015,38(3): 344–347.

[36] 张兴华,郑连斌,宇克莉,等.闽南人的体质特征[J].人类学学报,2015,34(4): 516–527.

[37] 李咏兰,郑连斌,席焕久,等.中国北方、南方汉族体重的差异[J].解剖学报,2015,46(2): 270–274.

[38] 包金萍,郑连斌,张兴华,等.海南琼海汉族成人皮褶厚度的年龄变化[J].人类学学报,2015,34(1): 97–104.

[39] Zhao D P, Zheng L B, Li Y L, et al. Area-level socioeconomic disparities impact adult overweight and obesity risks of Han ethnicity in China.[J]. American Journal of Human Biology, 2015, 27(1): 129–32.

[40] 李咏兰,郑连斌,席焕久,等.中国汉族城市成年人身体肥胖指数值的纬度性分布[J].解剖学报,2015,46(4): 572–576.

[41] 李咏兰,郑连斌,旺庆.鄂尔多斯蒙古族头面部的人体测量学[J].解剖学报,2015,46(5): 684–689.

[42] 张兴华,郑连斌,宇克莉,等.四川凉山彝族身体围度特征分析[J].解剖学杂志,2015,38(5): 596–598.

[43] 李咏兰，郑连斌，冯晨露，等．革家人的体质特征［ J ］．人类学学报，2015，34（2）：234–244.

[44] Zhao D P, Li Y L, Zheng L B. Ethnic inequalities and sex differences in body mass index among Tibet minorities in China: Implication for overweight and obesity risk//Chinese Society for Anatomical Sciences. Abstracts of the 18th Congress of the International Federation of Associations of Anatomists (IFAA 2014). 2014：1.

[45] 刘燕，陆舜华，李玉玲，等．湖南宁乡地区城市汉族体型分析［ J ］．解剖学杂志，2014，37（2）：238–242.

[46] 包金萍，郑连斌，宇克莉，等．云南汉族成人皮褶厚度［ J ］．解剖学杂志，2014，37（4）：533–536+547.

[47] Li Y L, Zheng L B, Xi H J, et al. Body weights in Han Chinese populations［ J ］. Chinese Science Bulletin, 2014, 59(35): 5096–5101.

[48] 宇克莉，郑连斌，赵大鹏，等．安徽汉族成人皮褶厚度的研究［ J ］．人类学学报，2014，33（2）：214–220.

[49] 郑连斌，陆舜华，包金萍，等．广东客家人 5 项舌运动类型的人类学研究［ J ］．人类学学报，2014，33（1）：109–117.

[50] 李咏兰，宇克莉，陆舜华，等．中国南方汉族群体的头面部特征［ J ］．人类学学报，2014，33（1）：101–108.

[51] 张兴华，郑连斌，宇克莉，等．江苏汉族成人皮褶厚度［ J ］．解剖学报，2014，45（4）：578–581.

[52] 郑明霞，郑连斌，陆舜华，等．江西客家人乡村成人围度值及其年龄变化［ J ］．解剖学报，2014，45（1）：134–139.

[53] 刘海萍，于会新，陆舜华，等．云南省蒙古族 9 项头面部群体遗传学特征［ J ］．基础医学与临床，2014，34（2）：176–178.

[54] Zhao D P, Li Y L, Zheng L B. Ethnic inequalities and sex differences in body mass index among Tibet minorities in China: Implicat［ J ］. American Journal of Human Biology, 2014, 26(6): 856–858.

[55] 宋雪，郑连斌，李咏兰，等．西双版纳傣族成人的体型［ J ］．解剖学杂志，2014，37（6）：795–799.

[56] 包金萍，郑连斌，张兴华，等．海南文昌城市汉族成人 Heath-Carter 法体型观察［ J ］．解剖学报，2013，44（1）：114–119.

[57] 宇克莉，郑连斌，赵大鹏，等．汉族江淮方言族群的体质特征［ J ］．解剖学报，2013，44（1）：124–132.

[58] 李咏兰，陆舜华，郑连斌，等．江西汉族城市成人围度值及其年龄变化［ J ］．解剖学报，2013，44（1）：133–139.

[59] 张瑜珂，李咏兰，陆舜华，等．浙江汉族 7 项不对称行为特征［ J ］．解剖学报，2013，44（2）：284–291.

[60] 李咏兰，陆舜华，郑连斌，等．浙江汉族的体质特征［ J ］．解剖学报，2013，44（5）：707–716.

[61] 宇克莉，郑连斌，胡莹，等. 福建汉族闽东语族群的体质特征[J]. 解剖学报，2013，44（6）：824-834.

[62] 宇克莉，郑连斌，包金萍，等. 屯堡人的体质特征[J]. 解剖学报，2013，44（6）：835-842.

[63] 郑连斌，宇克莉，包金萍，等. 云南汉族体质特征[J]. 云南大学学报（自然科学版），2013，35（5）：703-718.

[64] 李咏兰，陆舜华，郑连斌，等. 汉族湘语族群的7项不对称行为特征[J]. 人类学学报，2013，32（1）：101-109.

[65] 李咏兰，郑连斌，宇克莉，等. 南方汉族人头面部形态特征的年龄变化[J]. 科学通报，2013，58（4）：336-343.

[66] 郑连斌，李咏兰，陆舜华，等. 中国客家人体质特征[J]. 中国科学：生命科学，2013，43（3）：213-222.

[67] 郑连斌，李咏兰，宇克莉，等. 贵州屯堡人皮褶厚度的研究[J]. 南京医科大学学报（自然科学版），2013，33（7）：970-974.

[68] 倪晓璐，李咏兰，郑连斌，等. 贵州屯堡人头面部形态特征的年龄变化[J]. 解剖学杂志，2013，36（6）：1116-1121.

[69] 张兴华，郑连斌，宇克莉，等. 安徽滁州汉族体质特征[J]. 解剖学杂志，2013，36（1）：95-101.

[70] 李咏兰，陆舜华，张瑜珂，等. 浙江乡村汉族体型[J]. 解剖学杂志，2013，36（6）：1111-1115.

[71] Zhao D P, Li Y L, Zheng L B, et al. Brief communication: Body mass index, body adiposity index, and percent body fat in Asians[J]. American Journal of Physical Anthropology, 2013, 152(2): 294-299.

[72] Zheng L B, Li Y L, Lu S, et al. Physical characteristics of Chinese Hakka[J]. Science China Life Sciences, 2013, 56(6): 541-551.

[73] Li Y L, Zheng L B, Yu K, et al. Variation of head and facial morphological characteristics with increased age of Han in southern China[J]. Chinese Science Bulletin, 2013, 58(4-5): 517-524.

[74] 李玉玲，陆舜华，李咏兰，等. 湖南宁乡汉族成人体质特征[J]. 解剖学杂志，2013，36（3）：398-404.

[75] 包金萍，郑连斌，陆舜华，等. 江西赣州乡村客家人Heath-Carter法体型[J]. 解剖学杂志，2013，36（5）：982-985.

[76] 时蕊，郑连斌，胡莹，等. 汉族闽南语族群头面部形态特征的年龄变化[J]. 解剖学杂志，2013，36（2）：227-233.

[77] 旺庆，李咏兰，陆舜华，等. 绍兴地区汉族成人围度值年龄变化及城乡的比较[J]. 解剖学杂志，2013，36（4）：824-828.

[78] 傅媛，李玉玲，陆舜华，等. 湖南宁乡汉族成人皮褶厚度[J]. 解剖学杂志，2013，36（4）：829-833.

[79] 郑明霞，郑连斌，宋瓘兰，等. 海南文昌汉族成人头面部形态特征的年龄变化[J]. 解

剖学杂志,2013,36(3):387-391.

[80] 李咏兰,陆舜华,郑连斌,等.绍兴地区汉族成人头面部特征的年龄变化[J].解剖学杂志,2013,36(3):392-397.

[81] 张瑜珂,李咏兰,陆舜华,等.浙江地区汉族成人皮褶厚度的年龄变化[J].解剖学杂志,2013,36(1):102-105.

[82] 李咏兰,郑连斌,宇克莉,等.中国南方汉族成人头面部形态特征的年龄变化[J].科学通报,2013,58(4):336-343.

[83] 郑连斌,李咏兰,冯晨露,等.革家成年人头面部形态特征的年龄变化[J].解剖学报,2013,44(5):699-706.

[84] 郑连斌,陆舜华,包金萍,等.广东客家人头面部形态特征的年龄变化[J].浙江大学学报(医学版),2012,41(3):250-258.

[85] 郑连斌,包金萍,闫春燕,等.海南省汉族人头面部观察指标出现率的年龄变化[J].中国临床解剖学杂志,2012,30(4):389-392.

[86] 王杨,郑连斌,宇克莉,等.江苏淮安地区汉族Heath-Carter法体型研究[J].解剖学报,2012,43(1):123-129.

[87] Zheng L B, Lu S H, Ding B, et al. Comparison of somatotypes on 29 ethnic groups in China[J]. Acta Anatomica Sinica, 2012, 43(01): 130-134.

[88] 张晓瑞,郑连斌,陆舜华,等.江西赣州客家5项舌运动类型的研究[J].南京师大学报(自然科学版),2012,35(3):87-92.

[89] 王杨,郑连斌,陆舜华,等.广东客家人皮褶厚度特征[J].解剖学杂志,2012,35(4):506-509.

[90] 张兴华,郑连斌,包金萍,等.海南琼海城市汉族体型特征[J].解剖学杂志,2012,35(4):518-522.

[91] 李咏兰,陆舜华,郑连斌,等.湘语族群头面部特征的年龄变化[J].解剖学杂志,2012,35(2):223-228.

[92] 李咏兰,陆舜华,郑连斌,等.湘语族群城市成人围度值及其年龄变化[J].解剖学杂志,2012,35(2):217-220.

[93] 李咏兰,陆舜华,郑连斌,等.江西汉族人头面部形态特征的年龄变化[J].人类学学报,2012,31(2):193-201.

[94] 李咏兰,陆舜华,郑连斌,等.江西汉族体质特征[J].解剖学报,2012,43(1):114-122.

[95] 李咏兰,陆舜华,郑连斌,等.山西汉族成人皮褶厚度特点[J].解剖学报,2012,43(2):268-272.

[96] Li Y L, Zheng L B, Lu S H, et al. The Comparision of somatotypes of ethnic groups in China. Physical Anthropolop[M].北京:知识产权出版社,2012.

[97] 李咏兰,陆舜华,郑连斌,等.汉族湘语族群成人皮褶厚度[J].解剖学杂志,2012,35(4):501-505.

[98] 李传刚,李咏兰,陆舜华,等.湘语族群乡村汉族成人的体型特点[J].解剖学杂志,2012,35(4):514-517.

[99] 李咏兰，陆舜华，郑连斌，等. 汉族湘语族群体质特征[J]. 解剖学报，2012，43（5）：694–702.

[100] Zhao D P, Li B G, Yu K L, et al. Digit ratio (2D: 4D) and handgrip strength in subjects of Han ethnicity: Impact of sex and age[J]. American Journal of Physical Anthropology, 2012, 149(2): 266–271.

[101] 傅媛，李玉玲，陆舜华，等. 湖南宁乡汉族九项遗传学指标[J]. 解剖学杂志，2012，35（5）：660–664.

[102] 李玉玲，陆舜华，栗淑媛，等. 内蒙古兴安盟汉族体质调查[J]. 人类学学报，2012，31（1）：71–81.

[103] 李玉玲，陆舜华，陈琛，等. 广东粤语族群汉族体质特征[J]. 解剖学报，2012，43（6）：837–845.

[104] 李咏兰，陆舜华，郑连斌，等. 湘语族群乡村成人围度值年龄组的比较[J]. 解剖学报，2012，43（6）：826–831.

[105] 李传刚，李咏兰，陆舜华，等. 湘语族群汉族城市成人的体型特点[J]. 解剖学报，2012，43（6）：832–836.

[106] 郑连斌，张兴华，包金萍，等. 海南汉族体质特征[J]. 解剖学报，2012，43（6）：855–863.

[107] 郑连斌，陆舜华，丁博，等. 中国29个族群体型的比较（英文）[J]. 解剖学报，2012，43（1）：130–134.

[108] 李咏兰，陆舜华，郑连斌，等. 湘语族群头面部形态特征的年龄变化[J]. 解剖学杂志，2012，35（2）：223–228.

[109] 李咏兰，陆舜华，郑连斌，等. 江西城市汉族体型特征[J]. 解剖学杂志，2012，35（6）：826–829.

[110] 李传刚，李咏兰，陆舜华，等. 湘语族群乡村汉族成人的体型特征[J]. 解剖学杂志，2012，35（4）：514–517.

[111] 李咏兰，郑连斌，陆舜华，等. 中国布里亚特人的体质特征[J]. 人类学学报，2011，30（4）：357–367.

[112] 郑连斌，陆舜华，丁博，等. 云南蒙古族体质特征[J]. 人类学学报，2011，30（1）：74–85.

[113] 陆舜华，郑连斌，董其格其，等. 图瓦人和布里亚特人体型特点[J]. 解剖学杂志，2011，34（4）：544–547.

[114] 廖颖，黎霞，熊丽丹，等. 四川凉山彝族9项头面部群体遗传学特征[J]. 基础医学与临床，2011，31（8）：875–878.

[115] 杨建辉，陈利红，郑连斌. 仫佬族7项不对称行为的研究[J]. 现代预防医学，2011，38（8）：1470–1472+1475.

[116] 杨建辉，陈利红，郑连斌. 仫佬族4项人类学特征的研究[J]. 医学理论与实践，2011，24（11）：1347–1348.

[117] 李咏兰，陆舜华，国海，等. 山西城市汉族体型特点[J]. 解剖学报，2011，42（5）：707–712.

[118] 李咏兰，陆舜华，国海，等.山西汉族人头面部形态特征的年龄变化[J].解剖学报，2011,42(6): 840-845.

[119] 包金萍、郑连斌、张兴华，等.山东城市汉族体型[J].解剖学杂志，2011，34(1): 114-117.

[120] 张兴华，郑连斌，宇克莉，等.山东寿光汉族体质特征[J].人类学学报，2011，30(2): 206-217.

[121] 郑连斌，武亚文，张兴华，等.四川汉族体质特征[J].解剖学报，2011，42(5): 695-702.

[122] 郑连斌，张兴华，胡莹，等.四川邛崃汉族头面部形态特征的年龄变化[J].中山大学学报(医学科学版)，2011，32(6): 729-734，763.

[123] 张晓瑞，郑连斌，宇克莉，等.安徽汉族7项不对称行为特征的研究[C]//中国解剖学会.中国解剖学会2011年年会论文文摘汇编.天津: 天津师范大学生命科学学院，2011: 1.

[124] 龚忱，郑连斌，胡莹，等.四川资阳城市汉族体型的特点[J].解剖学杂志，2011，34(5): 695-698.

[125] 郑连斌，黎霞，张兴华，等.四川汉族人头面部形态特征的年龄变化[J].解剖学杂志，2011，34(5): 699-705.

[126] 杨建辉，郑连斌，陈利红.仫佬族9项头面部群体遗传学特征[J].基础医学与临床，2010，30(10): 1025-1028.

[127] 郑连斌，陆舜华，张兴华，等.中国莽人、僜人、珞巴族与门巴族Heath-Carter法体型研究[J].人类学学报，2010，29(2): 176-181.

[128] 廖彦博，李坤，郑连斌，等.广西京族体质人类学研究[J].人类学学报，2010，29(1): 100-102.

[129] 郑连斌，陆舜华，张兴华，等.珞巴族与门巴族的体质特征[J].人类学学报，2009，28(4): 401-407.

[130] 郑连斌，陆舜华，于会新，等.中国僜人体质特征[J].人类学学报，2009，28(2): 162-171.

[131] 郑连斌，陆舜华，许渤松，等.中国独龙族与莽人的体质特征[J].人类学学报，2008，4: 350-358.

[132] 郑连斌，陆舜华，罗东梅，等.怒族的体质调查[J].人类学学报，2008，2: 158-166.

[133] 梁明康，郑连斌，朱芳武，等.广西京族成人的体型特点[J].解剖学杂志，2008，31(2): 249-252.

[134] 谢宾，陆舜华，郑连斌，等.内蒙古地区俄罗斯族成人身体围度特征分析[J].解剖学杂志，2007，30(4): 483-486.

[135] 郑连斌，陆舜华，于会新，等.佤族的体质特征[J].人类学学报，2007，3: 249-258.

[136] Ding B, Zheng L B, Lu S H, et al. The variation of skinfold thickness of Mulam adults in China[J]. Journal of Life Sciences, 2007, 1(1): 55-59.

[137] 郑连斌，陆舜华，陈媛媛，等.中国克木人的体质特征[J].人类学学报，2007，1: 45-53.

[138] 郑连斌,陆舜华,丁博.仫佬族体质特征研究[J].人类学学报,2006,3：242–250.

[139] 陆舜华, 郑连斌, 索利娅, 等.俄罗斯族体质特征分析[J].人类学学报, 2005, 4：291–300.

[140] 杨建辉,郑连斌,陆舜华,等.布依族成人Heath-Carter法体型研究[J].人类学学报,2005,3：198–203.

[141] 郑明霞, 郑连斌.中国17个人群体部指数的多元分析[J].解剖学研究, 2005, 2：129–132+135.

[142] 郑明霞,郑连斌.体部测量项目的相关研究[J].解剖学杂志,2005,3：355–357.

[143] 郑连斌, 张淑丽, 陆舜华, 等.布依族体质特征研究[J].人类学学报, 2005, 2：137–144.

[144] 张淑丽, 郑连斌, 陆舜华, 等.布依族成人皮褶厚度的年龄变化[J].人类学学报, 2005, 1：58–63.

[145] 崔静, 郑连彬, 沈新生.新疆塔塔尔族体质特征调查[J].人类学学报, 2004, 1：47–54.

[146] 栗淑媛, 郑连斌, 陆舜华.中国36个人群头面部指数的主成分分析[J].解剖学杂志,2004,27(5)：555–558.

[147] 郑连斌, 栗淑媛, 陆舜华, 等.乌孜别克成人皮褶厚度的年龄变化[J].解剖学杂志, 2004,27(4)：438–440.

[148] 陆舜华, 郑连斌, 栗淑媛, 等.乌孜别克族成人的体型特点[J].人类学学报, 2004, 3：224–228.

[149] 郑连斌, 崔静, 陆舜华, 等.乌孜别克族体质特征研究[J].人类学学报, 2004, 1：35–45.

[150] 刘燕,陆舜华,郑连斌,等.内蒙古西部地区蒙古族、汉族4项人类群体遗传学特征的研究[J].遗传,2004,1：35–39.

[151] 郑明霞, 郑连斌, 陆舜华, 等.阿拉善盟蒙古族、汉族的舌运动类型[J].解剖学杂志, 2003,26(6)：606–608.

[152] 李玉玲,陆舜华,郑连斌.三种舌运动类型遗传方式的研究[J].遗传,2003,5：552–554.

[153] Zheng L B, Zheng Q X, Lu S H, et al. Study on seven asymmetric behavioral traits in three Mongolian groups[J]Anthropological Science, 2003, 111(2): 231–244.

[154] 郑连斌, 陆舜华, 栗淑媛.内蒙古6个人群舌运动类型研究[J].人类学学报, 2003, 3：241–245.

[155] 栗淑媛, 郑连斌, 朱钦, 等.达斡尔族学生体表面积研究[J].人类学学报, 2003, 1：51–55.

[156] 郑连斌, 朱钦, 王树勋, 等.达斡尔族成人的皮褶厚度及其年龄变化[J].人类学学报,2003,1：45–50.

[157] Zheng L B, Han Z Z, Lu S H, et al. Morphological traits in peoples of Mongolian nationality of the Hulunbuir league, Inner Mongolia, China[J]. Anthrop Anz, 2002, 60(2): 175–185.

[158] Lu S H, Han Z Z, Zheng L B, et al. Lateral functional dominance in behavioral traits observed in five populations of Inner Mongolia[J]. Anthropological Science, 2002, 110(3): 267–278.

[159] 郑连斌,陆舜华,郑琪,等.中国人群肤纹的主成分分析[J].人类学学报,2002,3: 231–238.

[160] 陆舜华,李咏兰,郑连斌,等.内蒙古5个民族12对性状的基因频率[J].遗传,2002, 2: 140–142.

[161] 李咏兰,郑连斌,陆舜华,等.达斡尔族、鄂温克族、鄂伦春族13项形态特征的研究[J].人类学学报,2001,3: 217–223.

[162] 郑连斌,谢宾,陆舜华,等.内蒙古呼伦贝尔盟蒙古族3个群体5项舌运动类型的研究[J].人类学学报,2001,2: 130–136.

[163] 韩在柱,陆舜华,郑连斌,等.兴安盟3个民族7种不对称行为特征的研究[J].人类学学报,2001,2: 137–143.

[164] 栗淑媛,郑连斌,陆舜华,等.人体头面部测量项目相关分析[J].解剖学杂志,2001, 2: 176–178.

[165] 栗淑媛,韩在柱,郑连斌,等.兴安盟3个民族5种舌运动类型的研究[J].人类学学报,2001,1: 76–78.

[166] 王树勋,郑连斌,朱钦,等.达斡尔族青少年体型的Heath-Carter人体测量法研究[J].人类学学报,2001,1: 45–51.

[167] 陆舜华,郑连斌,李咏兰,等.鄂伦春、鄂温克、达斡尔族一侧优势功能特征研究[J].遗传,2000,5: 287–291.

[168] 韩在柱,郑连斌,陆舜华,等.兴安盟3个民族10对性状的基因频率[J].遗传,2000, 4: 241–242.

[169] 朱钦,王树勋,阎桂彬,等.鄂伦春族成人的体型[J].解剖学杂志,2000,3: 208–212.

[170] 郑连斌,郑明霞,陆舜华,等.亚洲21个人群体部特征的比较研究[J].人类学学报, 2000,1: 49–56.

[171] 朱钦,王树勋,阎桂彬,等.鄂伦春族体质现状及其与60年前资料的比较[J].人类学学报,1999,4: 296–306.

[172] 李咏兰,郑连斌,陆舜华,等.内蒙古达斡尔族舌运动类型的遗传学研究[J].遗传, 1999,5: 20–22.

[173] 郑明霞,李咏兰,栗淑媛,等.内蒙古7个群体优势眼的调查[J].遗传,1999,4: 19–21.

[174] Zheng L B, Ao Z Y, Wo J Y. Study on pottical type, palmar and plantar digital formulae, hand clasping, arm folding, handedness, leg folding and stride type in the Daur population, China[J]. Anthropo Anz, 1999, 57(4): 361–369.

[175] 朱钦,郑连斌,金寅淳,等.現在の内蒙古哲里木地域蒙古族の形態特徴とこの60年間の動向[J]. Anthropological Science,1999,106(2): 143–150.

[176] 陆舜华,郑连斌,韩莉,等.蒙古族和汉族新生儿及幼儿蒙古斑的研究[J].人类学学报,1999,1: 48–54.

[177] 郑连斌，陆舜华，李晓卉，等.汉、回、蒙古族拇指类型、环食指长、扣手、交叉臂及惯用手的研究[J].遗传，1998，4：14-19.
[178] 郑连斌，朱钦，阎桂彬，等.达斡尔族成人体型研究[J].人类学学报，1998，2：72-78.
[179] 韩在柱，郑连斌，陆舜华.达斡尔族学生皮下脂肪发育的研究[J].人类学学报，1998，2：79-80+82-85.
[180] 陆舜华，郑连斌，张炳文，等.习舞青少年的体型初探[J].人类学学报，1998，1：46-50.
[181] 朱钦，郑连斌，王巧玲，等.回族成人的Heath-Carter法体型研究[J].解剖学杂志，1997，6：600-604.
[182] 郑连斌，陆舜华，李晓卉，等.内蒙古三个民族舌运动类型的遗传学研究[J].遗传，1997，3：23-29.
[183] 郑连斌，陆舜华.我国23个群体体质的聚类分析与主成分分析[J].人类学学报，1997，2：66-73.
[184] 郑连斌，朱钦，王巧玲，等.宁夏回族体质特征研究[J].人类学学报，1997，1：12-22.
[185] 陆舜华，郑连斌，李咏兰，等.内蒙地区蒙古、汉、回、朝鲜族指纹白线分析[J].人类学学报，1996，4：356-361.
[186] 郑连斌，陆舜华，阎桂彬，等.蒙古族体型的Heath-Carter人体测量法研究[J].人类学学报，1996，3：218-224.
[187] 郑连斌，栗淑媛，陆舜华，等.内蒙古8个群体女生月经初潮的月分布[J].人类学学报，1996，1：74-79.
[188] 陆舜华，郑连斌，张炳文.内蒙地区蒙古、汉、回、朝鲜族肤纹特征比较研究[J].人类学学报，1995，3：240-246.
[189] 郑连斌，栗淑媛，顾捷，等.乌兰浩特市蒙古族学生体质发育营养状况分析[J].中国学校卫生，1995，16(3)：165-167.
[190] 郑连斌，栗淑媛，陆舜华，等.呼和浩特地区蒙、汉族女生月经初潮年龄的变化[J].人类学学报，1995，14(2)：169-175.
[191] 郑连斌，李咏兰，陆舜华.内蒙古四个民族耳垂基因频率[J].遗传，1995，17(2)：12-13.
[192] 陆舜华，栗淑媛，郑连斌，等.内蒙古朝鲜族和回族苯硫脲味盲基因频率[J].遗传，1994，16(6)：29.
[193] 郑连斌，陆舜华，马小林，等.乌兰浩特市朝鲜族、汉族学生体质发育的比较分析[J].人类学学报，1994，13(1)：72-77.
[194] 郑连斌，陆舜华，王双喜，等.内蒙古蒙、汉民族PTC味盲基因频率[J].遗传，1991，13(4)：34.
[195] 郑连斌，桂花，王爱珑.蒙古族发旋和手指毛的调查研究[J].内蒙古大学学报(自然科学版).1989，20(1)：97-102.
[196] 郜丽华，王爱珑，郑连斌.蒙古族的蒙古褶、上眼睑褶皱的分析研究[J].内蒙古大学学报(自然科学版)，1989，20(1)：103-107.
[197] 郑连斌，呼尔查毕力格，苏尤拉图.蒙古族人手纹分析研究[J].内蒙古师范大学学

报（自然科学汉文版），1986，2：8-13.
[198] 郑连斌，郃晓春，赛娜，等.关于蒙族、达斡尔族ABO血型的调查[J].内蒙古师范大学学报（自然科学汉文版），1984，2：44-49.
[199] 李咏兰，郑连斌.中国蒙古族体质人类学研究[M].北京：科学出版社，2018.
[200] 郑连斌，李咏兰，席焕久.中国汉族体质人类学研究[M].北京：科学出版社，2017.
[201] 郑连斌.青春期教育概论[M].呼和浩特：内蒙古教育出版社，1996.

获奖情况

[1] 1993年获内蒙古自治区普通高等学校优秀教学成果二等奖。
[2] 1999年获内蒙古自治区医药卫生科技进步一等奖。
[3] 1999年获曾宪梓教育基金会授予的高等师范院校优秀教师三等奖。
[4] 2002年获内蒙古自治区科技进步三等奖。
[5] 2003年获天津师范大学教学名师称号。
[6] 2010获天津市自然科学奖三等奖。
[7] 2019年获天津市道德模范提名奖。

代表作

汉族的身高、体重和身体比例*

已有对中国人身高的报道[1-3]。现有的中国成人体质测量多将中国人按照南、北方和性别分为4个族群[4]，或按照城乡、性别划分4个族群[5]，进行比较的各族群测量年代相隔较远，或由不同的研究组测量完成[2]。身高、体重是体质人类学最重要的测量指标。近年来社会上流传着有关中国人身高、体重的各种数据。这些数据来源往往不清楚，被测量者也不分民族，甚至不知道被测量者的年龄范围。目前亟需一份可靠的、完整的各地中国人的身高、体重、身体比例的基本数据。

第一节　汉族的身高

一、各地区汉族的身高

（一）汉族身高均数

1. 乡村汉族身高均数

36个乡村男性族群中保定、富平、嘉兴、哈尔滨、张家口、滁州、榆树、平凉8个族群身高均数依次分居前1～8位，前8个族群主要属于北方汉族。而赣州、娄底、景德镇、长沙、安顺、成都、资阳、宜春8个南方族群身高均数依次分居29～36位（表1）。

36个乡村女性族群中保定、淮安、富平、荆门、潍坊、荆州、滁州、锦州8个族群身高均数依次分居前1～8位。荆门、荆州、滁州汉族分布于长江两岸的江淮平原、江汉平原，均进入前8位。娄底、赣州、化州、景德镇、安顺、成都、资阳、宜春8个族群依次分居29～36位（表2）。女性后8位的族群与男性后8位族群基本相同。

---

* 节选自《中国汉族体质人类学研究》（郑连斌，李咏兰，席焕久，等著，科学出版社，2017）中关于汉族身高的内容。

表1 汉族乡村男性36个族群的身高均数以及与年龄的相关分析与方差分析(mm,Mean ± SD)

地区	20 ~ 44 岁组	45 ~ 59 岁组	60 ~ 80 岁组	合计	相关分析		方差分析	
					r	*P*	*F*	*P*
哈尔滨	1 696.6 ± 58.7	1 655.6 ± 65.0	1 657.9 ± 48.3	1 676.9 ± 61.7	−0.41**	0.00	11.62**	0.01
榆树	1 703.4 ± 58.6	1 657.9 ± 54.1	1 648.0 ± 61.6	1 673.9 ± 61.8	−0.45**	0.00	16.28**	0.00
乌兰浩特	1 674.4 ± 52.6	1 642.2 ± 55.5	1 634.0 ± 45.3	1 657.3 ± 55.2	−0.32**	0.00	13.35**	0.00
昌图	1 680.6 ± 54.4	1 663.9 ± 58.0	1 645.6 ± 58.5	1 667.7 ± 57.8	−0.28**	0.00	7.38**	0.00
锦州	1 686.8 ± 61.1	1 662.4 ± 47.6	1 626.7 ± 57.1	1 666.6 ± 60.7	−0.44**	0.00	20.42**	0.00
张家口	1 695.8 ± 66.6	1 660.7 ± 59.1	1 652.5 ± 48.7	1 676.8 ± 54.9	−0.39**	0.00	25.25**	0.00
保定	1 720.8 ± 55.2	1 669.7 ± 50.0	1 667.3 ± 53.7	1 700.2 ± 57.5	−0.30**	0.00	16.23**	0.00
晋中	1 679.3 ± 59.6	1 655.0 ± 58.3	1 626.5 ± 54.2	1 660.5 ± 62.6	−0.41**	0.00	14.95**	0.00
潍坊	1 687.9 ± 58.4	1 664.9 ± 54.5	1 659.1 ± 60.9	1 672.0 ± 58.4	−0.22**	0.00	4.76**	0.01
南阳	1 678.8 ± 64.0	1 649.8 ± 61.5	1 641.7 ± 77.6	1 663.4 ± 68.0	−0.36**	0.00	26.91**	0.00
新野	1 692.0 ± 64.0	1 653.1 ± 50.9	1 632.5 ± 50.1	1 668.3 ± 62.5	−0.42**	0.00	22.79**	0.00
蒲城	1 686.8 ± 53.9	1 661.4 ± 51.2	1 637.0 ± 57.8	1 669.6 ± 57.0	−0.20**	0.00	16.02**	0.00
富平	1 698.8 ± 52.5	1 701.1 ± 60.9	1 661.6 ± 60.2	1 691.7 ± 58.9	−0.33**	0.00	8.63**	0.00
平凉	1 680.9 ± 54.6	1 673.4 ± 63.6	1 653.9 ± 55.2	1 672.5 ± 59.4	−0.23**	0.00	5.39**	0.005

（续　表）

地区	20 ～ 44 岁组	45 ～ 59 岁组	60 ～ 80 岁组	合计	相关分析		方差分析	
					r	*P*	*F*	*P*
武威	1 670.2 ± 54.9	1 644.6 ± 61.6	1 629.3 ± 56.6	1 655.1 ± 59.4	−0.33**	0.00	10.59**	0.00
荆门	1 678.7 ± 52.4	1 650.9 ± 53.6	1 620.5 ± 68.3	1 657.9 ± 60.4	−0.42**	0.00	20.27**	0.00
荆州	1 682.4 ± 64.1	1 646.3 ± 65.6	1 622.5 ± 63.5	1 658.2 ± 69.4	−0.40**	0.00	13.32**	0.00
成都	1 648.2 ± 46.3	1 606.5 ± 41.3	1 583.7 ± 47.1	1 624.4 ± 63.0	−0.46**	0.00	22.73**	0.00
资阳	1 661.4 ± 53.4	1 605.3 ± 56.0	1 561.0 ± 44.7	1 624.2 ± 69.0	−0.67**	0.00	57.00**	0.00
安顺	1 644.4 ± 59.3	1 617.7 ± 57.7	1 593.1 ± 69.0	1 626.1 ± 63.8	−0.36**	0.00	12.09**	0.00
昆明	1 678.2 ± 56.8	1 644.5 ± 46.6	1 632.8 ± 43.4	1 660.2 ± 66.5	−0.38**	0.00	9.71**	0.00
滁州	1 693.4 ± 63.2	1 662.7 ± 60.4	1 636.3 ± 60.3	1 676.3 ± 65.3	−0.44**	0.00	11.77**	0.00
淮安	1 691.1 ± 56.2	1 670.4 ± 61.6	1 628.3 ± 56.2	1 670.7 ± 63.0	−0.43**	0.00	20.88**	0.00
嘉兴	1 707.5 ± 51.9	1 674.5 ± 48.1	1 629.5 ± 49.5	1 677.0 ± 69.7	−0.45**	0.00	23.99**	0.00
绍兴	1 686.5 ± 47.9	1 645.6 ± 64.6	1 630.2 ± 56.0	1 657.5 ± 67.6	−0.52**	0.00	15.29**	0.00
景德镇	1 656.1 ± 62.5	1 618.7 ± 63.8	1 601.9 ± 52.8	1 633.3 ± 64.9	−0.39**	0.00	13.57**	0.00
宜春	1 645.5 ± 64.2	1 620.4 ± 57.7	1 573.8 ± 50.9	1 624.0 ± 65.5	−0.46**	0.00	20.81**	0.00
长沙	1 653.1 ± 58.4	1 612.7 ± 54.2	1 598.4 ± 50.9	1 630.4 ± 60.2	−0.44**	0.00	17.69**	0.00
娄底	1 666.0 ± 63.4	1 614.1 ± 50.9	1 605.6 ± 68.6	1 640.2 ± 67.1	−0.46**	0.00	20.08**	0.00

（续　表）

地区	20～44岁组	45～59岁组	60～80岁组	合计	相关分析		方差分析	
					r	*P*	*F*	*P*
赣州	1 653.0 ± 60.4	1 640.2 ± 59.6	1 612.2 ± 51.5	1 641.0 ± 60.2	−0.26**	0.00	6,43**	0.00
梅州	1 672.1 ± 60.7	1 630.1 ± 67.4	1 633.9 ± 52.7	1 649.6 ± 64.6	−0.32**	0.00	8.85**	0.00
福州	1 704.7 ± 57.7	1 647.7 ± 54.5	1 631.7 ± 74.8	1 669.9 ± 67.9	−0.47**	0.00	25.14**	0.00
漳州	1 678.1 ± 60.6	1 642.9 ± 56.0	1 624.4 ± 59.3	1 657.2 ± 62.9	−0.43**	0.00	12.39**	0.00
文昌	1 682.7 ± 61.4	1 648.5 ± 72.6	1 620.0 ± 66.3	1 656.3 ± 70.3	−0.39**	0.00	16.54**	0.00
琼海	1 678.1 ± 62.2	1 637.3 ± 63.9	1 621.7 ± 46.1	1 653.6 ± 64.0	−0.37**	0.00	13.19**	0.00
化州	1 690.5 ± 60.3	1 641.6 ± 59.4	1 621.4 ± 58.8	1 656.2 ± 58.2	−0.52**	0.00	47.36**	0.00

表2　汉族乡村女性36个族群的身高均数以及与年龄的相关分析与方差分析（mm，Mean ± SD）

地区	20～44岁组	45～59岁组	60～80岁组	合计	相关分析		方差分析	
					r	*P*	*F*	*P*
哈尔滨	1 571.4 ± 57.2	1 554.9 ± 52.1	1 535.7 ± 51.4	1 556.0 ± 64.0	−0.26**	0.00	6.55**	0.00
榆树	1 582.5 ± 59.5	1 549.3 ± 49.2	1 527.9 ± 46.5	1 558.4 ± 57.0	−0.48**	0.00	15.70**	0.00
乌兰浩特	1 557.3 ± 43.4	1 535.7 ± 52.0	1 512.1 ± 66.3	1 541.2 ± 54.0	−0.37**	0.00	14.88**	0.00
昌图	1 557.7 ± 55.1	1 543.1 ± 53.6	1 526.4 ± 60.0	1 545.1 ± 56.6	−0.20**	0.00	5.63**	0.00
锦州	1 573.4 ± 45.2	1 559.0 ± 49.5	1 524.6 ± 49.7	1 559.6 ± 50.7	−0.36**	0.00	19.94**	0.00

（续　表）

地区	20 ～ 44 岁组	45 ～ 59 岁组	60 ～ 80 岁组	合计	相关分析		方差分析	
					r	P	F	P
张家口	1 567.2 ± 54.4	1 567.2 ± 54.4	1 535.7 ± 53.8	1 554.2 ± 56.6	−0.31**	0.00	13.68**	0.00
保定	1 587.0 ± 50.9	1 570.0 ± 44.5	1 548.1 ± 48.4	1 574.2 ± 50.7	−0.31**	0.00	9.51**	0.00
晋中	1 552.5 ± 53.1	1 533.7 ± 53.6	1 515.2 ± 60.0	1 539.0 ± 56.2	−0.29**	0.00	8.81**	0.00
潍坊	1 579.9 ± 58.3	1 559.0 ± 56.3	1 546.2 ± 50.3	1 566.4 ± 57.5	−0.26**	0.00	5.26**	0.00
南阳	1 574.3 ± 48.3	1 552.7 ± 58.1	1 519.0 ± 53.5	1 558.2 ± 56.1	−0.32**	0.00	19.19**	0.00
新野	1 573.9 ± 52.0	1 556.4 ± 59.8	1 523.8 ± 54.1	1 558.6 ± 57.9	−0.30**	0.00	14.53**	0.00
蒲城	1 559.1 ± 58.7	1 551.2 ± 53.2	1 529.7 ± 39.2	1 551.1 ± 54.7	−0.27**	0.00	4.83**	0.00
富平	1 576.8 ± 55.9	1 571.5 ± 66.4	1 533.3 ± 55.9	1 566.6 ± 61.4	−0.20**	0.00	9.98**	0.00
平凉	1 572.9 ± 56.2	1 548.3 ± 55.2	1 510.0 ± 59.8	1 553.3 ± 61.3	−0.46**	0.00	22.88**	0.00
武威	1 555.3 ± 60.1	1 525.1 ± 51.7	1 502.9 ± 48.2	1 538.0 ± 59.5	−0.37**	0.00	16.80**	0.00
荆门	1 580.2 ± 54.2	1 564.9 ± 46.5	1 530.4 ± 48.9	1 566.5 ± 54.7	−0.40**	0.00	14.80**	0.00
荆州	1 571.1 ± 47.0	1 570.3 ± 55.1	1 540.6 ± 61.3	1 564.6 ± 60.0	−0.18**	0.00	5.33**	0.00
成都	1 542.2 ± 46.1	1 500.7 ± 56.6	1 444.6 ± 50.8	1 517.6 ± 60.8	−0.61**	0.00	53.29**	0.00
资阳	1 543.5 ± 49.1	1 504.7 ± 49.3	1 460.3 ± 37.5	1 517.2 ± 57.1	−0.64**	0.00	45.52**	0.00
安顺	1 543.0 ± 57.6	1 510.5 ± 52.1	1 475.3 ± 54.8	1 518.1 ± 60.5	−0.46**	0.00	27.53**	0.00
昆明	1 557.9 ± 54.4	1 544.3 ± 47.4	1 525.0 ± 54.1	1 546.8 ± 53.2	−0.30**	0.00	5.81**	0.00

（续　表）

地区	20 ～ 44 岁组	45 ～ 59 岁组	60 ～ 80 岁组	合计	相关分析		方差分析	
					r	P	F	P
滁州	1 570.0 ± 45.3	1 567.6 ± 54.8	1 525.4 ± 59.4	1 561.8 ± 53.3	-0.27^{**}	0.00	9.21^{**}	0.00
淮安	1 586.5 ± 53.0	1 560.3 ± 49.7	1 523.3 ± 57.5	1 566.9 ± 58.0	-0.46^{**}	0.00	21.44^{**}	0.00
嘉兴	1 576.3 ± 53.7	1 558.6 ± 53.6	1 525.1 ± 49.3	1 559.3 ± 56.6	-0.37^{**}	0.00	15.20^{**}	0.00
绍兴	1 572.7 ± 57.7	1 552.4 ± 56.4	1 515.4 ± 51.9	1 552.7 ± 60.5	-0.42^{**}	0.00	15.57^{**}	0.00
景德镇	1 544.6 ± 47.1	1 514.5 ± 43.6	1 466.8 ± 47.4	1 518.9 ± 56.7	-0.59^{**}	0.00	44.98^{**}	0.00
宜春	1 543.1 ± 49.0	1 513.3 ± 47.4	1 473.9 ± 48.1	1 517.0 ± 54.7	-0.48^{**}	0.00	28.22^{**}	0.00
长沙	1 555.8 ± 51.5	1 520.2 ± 50.5	1 508.0 ± 48.2	1 534.6 ± 54.7	-0.36^{**}	0.00	19.11^{**}	0.00
娄底	1 552.0 ± 53.8	1 523.8 ± 63.6	1 500.5 ± 34.6	1 533.6 ± 57.5	-0.46^{**}	0.00	27.53^{**}	0.00
赣州	1 549.7 ± 48.5	1 518.1 ± 53.4	1 502.4 ± 42.6	1 532.2 ± 52.5	-0.38^{**}	0.00	28.93^{**}	0.00
梅州	1 567.9 ± 54.4	1 529.8 ± 52.9	1 520.2 ± 67.4	1 543.1 ± 60.0	-0.38^{**}	0.00	11.66^{**}	0.00
福州	1 578.0 ± 56.5	1 553.8 ± 55.4	1 530.6 ± 54.2	1 559.2 ± 67.9	-0.36^{**}	0.00	9.60^{**}	0.00
漳州	1 560.8 ± 50.0	1 548.7 ± 51.4	1 534.0 ± 41.3	1 551.3 ± 49.6	-0.32^{**}	0.00	4.31^{*}	0.02
文昌	1 555.4 ± 53.4	1 537.1 ± 51.6	1 480.9 ± 47.9	1 535.0 ± 60.1	-0.49^{**}	0.00	30.38^{**}	0.00
琼海	1 554.3 ± 44.2	1 536.1 ± 54.1	1 508.1 ± 48.9	1 540.8 ± 50.8	-0.34^{**}	0.00	9.95^{**}	0.00
化州	1 549.2 ± 50.9	1 524.7 ± 48.1	1 501.0 ± 53.6	1 530.6 ± 53.7	-0.39^{**}	0.00	32.91^{**}	0.00

2. 城市汉族身高均数

31个城市男性族群中，锦州、榆树、哈尔滨、潍坊、保定、淮安、滁州、西安汉族族群依次进入身高的1～8位。化州、娄底、赣州、资阳、长沙、成都、宜春、景德镇8个南方族群身高均数依次分居24～31位。值得注意的是中国南方的福州、绍兴、荆州、漳州汉族人身高进入前16位，而中国北方的晋中汉族身高排在第18位（表3）。

31个城市女性族群中，锦州、潍坊、保定、西安、哈尔滨、福州、乌兰浩特、荆门汉族身高均数依次分居前1～8位。娄底、琼海、宜春、长沙、化州、景德镇、成都、资阳汉族依次分居24～31位（表4）。

（二）城市、乡村汉族身高比较

31个城市族群都各有一个邻近的相对应的乡村族群。将31个城市汉族与各自对应的乡村族群身高进行比较（表3，表4），发现有多数城市族群（21个男性族群、18个女性族群）身高大于邻近的乡村族群（$P < 0.01$或$0.01 < P < 0.05$）。总体上目前中国城市和乡村之间体质上尚存在差距。河北保定、浙江嘉兴、海南文昌城、乡族群的身高差异均无统计学意义（$P > 0.05$）。这3个地区都位于中国东部，提示中国部分东部地区城、乡间的差异已经很小。

综合男性、女性身高排序结果表明，中国北方乡村和城市汉族身高大于南方乡村、城市汉族。但江淮平原、江汉平原的男、女性身高已经高于部分北方汉族族群。

（三）北方汉族、南方汉族身高的比较

在地理学上，中国南方、北方以秦岭山脉、淮河为界。我们按照地理学的分界方法，将36个族群分为南方、北方族群。北方汉族乡村男性、女性身高大于南方汉族（$P < 0.01$）。u检验显示北方乡村男性、女性的耳上头高、形态面高、坐高、下肢全长均大于南方汉族（$P < 0.01$），即头、面、躯干、下肢（身高的主要组成部分）的高度北方汉族都大于南方汉族。此外，北方乡村汉族的身高下肢长指数大于南方汉族（$P < 0.01$），但下身长坐高指数小于南方汉族（$P < 0.01$或$0.01 < P < 0.05$）（表5），表明北方乡村汉族下肢高度在身高中的比例超过了南方汉族，而躯干与腿的比例小于南方汉族。

北方、南方城市汉族的比较结果与乡村汉族的比较结果一致。北方城市汉族的身高及头部、面部、躯干、下肢高度的均数大于南方城市汉族（$P < 0.01$）（表6），在整个身材比例上，北方汉族腿显得更长些，上身显得更短些。

表3　汉族31个城市男性族群的身高均数以及与年龄的相关分析与方差分析（mm，Mean ± SD）

地　区	男　性				相关分析		方差分析		u检验
	20～44岁组	45～59岁组	60～80岁组	合　计	r	P	F	P	u
哈尔滨	1 720.5 ± 66.4	1 670.7 ± 52.8	1 660.2 ± 55.3	1 700.9 ± 65.5	−0.45**	0.00	15.63**	0.00	3.52**
榆树	1 721.9 ± 61.3	1 670.9 ± 63.7	1 644.6 ± 53.2	1 703.8 ± 67.4	−0.58**	0.00	20.33**	0.00	4.17**
乌兰浩特	1 700.2 ± 62.4	1 678.8 ± 46.8	1 664.1 ± 70.6	1 687.5 ± 63.5	−0.23**	0.00	3.86*	0.02	4.78**
锦州	1 734.5 ± 55.2	1 700.6 ± 56.0	1 663.1 ± 53.8	1 708.1 ± 68.4	−0.47**	0.00	30.73**	0.00	7.05**
张家口	1 685.3 ± 48.2	1 668.4 ± 60.0	1 670.3 ± 60.2	1 677.1 ± 54.9	−0.19**	0.00	2.77	0.07	0.14
保定	1 718.8 ± 56.1	1 696.7 ± 61.4	1 654.4 ± 41.9	1 699.8 ± 60.1	−0.44**	0.00	14.84**	0.00	0.06
晋中	1 689.1 ± 64.2	1 663.6 ± 52.3	1 658.9 ± 62.0	1 675.2 ± 61.4	−0.28**	0.00	4.50*	0.01	2.30*
潍坊	1 718.2 ± 66.1	1 695.0 ± 63.1	1 665.5 ± 52.2	1 700.9 ± 65.5	−0.36**	0.00	8.15**	0.00	4.57**
南阳	1 674.8 ± 65.3	1 687.0 ± 64.0	1 674.2 ± 58.7	1 677.9 ± 64.1	−0.05	0.51	0.74	0.48	2.32*
西安	1 703.8 ± 54.9	1 682.3 ± 68.1	1 653.0 ± 63.5	1 687.6 ± 63.4	−0.36**	0.00	10.49**	0.00	0.71
兰州	1 693.8 ± 51.5	1 671.7 ± 68.0	1 671.3 ± 64.9	1 682.0 ± 60.8	−0.17*	0.01	3.39*	0.04	4.70**
荆门	1 692.9 ± 61.3	1 662.7 ± 48.6	1 645.7 ± 42.1	1 674.2 ± 57.5	−0.414**	0.00	9.85**	0.00	2.56*
荆州	1 704.5 ± 57.2	1 660.2 ± 67.4	1 652.4 ± 45.5	1 679.0 ± 63.7	−0.42**	0.00	10.99**	0.00	2.84**
成都	1 665.9 ± 64.2	1 636.3 ± 67.6	1 608.1 ± 58.9	1 648.1 ± 67.6	−0.43**	0.00	8.21**	0.00	3.38**

（续 表）

地 区	男 性				相关分析		方差分析		u检验
	20～44岁组	45～59岁组	60～80岁组	合 计	r	P	F	P	u
资阳	1 672.8 ± 67.3	1 631.5 ± 51.4	1 588.8 ± 65.8	1 652.4 ± 69.8	−0.50**	0.00	14.74**	0.00	3.68**
昆明	1 694.7 ± 65.0	1 656.2 ± 54.1	1 650.7 ± 52.6	1 674.6 ± 62.5	−0.34**	0.00	8.81**	0.00	2.09*
滁州	1 709.5 ± 61.4	1 670.5 ± 62.1	1 666.0 ± 54.6	1 690.5 ± 63.0	−0.37**	0.00	9.17**	0.00	2.06*
淮安	1 721.9 ± 53.7	1 674.2 ± 46.0	1 650.6 ± 60.0	1 693.1 ± 60.1	−0.54**	0.00	24.52**	0.00	3.47**
嘉兴	1 696.8 ± 59.4	1 668.9 ± 63.7	1 630.6 ± 53.4	1 675.2 ± 64.5	−0.43**	0.00	13.54**	0.00	0.24
绍兴	1 700.6 ± 58.5	1 679.0 ± 50.9	1 657.8 ± 55.9	1 683.9 ± 57.2	−0.32**	0.00	8.52**	0.00	3.84**
景德镇	1 655.9 ± 58.9	1 639.1 ± 54.9	1 607.8 ± 59.2	1 638.5 ± 60.2	−0.34**	0.00	8.14**	0.00	0.77
宜春	1 670.1 ± 61.8	1 631.3 ± 53.4	1 586.3 ± 52.7	1 641.9 ± 65.9	−0.55**	0.00	23.74**	0.00	2.53*
长沙	1 668.3 ± 51.5	1 632.9 ± 59.7	1 629.0 ± 50.3	1 650.4 ± 56.5	−0.34**	0.00	8.96**	0.00	3.17**
娄底	1 673.5 ± 59.1	1 656.5 ± 50.5	1 606.0 ± 53.0	1 654.0 ± 60.3	−0.43**	0.00	16.28**	0.00	2.03*
赣州	1 659.0 ± 56.9	1 663.5 ± 65.6	1 623.5 ± 58.9	1 652.5 ± 61.6	−0.26**	0.00	5.19**	0.01**	1.73
梅州	1 678.4 ± 62.8	1 663.4 ± 57.8	1 629.1 ± 54.9	1 664.2 ± 62.3	−0.33**	0.00	7.31**	0.00	2.04*
福州	1 706.2 ± 54.3	1 678.0 ± 59.1	1 649.2 ± 53.4	1 685.2 ± 59.6	−0.40**	0.00	11.53**	0.00	2.21*
漳州	1 692.0 ± 53.3	1 670.2 ± 54.3	1 648.4 ± 63.5	1 676.5 ± 57.9	−0.36**	0.00	6.79**	0.00	2.87*

（续　表）

地　区	男　　性				相关分析		方差分析		u检验
	20～44岁组	45～59岁组	60～80岁组	合　计	r	P	F	P	u
文昌	1 683.0 ± 53.4	1 666.4 ± 54.3	1 637.4 ± 54.7	1 668.9 ± 56.3	−0.36**	0.00	7.75**	0.00	1.90
琼海	1 677.6 ± 60.7	1 646.8 ± 67.6	1 643.3 ± 44.4	1 661.7 ± 61.9	−0.27**	0.00	5.77**	0.00	1.18
化州	1 667.9 ± 51.0	1 666.6 ± 51.2	1 608.2 ± 57.9	1 657.0 ± 56.7	−0.37**	0.00	18.00**	0.00	2.51**

注：u值为城、乡男性间u检验值

表4　汉族31个城市女性族群的身高均数以及与年龄的相关分析与方差分析（mm，Mean ± SD）

地　区	女　　性				相关分析		方差分析		u检验
	20～44岁组	45～59岁组	60～80岁组	合　计	r	P	F	P	u
哈尔滨	1 603.4 ± 53.1	1 565.7 ± 57.1	1 539.3 ± 49.5	1 577.9 ± 59.2	−0.44**	0.00	10.48**	0.00	3.43**
榆树	1 597.5 ± 60.6	1 567.1 ± 55.2	1 527.4 ± 49.5	1 572.1 ± 62.6	−0.48**	0.00	10.83**	0.00	2.17*
乌兰浩特	1 588.2 ± 53.1	1 572.0 ± 39.7	1 543.0 ± 63.7	1 574.7 ± 54.4	−0.35**	0.00	9.48**	0.00	6.33*
锦州	1 605.2 ± 57.7	1 594.7 ± 62.7	1 547.1 ± 60.3	1 589.7 ± 63.6	−0.41**	0.00	17.92**	0.00	5.89**
张家口	1 592.4 ± 45.8	1 572.3 ± 60.2	1 529.7 ± 48.0	1 572.7 ± 56.6	−0.40**	0.00	25.32**	0.00	4.23**
保定	1 593.6 ± 50.4	1 568.0 ± 35.3	1 554.6 ± 40.4	1 579.0 ± 47.4	−0.40**	0.00	10.01**	0.00	0.93

（续　表）

地　区	女　性				相关分析		方差分析		u检验
	20～44岁组	45～59岁组	60～80岁组	合　计	r	P	F	P	u
晋中	1 575.5 ± 54.3	1 561.5 ± 42.9	1 514.2 ± 60.8	1 558.1 ± 57.5	−0.46**	0.00	15.56**	0.00	3.26**
潍坊	1 596.3 ± 54.4	1 572.1 ± 56.8	1 550.5 ± 49.8	1 584.4 ± 56.6	−0.32**	0.00	6.86**	0.00	2.92**
南阳	1 581.9 ± 67.7	1 581.5 ± 59.7	1 525.1 ± 51.7	1 570.4 ± 66.3	−0.31**	0.00	13.09**	0.00	2.08*
西安	1 593.1 ± 56.5	1 581.0 ± 52.7	1 536.3 ± 49.7	1 578.1 ± 57.7	−0.35**	0.00	16.65**	0.00	2.10*
兰州	1 577.6 ± 52.2	1 558.2 ± 54.7	1 532.5 ± 49.3	1 563.2 ± 54.9	−0.32**	0.00	10.19**	0.00	4.69**
荆门	1 579.5 ± 45.2	1 573.2 ± 44.9	1 558.8 ± 39.5	1 573.0 ± 44.4	−0.21**	0.01	2.71	0.07	1.29
荆州	1 578.6 ± 52.9	1 571.6 ± 51.0	1 551.3 ± 67.8	1 570.1 ± 56.4	−0.17*	0.04	2.70	0.07	0.89
成都	1 553.2 ± 46.0	1 530.2 ± 51.3	1 482.9 ± 51.7	1 530.3 ± 55.6	−0.51**	0.00	26.28**	0.00	2.13*
资阳	1 559.5 ± 49.6	1 503.7 ± 49.6	1 482.1 ± 51.5	1 529.5 ± 59.7	−0.60**	0.00	32.23**	0.00	1.96*
昆明	1 584.0 ± 47.6	1 552.3 ± 46.7	1 545.3 ± 40.7	1 565.0 ± 49.3	−0.36**	0.00	12.53**	0.00	3.31**
滁州	1 579.8 ± 47.3	1 574.7 ± 53.7	1 543.0 ± 56.7	1 570.6 ± 53.2	−0.24*	0.02	5.91**	0.00	1.52
淮安	1 595.1 ± 60.5	1 562.6 ± 40.4	1 529.6 ± 49.6	1 571.8 ± 58.7	−0.47**	0.00	18.18**	0.00	0.79
嘉兴	1 587.7 ± 56.6	1 560.9 ± 58.5	1 539.1 ± 61.7	1 567.4 ± 60.9	−0.33**	0.00	7.56**	0.00	1.29
绍兴	1 575.7 ± 53.0	1 571.8 ± 53.6	1 548.2 ± 53.8	1 568.8 ± 54.1	−0.26**	0.00	2.87	0.06	2.64**

（续　表）

地　区	女　性				相关分析		方差分析		u检验
	20～44岁组	45～59岁组	60～80岁组	合　计	r	P	F	P	u
景德镇	1 550.6 ± 43.6	1 530.9 ± 44.0	1 519.9 ± 57.3	1 536.6 ± 48.0	-0.32^{**}	0.00	5.04^{**}	0.01	3.16^{**}
宜春	1 558.3 ± 49.9	1 535.7 ± 45.0	1 509.4 ± 53.7	1 541.7 ± 52.1	-0.38^{**}	0.00	10.92^{**}	0.00	4.32^{**}
长沙	1 553.5 ± 43.4	1 534.4 ± 40.8	1 501.2 ± 64.1	1 537.1 ± 51.0	-0.45^{**}	0.00	14.52^{**}	0.00	0.47
娄底	1 559.9 ± 42.5	1 540.4 ± 51.6	1 506.0 ± 45.5	1 543.2 ± 50.2	-0.37^{**}	0.00	15.20^{**}	0.00	1.73
赣州	1 571.3 ± 49.7	1 546.4 ± 55.2	1 521.7 ± 50.6	1 553.4 ± 54.9	-0.37^{**}	0.00	10.35^{**}	0.00	3.63^{**}
梅州	1 560.9 ± 54.6	1 552.3 ± 65.5	1 510.8 ± 57.7	1 548.2 ± 61.5	-0.29^{**}	0.00	9.44^{**}	0.00	0.79
福州	1 600.3 ± 50.3	1 573.2 ± 62.2	1 538.8 ± 53.1	1 577.1 ± 60.1	-0.40^{**}	0.00	12.66^{**}	0.00	2.63^{**}
漳州	1 570.9 ± 50.0	1 559.7 ± 50.9	1 525.1 ± 53.6	1 557.7 ± 53.7	-0.40^{**}	0.00	8.66^{**}	0.00	1.16
文昌	1 559.8 ± 52.0	1 545.2 ± 47.2	1 530.6 ± 47.8	1 550.6 ± 51.0	-0.20^{**}	0.01	4.19^{*}	0.02	2.65^{**}
琼海	1 557.1 ± 50.5	1 531.4 ± 42.0	1 512.0 ± 51.9	1 541.9 ± 51.3	-0.37^{**}	0.00	9.91^{**}	0.00	0.19
化州	1 553.9 ± 49.1	1 537.2 ± 47.6	1 490.2 ± 61.0	1 536.8 ± 56.0	-0.44^{**}	0.00	22.47^{**}	0.00	1.46

注：u值为城、乡男性间u检验值

表5　北方汉族、南方汉族乡村成年人身高及相关指标、指数的比较（mm，Mean ± SD）

变　量	男　性			女　性		
	北方汉族（n=3 841）	南方汉族（n=4 333）	u	北方汉族（n=3 810）	南方汉族（n=4 517）	u
身　高	1 671.2 ± 61.5	1 649.5 ± 67.4	15.22^{**}	1 554.7 ± 57.3	1 539 ± 61.8	12.05^{**}
耳上头高	132.3 ± 25.3	121.7 ± 27	18.32^{**}	125.6 ± 19.7	123.4 ± 9.4	6.29^{**}
形态面高	126.2 ± 16.7	123.5 ± 8.7	9.00^{**}	118.6 ± 10.6	115.9 ± 7.9	13.26^{**}
坐　高	893.9 ± 48.2	887.5 ± 39.0	6.55^{**}	841.8 ± 44.8	836.0 ± 35.1	6.39^{**}
躯干前高	586.8 ± 52.2	582.7 ± 34.7	4.41^{**}	556.6 ± 48.7	552.4 ± 33.8	4.48^{**}
下肢全长	888.1 ± 41.7	863.7 ± 38.9	27.25^{**}	835.6 ± 39.4	819.4 ± 40.6	18.49^{**}
身高下肢长指数	53.2 ± 2.0	52.4 ± 1.8	18.91^{**}	53.8 ± 1.8	53.2 ± 1.7	13.15^{**}
下身长坐高指数	115.5 ± 16.0	116.1 ± 6.4	2.17^{*}	118.5 ± 9.8	119.4 ± 10.0	4.21^{**}

表6　北方汉族、南方汉族城市成年人指标、指数的比较（mm，Mean ± SD）

变　量	男　性			女　性		
	北方汉族（n=2 002）	南方汉族（n=3 046）	u	北方汉族（n=2 105）	南方汉族（n=3 298）	u
身　高	1 688.5 ± 63	1 663.5 ± 64.6	13.63**	1 574.9 ± 58.9	1 552.9 ± 56.4	13.58**
耳上头高	133.4 ± 20.5	126.6 ± 18.8	11.75**	129.7 ± 33.1	123.5 ± 9.6	8.35**
形态面高	128 ± 12.1	124.1 ± 10	11.78**	120.1 ± 11.2	117 ± 9.6	10.35**
坐　高	906.8 ± 38.2	893.2 ± 48.4	11.12**	854.2 ± 36.1	844.8 ± 33.2	9.61**
躯干前高	599.6 ± 43.8	588.8 ± 37.9	9.00**	564.0 ± 39.7	556.6 ± 31.7	7.26**
下肢全长	910.1 ± 20.4	873.7 ± 40.0	7.89**	861.4 ± 70.9	825.4 ± 37.0	21.44**
身高下肢长指数	53.9 ± 11.8	52.6 ± 2.4	4.96**	54.7 ± 4.6	53.2 ± 1.7	15.14**
下身长坐高指数	116.3 ± 8.2	116.2 ± 9.5	0.40	118.8 ± 10.0	119.5 ± 6.5	2.84**

有学者认为高身材的人大致都是腿较长，身材的高矮在很大程度上取决于下肢的长短，而不是坐高[8]。本文研究支持这一观点。

生活在中国南方气温高的地区的族群体型上普遍趋于瘦小，生活在中国北方气温低的地区的族群体型上总是显得高大。中国各地汉族的身高反映了他们各自对环境的适应，符合贝格曼法则[9]。

（四）各地区汉族身高的分类

据Martin的身高分类标准，男性1 700 ～ 1 799 mm被认为是高身材，1 670 ～ 1 699 mm为超中等身材，1 640 ～ 1 669 mm为中等的身材，1 600 ～ 1 639 mm为亚中等身材。对于女性，1 590 ～ 1 679 mm为高身材，1 560 ～ 1 589 mm为超中等身材，1 530 ～ 1 559 mm为中等身材，1 490 ～ 1 529 mm为亚中等身材[10]。

1. 男性身高的分类

根据身高的合计资料，乡村男性中，北方的保定族群为高身材，富平、湖州、哈尔滨、张家口、滁州、榆树、平凉、潍坊、淮安等北方、江淮地区、杭嘉湖地区的9个族群为超中等身材，景德镇、长沙、安顺、成都、资阳、宜春6个南方族群为亚中等身材。其余20个族群（既包括北方族群，又包括南方族群）为中等身材。目前汉族乡村男性族群多为中等身材。

城市男性中，锦州、榆树、哈尔滨、潍坊4个北方族群为高身材，文昌、梅州、琼海、化州、娄底、赣州、资阳、长沙、成都、宜春10个南方族群为中等身材，南方的景德镇族群为亚中等身材。超中等身材共16个族群（既包括北方族群，又包括南方族群），约占城市男性族群的一半。

目前乡村男性族群多为中等身材，约有一半的城市男性族群为超中等身材。

2. 女性身高的分类

根据身高的合计资料，乡村女性中，保定、淮安、富平、荆门、潍坊、荆州、滁州等北方及长江中下游地区的7个族群为超中等身材，景德镇、安顺、成都、资阳、宜春5个南方族群为亚中等身材，其余24个族群为中等身材。目前乡村女性族群多为中等身材。

城市女性中，资阳族群为亚中等身材，晋中、漳州、赣州、文昌、梅州、娄底、琼海、宜春、长沙、化州、景德镇、成都12个族群为中等身材，其余18个族群为超中等身材。中等、亚中等身材的族群中，除晋中族群外，均为南方族群。

综合女性分析结果，可以认为，目前乡村族群以中等身材为主，城市族群以超中等身材为主。北方城市族群主要为超中等身材，北方乡村族群、南方城市族群为超中等身材或中等身材。南方乡村族群为中等身材或亚中等身材。

3. 汉族各族群20～44岁组的身高分类

20～44岁组的身高反映了目前中青年人的身高水平，也预示今后一些年汉族人身高的变化趋势。在3个年龄组中，20～44岁组身高最高。

乡村男性20～44岁组中，安顺、宜春、成都、赣州、长沙、景德镇、资阳、娄底8个族群为中等身材，榆树、福州、嘉兴、保定4个族群为高身材，其余24个族群为超中等身材。

城市男性20～44岁组中，景德镇、赣州、成都、化州、长沙5个族群为中等身材，宜春、资阳、娄底、南阳、琼海、梅州、文昌、张家口、晋中、漳州、荆门、兰州、昆明、嘉兴14个族群为超中等身材，其余12个族群为高身材。

乡村女性20～44年龄组中，成都、安顺、宜春、资阳、景德镇、化州、赣州、娄底、晋中、琼海、武威、文昌、长沙、乌兰浩特、昌图、昆明、蒲城17个族群为中等身材，其余19个族群为超中等身材。目前汉族乡村女性20～44岁组多为超中等身材。

城市女性20～44岁组中，景德镇、成都、长沙、化州、琼海、宜春、资阳、文昌、娄底9个族群为中等身材，梅州、漳州、赣州、晋中、绍兴、兰州、荆州、荆门、滁州、南阳、昆明、嘉兴、乌兰浩特13个族群为超中等身材，其余9个族群为高身材。

20～44岁组中接近11%的乡村男性、40%的城市男性、30%的城市女性族群身高已经进入高身材族群，而乡村女性尚无高身材族群。男性、女性20～44岁组均进入高身材族群的有北方的锦州、榆树、哈尔滨、潍坊、保定、西安族群和南方的淮安、福州族群。

4. 近年来汉族身高的变化

相关分析显示，乡村男性、乡村女性、城市男性、城市女性的所有族群身高与年龄的相关系数均为负值，且除南阳城市男性外，均为$0.01 < P < 0.05$或$P < 0.01$，说明随年龄增长，汉族各族群的身高呈现线性下降变化规律。

方差分析显示，除张家口、南阳的城市男性和荆门、荆州、绍兴的城市女性外，其余乡村男性、乡村女性、城市男性、城市女性的年龄组间身高差异均具有统计学意义（$0.01 < P < 0.05$或$P < 0.01$）。这表明近些年来，汉族人身材出现了明显的变化。

本资料为横断面调查，不是纵向追踪调查（纵向调查在大样本研究中是不可能的）。由于本次研究样本量大，且由同一个研究组用同样方法在短时间完成，因此可以弥补非追踪调查的不足。

20～44岁组的组中年龄为32岁，60～80岁组的组中年龄为70岁，可以大致认为这两个年龄组的年龄相距38岁左右。总体说来，60～80岁组中身材较矮的族群38年来身高增加幅度较大，而身材较高的族群身高增加幅度较小，如乡村男性中，38年间原身材较高的榆树族群由1 648.0 mm增加到1 703.4 mm，增加了55.4 mm；蒲城族群由1 637.0 mm增加到1 686.8 mm，增加了49.8 mm。而原身材较矮的资阳族群由1 561.0 mm增加到1 661.4 mm，增加了100.4 mm；宜春族群由1 573.8 mm增加到1 645.5 mm，增加了71.7 mm。

应该说明，随年龄增长，由于人椎间软骨失去弹性和可塑性而变得扁平，身高也逐渐降低。一般认为身高稳定期为45～50岁，55岁以前每5年身高平均减少5 mm，以后每5年减少7 mm[8]。考虑到这种生理性变化，汉族人38年来的身高实际增加值比上述计算的增加值小一些。

标准差的大小，反映了一组数据的离散程度。将36个乡村男性族群的20～44岁组、45～59岁组、60～80岁组的身高均数再求均数、标准差。乡村男性3个年龄组依次为1 680.1 mm ± 18.1 mm、1 647.1 mm ± 21.2 mm、1 627.3 mm ± 25.6 mm，同样方法可得到乡村女性依次为1 563.8 mm ± 13.5 mm、1 542.5 mm ± 20.3 mm、1 513.4 mm ± 24.7 mm，城市男性依次为1 692.0 mm ± 20.8 mm、1 665.8 mm ± 18.3 mm、1 641.9 mm ± 24.4 mm，城市女性依次为1 577.2 mm ± 16.9 mm、1 557.5 mm ± 19.8 mm、1 527.3 mm ± 20.5 mm。各族群的均数均是20～44岁组最大，45～59岁组次之，60～80岁组最小，20～44岁组标准差均小于60～80岁组。

这提示中国20～44岁组各族群身高均数彼此差距较小，而60～80岁组各族群身高均数彼此差距较大，表明目前各地年轻人的身高差距已经在缩小。

二、汉族方言族群的身高

汉族各方言族群身高均数间的差异多具有统计学意义（表8）。乡村男性族群身高均数在1 628.0～1 674.0 mm，城市族群男性身高均数在

1 640.0 ～ 1 696.0 mm。乡村女性族群身高均数在1 518.0 ～ 1 565.0 mm，城市女性族群身高均数在1 536.0 ～ 1 580.0 mm。

对11个汉族方言族群的身高分3个年龄组进行统计。对60 ～ 80岁组（老年人，在1933~1953年出生）、40 ～ 49岁组（中年人，在1973 ～ 1963年出生）、20 ～ 29岁组（青年人，在1983 ～ 1993年出生），以及合计资料按照身高值的大小进行排序比较，以比较各汉族方言族群的老年人、中年人、青年人的相对身高。

（一）汉族方言族群的身高

1. 汉族方言族群乡村成年人的身高

乡村男性中，中部汉族（江淮方言汉族、吴语族群）、北方汉族身材较高，彼此接近（$P > 0.05$），高于南方其他6个族群（$0.01 < P < 0.05$或$P < 0.01$）；在南方6个族群中，闽语族群、粤语族群相对较高，高于西南方言汉族、客家人、湘语族群、赣语族群（$P < 0.01$）；西南方言汉族、赣语族群、湘语族群、客家人男性身材较矮小（表7，表8）。

不同年龄组汉族乡村男性身高排序有一些变化：在60 ～ 80岁组、40 ～ 49岁组、20 ～ 29岁组中吴语族群身高分别为第5、5、3位，江淮方言汉族分别为第4、2、1位，西北方言族群分别为第2、1、5位。在当代汉族乡村成人中，江淮方言族群、吴语族群身高排序提前，已是明显的事实，而西北方言族群的身高排序推后，也是很明显的。江淮方言族群的身高变化自20世纪50年代即开始，而吴语族群、西北方言族群变化分别自70 ～ 80年代开始。

乡村女性中，位于长江下游的江淮地区、吴语地区已经逐渐成为汉族乡村女性身材最高的地区。目前50 ～ 59岁组（出生于1953~1963年）中，江淮方言族群女性身材已经是全国第一，其后依次是华北方言族群、吴语族群、东北方言族群、西北方言族群。上述族群身高值高于南方6个族群（$0.01 < P < 0.05$或$P < 0.01$）。闽语族群是女性中身材居中的族群，低于北方汉族（$0.01 < P < 0.05$或$P < 0.01$），但高于其他5个南方族群（$P < 0.01$）。赣语族群、粤语族群女性最矮（表9，表10）。

不同年龄组汉族乡村女性身高排序有一些变化。在60 ～ 80岁组、40 ～ 49岁组、20 ～ 29岁组中江淮方言汉族分别为第2、1、1位，吴语族群身高分别为第5、2、2位，西北方言族群分别为第2、1、5位。

江淮方言族群乡村成人是当今11个汉族方言族群中身材最高的族群，

表 7　汉族各方言族群乡村男性不同年龄组的身高均数（mm）

族群	20～29岁组		30～39岁组		40～49岁组		50～59岁组		60～80岁组		合计	
	n	Mean ± SD	*n*	Mean ± SD	*n*	Mean ± SD	*n*	Mean ± SD	*n*	Mean ± SD	*n*	Mean ± SD
东北方言族群	195	170.8 ± 5.7	234	167.5 ± 5.3	238	166.6 ± 5.6	233	165.4 ± 5.7	230	164.2 ± 5.5	1 130	166.8 ± 6.0
华北方言族群	318	171.5 ± 6.5	330	168.1 ± 5.7	362	166.7 ± 5.8	351	165.7 ± 5.7	347	164.9 ± 6.0	1 708	167.3 ± 6.4
西北方言族群	204	169.3 ± 5.2	198	168.0 ± 5.5	207	167.6 ± 6.1	202	166.7 ± 6.3	196	164.4 ± 5.9	1 007	167.2 ± 6.0
西南方言族群	250	168.4 ± 6.1	250	165.5 ± 6.0	264	164.4 ± 5.8	259	161.2 ± 5.9	250	160.2 ± 6.7	1 273	164.1 ± 6.7
江淮方言族群	106	171.6 ± 5.4	59	167.8 ± 5.6	96	167.0 ± 5.7	75	166.4 ± 6.6	78	163.0 ± 5.8	414	167.3 ± 6.4
吴语族群	76	170.9 ± 6.1	67	168.7 ± 6.2	70	166.6 ± 5.7	80	165.7 ± 6.9	81	162.4 ± 6.5	374	166.7 ± 6.9
闽语族群	167	169.4 ± 5.8	138	168.5 ± 6.0	140	165.2 ± 6.1	149	164.1 ± 6.6	157	162.4 ± 6.2	751	165.9 ± 6.7
赣语族群	73	166.6 ± 6.7	78	164.5 ± 5.9	91	163.6 ± 6.1	77	160.9 ± 5.9	79	158.8 ± 5.3	398	162.9 ± 6.5
客家人	69	168.3 ± 5.5	70	165.0 ± 6.2	69	163.5 ± 6.0	69	163.5 ± 6.5	68	162.2 ± 5.2	345	164.5 ± 6.2
粤语族群	67	171.0 ± 5.4	64	165.0 ± 4.3	77	164.4 ± 6.1	79	164.1 ± 5.9	98	162.1 ± 5.8	395	165.6 ± 6.6
湘语族群	81	168.2 ± 6.1	80	165.3 ± 6.1	80	163.4 ± 5.2	75	160.3 ± 4.8	77	160.2 ± 6.0	393	163.5 ± 6.4

表8　汉族方言族群乡村男性间身高值的u检验

族　群	东北方言族群	华北方言族群	西北方言族群	西南方言族群	江淮方言族群	吴语族群	闽语族群	赣语族群	客家人	粤语族群	湘语族群
东北方言族群	—	2.22*	1.73	10.40**	1.58	0.10	2.79**	10.49**	5.98**	3.02**	8.82**
华北方言族群	2.22*	—	0.29	13.19**	0.14	1.44	4.72**	12.22**	7.55**	4.50**	10.54**
西北方言族群	1.73	0.29	—	11.74**	0.33	1.21	4.18**	11.50**	7.05**	4.13**	9.88**
西南方言族群	10.40**	13.19**	11.74**	—	8.88**	6.55**	6.00**	3.22**	1.09	3.98**	1.47
江淮方言族群	1.58	0.14	0.33	8.88**	—	1.28	3.54**	9.85**	6.17**	3.72**	8.46**
吴语族群	0.10	1.44	1.21	6.55**	1.28	—	1.85	7.97**	4.54**	2.26*	6.65**
闽语族群	2.79**	4.72**	4.18**	6.00**	3.54**	1.85	—	7.51**	3.45**	0.74	5.94**
赣语族群	10.49**	12.22**	11.50**	3.22**	9.85**	7.97**	7.51**	—	3.49**	5.86**	1.46
客家人	5.98**	7.55**	7.05**	1.09	6.17**	4.54**	3.45**	3.49**	—	2.35*	2.09*
粤语族群	3.02**	4.50**	4.13**	3.98**	3.72**	2.26*	0.74	5.86**	2.35*	—	4.48**
湘语族群	8.82**	10.54**	9.88**	1.47	8.46**	6.65**	5.94**	1.46	2.09*	4.48**	—

表9　汉族各方言族群乡村女性不同年龄组的身高均数(mm)

族　群	20～29岁组		30～39岁组		40～49岁组		50～59岁组		60～80岁组		合　计	
	n	Mean ± SD	*n*	Mean ± SD	*n*	Mean ± SD	*n*	Mean ± SD	*n*	Mean ± SD	*n*	Mean ± SD
东北方言族群	186	157.8 ± 5.4	238	156.2 ± 4.9	250	155.8 ± 5.0	233	154.4 ± 5.4	238	152.5 ± 5.5	1 145	155.2 ± 5.5
华北方言族群	320	158.1 ± 4.9	329	156.2 ± 5.7	341	156.3 ± 5.6	338	154.8 ± 5.6	320	153.1 ± 5.5	1 648	155.7 ± 5.7
西北方言族群	196	157.4 ± 5.5	207	156.9 ± 5.7	205	155.1 ± 6.1	205	154.7 ± 5.8	193	152.0 ± 5.3	1 006	155.2 ± 6.0
西南方言族群	258	157.0 ± 5.1	256	154.8 ± 5.3	274	154.5 ± 5.5	248	152.4 ± 5.8	242	149.8 ± 6.4	1 278	153.8 ± 6.1
江淮方言族群	85	158.8 ± 5.4	72	157.4 ± 4.8	87	157.4 ± 4.9	77	155.5 ± 5.0	70	152.4 ± 5.8	391	156.5 ± 5.6
吴语族群	81	158.5 ± 4.9	79	156.7 ± 6.1	85	156.9 ± 4.8	88	154.4 ± 5.8	81	151.7 ± 5.2	414	155.6 ± 5.9
闽语族群	159	157.0 ± 5.2	142	155.7 ± 4.9	149	154.7 ± 5.1	147	154.1 ± 5.4	140	151.3 ± 5.4	737	154.7 ± 5.6
赣语族群	67	155.0 ± 4.7	86	154.1 ± 4.6	81	152.5 ± 4.9	82	151.0 ± 4.5	80	146.7 ± 5.1	396	151.8 ± 5.6
客家人	75	156.7 ± 5.0	76	155.3 ± 5.4	75	153.3 ± 4.9	75	152.1 ± 5.4	72	151.1 ± 5.7	373	153.7 ± 5.6
粤语族群	94	156.4 ± 4.7	92	154.0 ± 5.2	102	152.8 ± 5.1	104	152.4 ± 4.6	100	150.1 ± 5.4	492	153.1 ± 5.4
湘语族群	82	155.9 ± 5.1	88	155.1 ± 5.8	90	154.4 ± 5.5	89	151.5 ± 5.3	86	150.3 ± 4.3	435	153.4 ± 5.6

表10 汉族方言族群乡村女性间身高值的u检验

族 群	东北方言族群	华北方言族群	西北方言族群	西南方言族群	江淮方言族群	吴语族群	闽语族群	赣语族群	客家人	粤语族群	湘语族群
东北方言族群	—	2.18^{*}	0.00	6.14^{**}	3.76^{**}	1.12	2.06^{*}	10.59^{**}	4.45^{**}	7.42^{**}	5.79^{**}
华北方言族群	2.18^{*}	—	2.01^{*}	8.77^{**}	2.42^{*}	0.31	4.09^{**}	12.52^{**}	6.07^{**}	10.95^{**}	9.23^{**}
西北方言族群	0.00	2.01^{*}	—	5.69^{**}	3.61^{**}	1.07	1.93	10.16^{**}	4.28^{**}	7.92^{**}	6.49^{**}
西南方言族群	6.14^{**}	8.77^{**}	5.69^{**}	—	8.29^{**}	5.24^{**}	3.43^{**}	6.05^{**}	0.12	2.76^{**}	1.38
江淮方言族群	3.76^{**}	2.42^{*}	3.61^{**}	8.29^{**}	—	2.16^{**}	5.16^{**}	11.89^{**}	6.78^{**}	10.04^{**}	8.88^{**}
吴语族群	1.12	0.31	1.04	5.24^{**}	2.16^{**}	—	2.49^{*}	9.23^{**}	4.42^{**}	7.03^{**}	5.98^{**}
闽语族群	2.06^{*}	4.09^{**}	1.93	3.43^{**}	5.16^{**}	2.49^{*}	—	8.36^{**}	2.67^{**}	5.75^{**}	4.42^{**}
赣语族群	10.59^{**}	12.52^{**}	10.16^{**}	6.05^{**}	11.89^{**}	9.23^{**}	8.36^{**}	—	4.80^{**}	3.68^{**}	4.64^{**}
客家人	4.45^{**}	6.07^{**}	4.28^{**}	0.12	6.78^{**}	4.42^{**}	2.67^{**}	4.80^{**}	—	1.88	0.90
粤语族群	7.42^{**}	10.95^{**}	7.92^{**}	2.76^{**}	10.04^{**}	7.03^{**}	5.75^{**}	3.68^{**}	1.88	—	1.02
湘语族群	5.79^{**}	9.23^{**}	6.49^{**}	1.38	8.88^{**}	5.98^{**}	4.42^{**}	4.64^{**}	0.90	1.02	—

男性、女性身高为第1位。华北方言族群在11个汉族方言族群中，男性、女性身高均为第2位。东北方言族群男性身高为第4位、女性身高为第5位。西北方言族群男性身高为第3位、女性身高为第4位。吴语族群与江淮方言汉族有较为一致的基因结构，男性身高为第5位、女性身高为第3位。闽语族群男性、女性身高均为第6位。客家人男性、女性身高均为第8位。西南方言族群与东北、华北、西北地区的汉族都操汉语北方话，但其身高明显低于北方汉族，男性身高为第9位、女性身高为第7位。粤语族群男性身高为第7位，女性为第10位。湘语族群身材矮小，男性身高为第10位、女性身高为第9位。赣语族群是汉族中最矮的族群，男性、女性身高均为第11位。

有研究认为中国族群以环渤海地区青年平均身高最高[2]，本文研究发现，20～29岁青年男性和女性则是江淮方言族群最高，与上述研究结果不同。

2. 汉族方言族群城市成年人的身高

方言族群城市男性中，中部汉族（江淮方言汉族、吴语族群）、北方汉族3个族群身材较高，高于南方其他6个族群（$P < 0.01$）；在南方6个族群中，闽语族群相对较高，高于西南方言汉族、客家人、湘语族群、赣语族群（$P < 0.01$或$0.01 < P < 0.05$），赣语族群身材最矮（$P < 0.01$或$0.01 < P < 0.05$）（表11，表12）。

不同年龄组汉族城市男性身高排序有一些变化：在60～80岁组、40～49岁组、20～29岁组中华北方言族群身高分别为第1、3、3位，西北方言族群分别为第2、4、4位，东北方言族群分别为第4、2、1位。在北方三个族群中，东北族群身高增长速度最快，在40～岁组（1962年以后出生）已经超过了华北和西北方言族群。江淮方言汉族分别为第3、1、2位，自20世纪60年代以后其身高稳居前两位。

在11个方言族群城市女性合计资料中，东北方言族群身高最高，然后依次是华北方言族群、江淮方言族群、西北方言族群、吴语族群，粤语族群、赣语族群、湘语族群身高分居倒数第1、2、3位（表13）。

方言族群城市女性中，北方汉族、中部汉族（江淮方言族群、吴语族群）5个族群彼此间身高差异均无统计学意义（$P > 0.05$），均高于南方6个族群（$P < 0.01$）。南方族群中赣语族群、湘语族群、粤语族群彼此身高差异均无统计学意义（$P > 0.05$）（表14）。

东北方言族群在汉族各族群中身材最高，华北方言族群身材较高。江淮方言汉族和吴语族群属于中部汉族。江淮方言汉族男性身材仅矮于东北方

表11 汉族各方言族群城市男性身高均数(mm)

族群	20～29岁组		30～39岁组		40～49岁组		50～59岁组		60～80岁组		合计	
	n	Mean ± SD	*n*	Mean ± SD	*n*	Mean ± SD	*n*	Mean ± SD	*n*	Mean ± SD	*n*	Mean ± SD
东北方言族群	153	1 737.9 ± 60.5	134	1 705.8 ± 58.8	123	1 687.9 ± 60.4	125	1 684.9 ± 61.0	148	1 656.7 ± 64.0	689	1 695.2 ± 67.0
华北方言族群	181	1 714.9 ± 57.1	174	1 681.8 ± 61.1	192	1 682.6 ± 63.1	178	1 678.7 ± 60.1	180	1 665.9 ± 56.0	905	1 684.8 ± 61.6
西北方言族群	86	1 710.5 ± 60.0	73	1 694.8 ± 43.1	84	1 676.6 ± 60.5	84	1 679.9 ± 68.1	80	1 662.2 ± 64.5	407	1 684.9 ± 62.1
西南方言族群	153	1 709.8 ± 56.3	150	1 670.4 ± 64.1	150	1 666.9 ± 62.8	149	1 642.0 ± 58.1	124	1 633.4 ± 56.7	726	1 665.8 ± 65.3
江淮方言族群	68	1 736.3 ± 56.6	60	1 693.7 ± 50.9	55	1 696.4 ± 58.0	56	1 670.8 ± 53.1	70	1 659.4 ± 57.1	309	1 691.6 ± 61.5
吴语族群	69	1 704.3 ± 57.5	50	1 701.9 ± 50.9	54	1 673.5 ± 63.4	58	1 672.5 ± 56.8	59	1 644.0 ± 55.9	290	1 679.5 ± 61.0
闽语族群	122	1 706.9 ± 56.8	122	1 681.2 ± 56.5	124	1 663.6 ± 51.7	122	1 667.9 ± 60.9	120	1 644.5 ± 53.8	610	1 672.9 ± 59.5
赣语族群	61	1 678.9 ± 61.0	58	1 657.7 ± 55.6	57	1 633.8 ± 59.3	62	1 636.8 ± 52.1	67	1 598.2 ± 57.0	305	1 640.2 ± 63.1
客家人	61	1 688.4 ± 58.7	60	1 656.9 ± 62.2	58	1 666.3 ± 55.2	60	1 655.4 ± 61.9	64	1 626.1 ± 56.7	303	1 658.3 ± 62.1
粤语族群	49	1 676.2 ± 54.9	38	1 658.1 ± 46.6	34	1 678.8 ± 41.5	40	1 655.5 ± 54.6	34	1 608.2 ± 57.9	195	1 657.0 ± 56.7
湘语族群	62	1 685.8 ± 50.5	61	1 663.5 ± 59.2	60	1 652.3 ± 49.3	61	1 642.9 ± 58.3	62	1 617.5 ± 52.6	306	1 652.4 ± 58.4

表12　汉族各方言族群城市男性身高值间的 *u* 检验

族　群	东北方言族群	华北方言族群	西北方言族群	西南方言族群	江淮方言族群	吴语族群	闽语族群	赣语族群	客家人	粤语族群	湘语族群
东北方言族群	—	3.18**	2.58**	8.36**	0.83	3.57**	6.36**	12.45**	8.43**	7.97**	10.19**
华北方言族群	3.18**	—	0.03	5.99**	1.68	1.28	3.76**	10.75**	6.46**	6.11**	8.27**
西北方言族群	2.58**	0.03	—	4.88**	1.44	1.14	3.07**	9.43**	5.66**	5.48**	7.16**
西南方言族群	8.36**	5.99**	4.88**	—	6.06**	3.17**	2.08*	5.89**	1.74	1.86	3.25**
江淮方言族群	0.83	1.68	1.44	6.06**	—	2.42*	4.40**	10.23**	6.68**	6.46**	8.11**
吴语族群	3.57**	1.28	1.14	3.17**	2.42*	—	1.53	7.73**	4.20**	4.16**	5.53**
闽语族群	6.36**	3.76**	3.07**	2.08*	4.40**	1.53	—	7.54**	3.40**	3.37**	4.98**
赣语族群	12.45**	10.75**	9.43**	5.89**	10.23**	7.73**	7.54**	—	3.57**	3.09**	2.48**
客家人	8.43**	6.46**	5.66**	1.74	6.68**	4.20**	3.40**	3.57**	—	0.24	1.21
粤语族群	7.97**	6.11**	5.48**	1.86	6.46**	4.16**	3.37**	3.09**	0.24	—	0.88
湘语族群	10.19**	8.27**	7.16**	3.25**	8.11**	5.53**	4.98**	2.48*	1.21	0.88	—

表13　汉族各方言族群城市女性不同年龄组的身高均数（mm）

族　群	20～29岁组		30～39岁组		40～49岁组		50～59岁组		60～80岁组		合　计	
	n	Mean ± SD	*n*	Mean ± SD	*n*	Mean ± SD	*n*	Mean ± SD	*n*	Mean ± SD	*n*	Mean ± SD
东北方言族群	155	1 610.8 ± 73.4	148	1 588.6 ± 55.0	148	1 589.1 ± 52.7	151	1 567.5 ± 53.8	158	1 538.9 ± 57.7	760	1 579.6 ± 61.3
华北方言族群	182	1 594.8 ± 50.4	191	1 591.6 ± 50.8	188	1 578.0 ± 63.9	173	1 563.0 ± 49.5	172	1 532.3 ± 52.0	906	1 572.7 ± 58.1
西北方言族群	82	1 586.9 ± 58.1	79	1 586.0 ± 51.2	92	1 585.8 ± 54.9	97	1 560.3 ± 52.6	78	1 534.6 ± 49.2	428	1 570.9 ± 56.8
西南方言族群	152	1 573.2 ± 49.7	158	1 573.7 ± 48.4	159	1 562.8 ± 49.7	181	1 539.1 ± 54.7	167	1 523.2 ± 61.1	817	1 553.5 ± 56.6
江淮方言族群	61	1 598.4 ± 60.7	56	1 585.8 ± 47.7	69	1 569.9 ± 49.7	62	1 568.6 ± 48.3	64	1 536.3 ± 53.3	312	1 571.2 ± 55.9
吴语族群	60	1 589.8 ± 54.5	58	1 578.6 ± 53.6	64	1 574.5 ± 55.8	63	1 554.5 ± 55.2	60	1 543.7 ± 57.6	305	1 568.1 ± 57.5
闽语族群	136	1 580.8 ± 55.3	126	1 561.4 ± 47.7	128	1 565.2 ± 55.4	127	1 547.2 ± 52.3	121	1 527.2 ± 51.9	638	1 557.0 ± 55.5
赣语族群	61	1 567.3 ± 46.8	55	1 544.7 ± 44.3	61	1 540.5 ± 48.3	69	1 529.7 ± 41.8	59	1 514.8 ± 55.3	305	1 539.2 ± 50.2
客家人	66	1 576.0 ± 48.2	66	1 556.7 ± 50.8	67	1 563.3 ± 60.6	66	1 540.9 ± 60.2	66	1 515.8 ± 54.5	331	1 550.6 ± 58.6
粤语族群	50	1 566.6 ± 53.0	56	1 543.3 ± 36.1	52	1 548.3 ± 52.6	51	1 532.7 ± 49.0	48	1 490.2 ± 61.0	257	1 536.8 ± 56.0
湘语族群	64	1 560.8 ± 43.1	69	1 555.2 ± 43.8	70	1 545.4 ± 44.4	66	1 534.0 ± 46.1	64	1 503.5 ± 55.5	333	1 540.1 ± 50.6

表 14　汉族各方言族群城市女性身高值均数的 u 检验

族　群	东北方言族群	华北方言族群	西北方言族群	西南方言族群	江淮方言族群	吴语族群	闽语族群	赣语族群	客家人	粤语族群	湘语族群
东北方言族群	—	1.13	1.43	6.50**	1.27	1.94	5.33**	9.07**	5.96**	8.70**	9.01**
华北方言族群	1.13	—	0.54	6.96**	0.41	1.21	5.38**	9.69**	5.89**	9.01**	9.67**
西北方言族群	1.43	0.54	—	5.14**	0.07	0.65	3.95**	7.97**	4.80**	7.67**	7.89**
西南方言族群	6.50**	6.96**	5.14**	—	4.74**	3.80**	1.18	4.10**	0.77	4.16**	3.93**
江淮方言族群	1.27	0.41	0.07	4.74**	—	0.68	3.69**	7.48**	4.56**	7.30**	7.39**
吴语族群	1.94	1.21	0.65	3.80**	0.68	—	2.80**	6.61**	3.80**	6.52**	6.50**
闽语族群	5.33**	5.38**	3.95**	1.18	3.69**	2.80**	—	4.92**	1.64	4.89**	4.78**
赣语族群	9.07**	9.69**	7.97**	4.10**	7.48**	6.61**	4.92**	—	2.64**	0.53	0.23
客家人	5.96**	5.89**	4.80**	0.77	4.56**	3.80**	1.64	2.64**	—	2.90**	2.47*
粤语族群	8.70**	9.01**	7.67**	4.16**	7.30**	6.52**	4.89**	0.53	2.90**	—	0.74
湘语族群	9.01**	9.67**	7.89**	3.93**	7.39**	6.50**	4.78**	−0.23	2.47*	−0.74	—

言族群，女性矮于东北方言族群和华北方言族群。吴语族群男性、女性身高均排第5位，高于其他南方汉语方言族群。

在南方族群中客家人男性身高较矮。闽语族群在6个南方汉族族群中男性身材最高，女性身材较高。赣语族群是中国身材最矮的族群之一，男性身高最矮，女性与粤语族群、湘语族群接近。粤语族群男性身高在南方汉族中居中，女性身材较矮。在南方6个族群中，湘语族群身材较矮。

在城市男性、城市女性中，东北方言族群、江淮方言族群、华北方言族群、西北方言族群身材较高，赣语族群、湘语族群、粤语族群身材较矮。

城市女性在60～80岁组、40～49岁组、20～29岁组中，东北方言族群分别为第2、1、1位，其排序稳定。有些族群身高排序有一些变化，华北方言族群身高排序分别为第5、3、3位，身高增长速度较快。吴语族群身高分别为第1、4、4位，身高增长速度较慢。

20～29岁城市青年男性、女性均以东北方言族群最高，其次是江淮方言族群，不支持中国环渤海地区青年平均身高最高的观点。

半个多世纪以来，与华北、西北城市汉族男性比较，中国东北、西南、江淮城市汉族男性的身高都出现了较大的增长。与中国吴语族群、闽语族群、赣语族群城市女性比较，中国粤语族群、东北方言族群、华北方言族群、江淮方言族群、客家人城市女性身高都出现了较大的增长。

从农村和城市男性和女性的五个年龄组获得的数据，我们可以观察到与年龄增加的平均身高值逐步下降。这可能与医疗设施和卫生条件的改善，良好的营养和生理变化有关。

3. 汉语方言族群城市汉族与乡村汉族身高的比较

同性别城乡间身高均数比较（表15），多数族群城乡间差异具有统计学意义，均为城市人高于乡村人。男性间只有粤语族群城乡间身高差异无统计学意义。女性只有江淮方言族群、粤语族群、湘语族群城乡间身高差异无统计学意义。

（二）近半个世纪来汉族方言族群身高的变化

最近，英国埃塞克斯大学教授Hatton[12]发现，1870—1980年的110年间，欧洲21岁男性身高从167 cm增高到178 cm，1971—1975年21岁的英国男性为177.3 cm，1980年荷兰21岁男性平均身高为183 cm，是欧洲最高的族群，葡萄牙男人最矮，为173 cm。中国汉族近半个世纪以来的身高变化的全面研究

表15　城市汉族与乡村汉族身高的比较（mm，Mean ± SD）

族　群	乡村男性	城市男性	u	乡村女性	城市女性	u
东北方言族群	166.8 ± 6.0	169.5 ± 6.7	8.67**	155.2 ± 5.5	158.0 ± 6.1	10.20**
华北方言族群	167.3 ± 6.4	168.5 ± 6.2	4.65**	155.7 ± 5.7	157.3 ± 5.8	6.71**
西北方言族群	167.2 ± 6.0	168.5 ± 6.2	3.60**	155.2 ± 6.0	157.1 ± 5.7	5.69**
西南方言族群	164.1 ± 6.7	166.6 ± 6.5	8.18**	153.8 ± 6.1	155.4 ± 5.7	6.10**
江淮方言族群	167.3 ± 6.4	169.2 ± 6.2	4.02**	156.5 ± 5.6	157.1 ± 5.6	1.41
吴语族群	166.7 ± 6.9	168.0 ± 6.1	2.57*	155.6 ± 5.9	156.8 ± 5.8	2.72**
闽语族群	165.9 ± 6.7	167.3 ± 6.0	4.06**	154.7 ± 5.6	155.7 ± 5.6	3.30**
赣语族群	162.9 ± 6.5	164.0 ± 6.3	2.26*	151.8 ± 5.6	153.9 ± 5.0	5.23**
客家人	164.5 ± 6.2	165.8 ± 6.2	2.66**	153.7 ± 5.6	155.1 ± 5.9	3.22**
粤语族群	165.6 ± 6.6	165.7 ± 5.7	0.19	153.1 ± 5.4	153.7 ± 5.6	1.41
湘语族群	163.5 ± 6.4	165.2 ± 5.8	3.67**	153.4 ± 5.6	154.0 ± 5.1	1.55

尚未见报道。

1. 乡村汉族身高的变化

一般认为身高稳定期为45～50岁，55岁以前每5年身高平均减少5 mm，以后每5年减少7 mm[6]。从50岁到70岁（60～80岁的中间值），身高大约减少25 mm。

20～29岁组与60～80岁组身高之差大致反映了近50年间身高的变化，再考虑到正常人从青年（20～29岁）到老年（60～80岁）身高生理性减小25 mm左右，那么近50年间汉族男性身高增加幅度较大的依次为粤语族群（64.2 mm）、吴语族群（60.4 mm）、西南方言族群（57.8 mm）、江淮方言族群（57.4 mm）、湘语族群（54.2 mm）、赣语族群（53.1 mm）、闽语族群（44.6 mm）、华北方言族群（41.3 mm）、东北方言族群（40.7 mm）、客家人（36.0 mm）、西北方言族群（23.9 mm）。可以发现，尽管目前南方族群身高矮于北方族群，但是近50年来南方族群、中部族群男性身高增长的幅度大于北方族群。

近50年间女性身高增长幅度较大的依次为赣语族群（58.1 mm）、西南方言族群（47.0 mm）、吴语族群（43.4 mm）、江淮方言族群（38.6 mm）、粤语族群（37.6 mm）、闽语族群（35 mm）、客家人（31.2 mm）、湘语族群（30.2 mm）、西北方言族群（28.9 mm）、东北方言族群（28.6 mm）、华北方言族群（25.2 mm）。与男性一样，近50年来南方族群、中部族群女性身高增长的幅度也大于北方族群。

2. 城市汉族身高的变化

近50年间男性身高增加幅度较大的依次为东北方言族群（56.2 mm）、赣语族群（55.7 mm）、江淮方言族群（51.9 mm）、西南方言族群（51.4 mm）、湘语族群（43.3 mm）、粤语族群（43.0 mm）、闽语族群（37.4 mm）、客家人（37.3 mm）、吴语族群（35.3 mm）、华北方言族群（24.0 mm）、西北方言族群（23.3 mm）。可以发现，近50年来总体上南方族群、中部族群男性身高增长的幅度大于北方族群。

近50年间女性身高增加幅度较大的依次为粤语族群（51.4 mm）、东北方言族群（46.9 mm）、华北方言族群（37.5 mm）、西北方言族群（37.3 mm）、江淮方言族群（37.1 mm）、客家人（35.2 mm）、湘语族群（32.3 mm）、闽语族群（28.6 mm）、赣语族群（27.5 mm）、西南方言族群（25.0 mm）、吴语族群（21.1 mm）。总的说来，近50年来北方族群女性身高增长的幅度大于南方族群。

约110年欧洲男性身高增长了约11 cm，这相当于每10年增长约1 cm。根据我们所统计的11个民族在近50年的平均身高的计算（使用简单平均），我们发现农村男性的身高大约共增长5 cm。换句话说，他们的身高每10年增长了约1 cm。城市男性的总身高增加约4.2 cm，其对应于每10年大约增长0.84 cm。与欧洲男性相比，近50年中国男性身高的增长速度并不逊色。在过去50年身材显著增加的原因，与社会卫生资源、营养的改善有关。

三、汉族身高与纬度的相关分析

（一）城市汉族身高与纬度的相关分析

将10 451例城市汉族身高值与各测量地点的纬度进行线性相关分析，结果表明，男性身高与纬度的相关系数为0.180（$P < 0.01$），女性身高与纬度的

相关系数为0.189（$P < 0.01$）（表16）。汉族城市男性、女性的身高均与纬度显著正相关。总的说来，随纬度的升高，中国城市汉族身高呈线性增加，即总体上由南向北，中国城市汉族人身高增高。

有学者认为全世界人的身高的分布图呈现高矮驳杂，与地理位置和气候没有明显的关系[8]。身高的变化较小，变化幅度大约10%，并不符合某些特定的地理趋势[13]。本次研究证实人口达到6亿以上的中国城市汉族人身高与纬度呈正相关，表明在全球局部地区人的身高分布还是有一定规律的。

人的身高主要由头部、面部、躯干和下肢高度决定的。耳上头高反映头部高度，形态面高反映面部高度，坐高、躯干前高主要反映躯干的高度，下肢全长反映下肢的高度，身高下肢长指数反映了下肢长在整个身高中所占的比例，也间接反映躯干与下肢高度的比例。

相关分析显示，城市汉族的耳上头高、形态面高、坐高、躯干前高、下肢全长均与纬度呈显著正相关（表16），即随纬度增加，头部、面部、躯干、下肢的高度值均增大，共同导致身高与纬度的正相关。男性、女性的身高下肢长指数均与纬度呈显著正相关，说明随纬度增加，男性、女性的下肢全长增加的速度均超过身高增加的速度。从身高下肢长指数均数来说，有学者认为高身材的人大致都是腿较长，身材的高矮在很大程度上取决于下肢的长短，而不是坐高[8]。本文研究证实这一观点。

表16　城市汉族人体质指标、指数与纬度、身高的相关分析

变　量	男　性				女　性			
	纬　度		身　高		纬　度		身　高	
	r	*P*	*r*	*P*	*r*	*P*	*r*	*P*
身　高	0.585**	0.000	—	—	0.619**	0.000	—	—
耳上头高	0.286	0.119	0.382*	0.034	0.259	0.159	0.461**	0.009
形态面高	0.429**	0.000	0.133	0.475	0.453*	0.011	0.111	0.552
躯干前高	0.401*	0.026	0.633**	0.000	0.453*	0.011	0.659**	0.000
下肢全长	0.506**	0.004	0.614	0.000	0.408*	0.023	0.574**	0.001
身高下肢长指数	0.391*	0.032	0.380*	0.000	0.294	0.108	0.371*	0.040

（二）乡村汉族身高与纬度的相关分析

表17　乡村汉族指标、指数与纬度的相关分析

变量	纬度				变量	纬度			
	男性		女性			男性		女性	
	F	*P*	*F*	*P*		*F*	*P*	*F*	*P*
体重	0.689**	0.000	0.761**	0.000	下肢全长	0.606**	0.000	0.380*	0.022
身高	0.478**	0.003	0.403*	0.015	身高坐高指数	−0.025	0.886	−0.020	0.919
坐高	0.335*	0.046	0.354*	0.034	身高下肢长指数	0.474**	0.004	0.220	0.197

如表17所示，乡村男性体重、身高、坐高、下肢全长、身高下肢长指数与纬度呈显著正相关，身高坐高指数与纬度无显著相关。随纬度增加，男性的坐高、下肢全长值都呈线性增大。下肢全长与纬度的相关系数明显大于坐高与纬度的相关系数。这提示男性身高与纬度呈显著正相关的原因，主要是男性的下肢全长随纬度增加而线性增大造成的，其次才是坐高的线性增大。

乡村女性体重、身高、坐高、下肢全长与纬度呈显著正相关，身高下肢长指数、身高坐高指数与纬度无显著相关。随纬度增加，乡村女性的坐高、下肢全长均呈线性增大。坐高与纬度的相关系数为0.354，下肢全长与纬度的相关系数为0.380，这两个相关系数接近。可以说女性身高与纬度呈正相关的是女性的坐高值、下肢全长值随纬度增加而增大共同造成的。

（三）汉族身高与纬度呈显著正相关的原因初探

中国人体质特征的地区性分布研究已有报道。有些学者分别从宏观和微观的角度提出现代中国人可分成南北两个类群[14-16]。马立广等计算了中国族群身高与纬度的线性相关系数，认为中国族群随纬度增加，身高逐渐增高[3]。马立广等的研究不是对单一民族的研究，而且所涉及的族群的测量年代相差较大。

将36个汉族乡村族群身高均数与其生活地区的纬度进行线性相关分析

（表 17），结果表明中国汉族乡村男性、女性的身高均与纬度呈正相关。总体上，随纬度的升高，中国男性、女性身高呈线性增加。

人的身高主要是由上身高度和下肢高度决定的。坐高反映了上身的高度，下肢全长反映下身的高度，身高坐高指数与身高下肢长指数分别反映了上身、下身在整个身高中所占的比例。有研究证实，身高相同的女性与男性比较，女性的下肢长大于男性；而坐高相同的女性与男性比较，女性的下肢长小于男性[8]。

人的身体各部位的高度、长度有人种学的差异。有研究发现[17]，远东亚洲人身高比德国人低，胳膊和腿的长度比德国人短，但躯干高于德国人。Dewangan 等[18]报道，虽然韩国人身高一般比美国人或英国人矮，但坐高比他们高。Moss 等[19]认为，亚洲人和西方人人体测量的主要区别是腿的长度。Jung 等[20]研究表明，韩国人比美国人上半身长，但胳膊和腿短。坐高和下肢全长值均影响到身高，男性的坐高、下肢全长均与纬度呈正相关，下肢全长与纬度的相关系数明显大于坐高。这说明男性身高与纬度呈正相关的原因，主要是男性的下肢全长随纬度增加而增长造成的，其次才是坐高的增长。男性身高坐高指数与纬度不相关，说明随纬度增加坐高值与身高值增大的速度较为接近。身高下肢长指数与纬度呈正相关，即随纬度增加，汉族下肢长增长的速度超过了身高增长的速度。这说明，北方男性比南方男性高的原因，主要是北方男性腿更长些。有学者认为高身材的人大致都是腿较长，身材的高矮在很大程度上取决于下肢的长短，而不是坐高[8]。这一观点在本次男性资料中得到证实。

女性的坐高、下肢全长与身高、纬度均相关，说明坐高、下肢全长的增加引起了身高的线性增长，都是导致身高与纬度呈正相关的原因，即从中国南方到北方，上身高度、下身高度的增加共同导致女性身高的线性增长。从南方到北方，女性坐高、下肢全长增加的速度与身高增长的速度基本一致，所以女性的身高坐高指数、身高下肢长指数与纬度无关，提示北方女性与南方女性上身与下身的比例基本相同。

参考文献

[1] 季成叶.农村青年学生生长发育的环境差异[J].中国校医，1991，5(3)：125.
[2] 徐玖瑾，杜若甫.北京与重庆地区成人身高变化的研究[J].人类学学报，1985，4

(2): 151–159.

[3] 马立广，曹彦荣．徐玖瑾，等．中国102个人群的身高与地理环境相关性研究[J]．人类学学报，2008，27(3): 223–231.

[4] 江崇民，林莉萍．中日两国国民身高的比较研究[J]．北京体育大学学报，2002，24: 208–211.

[5] 杨晓光，李艳平，马冠生，等．中国2002年居民身高和体重水平及近10年变化趋势分析[J]．中华流行病学杂志，2005，26(7): 489–493.

[6] 長嶺晋吉．肥満の判定法[J]．医学のあゆみ，1977，101: 404.

[7] Brozek, J. Densitometric analysis of body composition: Revision of some quantitative assumptions[J]. Ann, N. Y. Acad. Sci, 1963, 110: 113–140.

[8] 雅·雅·罗金斯基，马·格·列文．人类学[M]．王培英，汪连兴，史庆礼，等译．北京：警官教育出版社，1993: 65–66，73–75.

[9] Roberts D F. Climate and Human Variability. 2nd edition[M]. Menlo Park, CA: The Benjamin-Cummings, 1978.

[10] Martin R, Saller K. Lehrbuch der Anthropologie[M]. Stuttgart: Gustav Fischer Verlag, 1956.

[11] Li Y L, Zheng L B, Xi H J, et al. Stature of Han Chinese dialect groups: a most recent survey[J]. Sci. Bull., 2015, 60(5): 565–569.

[12] Hatton T J. How have Europeans grown so tall?[M]. Oxford: Oxford University Press, 2013.

[13] Ruff C. Variation in human body size and shape[J]. Ann Rev Anthropol, 2002, 31: 211–232.

[14] 张振标．现代中国人体质特征及其类型的分析[J]．人类学学报，1988，7: 314–323.

[15] 刘武，杨茂有，王野城．现代中国人颅骨测量特征及其地区性差异的初步研究[J]．人类学学报，1991，10: 96–105.

[16] Wen B, Li H, Lu D R, et al. Genetic evidence supports demic diffusion of Han culture[J]. Nature, 2004. 431: 302–305.

[17] Lin Y C, Wang M J J, Wang E M. The comparisons of anthropometric characteristics among four peoples in East Asia[J]. Applied Ergonomics, 2004, 35(2): 173–178.

[18] Dewangan K N, Prasanna Kumar G V, Suja P L, et al. Anthropometric dimensions of farm youth of the north eastern region of India[J]. International Journal of Industrial Ergonomics, 2005, 35(11): 979–989.

[19] Moss S, Wang Z, Salloum M, et al. Anthropometry for World-SID: A world-harmonized midsize male side impact crash dummy[J]. SAE Technical Paper Series, 2000–01—2202.

[20] Jung S G, Kim G H, Roh W J. Comparison of basic body dimension between Korean and American for design application[J]. Korean Society of Basic Design and Art, 2000, 1(2): 65–75.

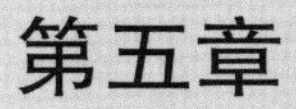

第五章

雪球滚积中华体

2021年人类学终身成就奖获得者

——徐杰舜

人类学生涯回顾

出生：河姆渡人的子孙

我祖籍浙江余姚，是河姆渡人的子孙。1943年12月24日，抗日战争期间的一个平安夜，我出生于湖南南部山区古城零陵（今为永州所辖之区），因零陵是舜陵所在之地，父亲给我取名“杰舜”，从此我拥有了一个在中国独一无二的姓名。

从1945年1月到1946年12月，我随父亲徐福尧，因其工作单位中央银行的迁移而颠沛流离，先后在云南下关（今红河州个旧市）、蒙自，广西南宁生活过短暂的日子。1946年始定居于汉口，从此记事开始，一直到1965年大学毕业才离开武汉。

求学：成为人类学家岑家梧先生的关门弟子

1948年9月入汉口黎黄陂路小学一年级。

1949年5月，目睹了父亲拒绝撤逃台湾，坚守中央银行，保卫国家财产的情景，迎来了武汉的解放。父亲因保卫中央银行有功，成为中国人民银行武汉分行金库科员，1951年被评为武汉市财贸银行系统乙等模范，这在我幼小的心灵中留下了深刻的印象。

此后，我先后求学于汉口江苏小学、中国人民银行武汉分行职工子弟小学（现为华中里小学），1955年小学毕业后考入武汉市28中学读初中，1958年考入武汉大学附属共青团中学读高一。1959年读高二时，我就读的学校与武汉市14中合并为武汉大学附属中学。1961年7月我高中毕业，考入中央民族学院分院（现为中南民族大学）历史系，成为著名人类学家、民族学家岑家梧教授的关门弟子。

在大学求学时最重要的四件事，影响了我终身的学术方向。

1945年两岁在湖南零陵

1946年三岁在云南蒙自

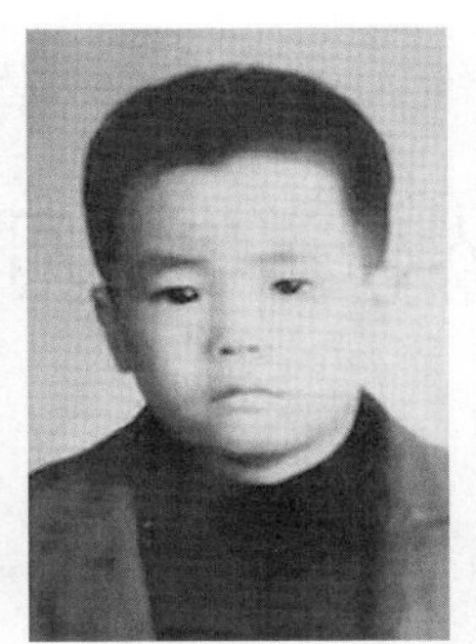
1948年五岁在湖北武汉

1948年冬五姊妹徐菊仙、徐湘芬、徐杰舜、徐滇蒙和徐杰芳合影

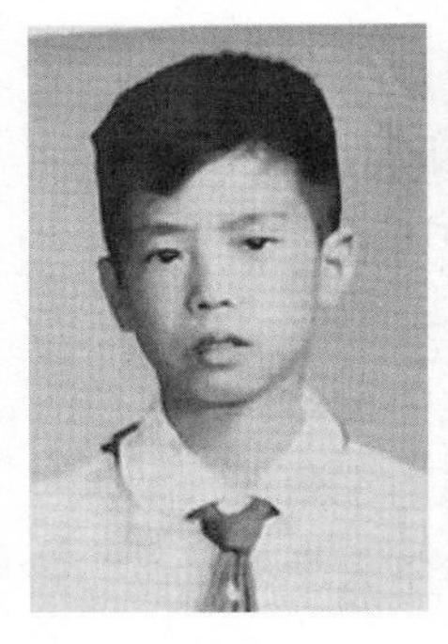
1955年小学毕业

1958年初中毕业

1961年高中毕业

一是与少数民族同学朝夕相处，从感性上认识了什么是民族。

二是听了岑家梧教授讲授的“原始社会史”课，精读了恩格斯的《家庭、私有制和国家的起源》，从理性上认识了什么是民族。

三是在学习《家庭、私有制和国家的起源》的基础上，参与了中国20世纪60年代关于民族概念的学术讨论，与师兄彭英明（后曾任中南民族学院院长）合作，在《江汉学报》1963年第5期上发表了处女作《试论从部落到民族发展的历史过程》，从此开启了我从事民族研究的学术之路。

四是1964年9月到1965年7月，大四时赴广西三江侗族自治县，参加了中央统战部组织的“三江四清工作队”（又称“社会主义教育工作队”），进驻程阳公社程阳桥畔的平寨，不自觉地开始了真正意义的人类学田野调查。十个月的田野经历，不仅锻炼了我的田野工作能力，也使我在不自觉中完成了人类学研究的“成年礼”。22年后我重返三江，于1992年才出版了关于三江

大四时在广西三江侗族自治县平寨村进行田野调查

平寨的民族志报告——《程阳桥风俗》。

从小学到大学，从城市到乡村，从汉族地区到少数民族地区，历时18年，我在不自觉中完成了非典型的人类学洗礼。

工作：相思湖畔"圆梦"民族研究

1965年7月，从广西三江经桂林回到武汉，参加了大学毕业时分配的工作。我父亲听说我被分配回了祖籍浙江，非常高兴，我也愉快地"打起背包就出发"，从汉口坐轮船，经上海，转火车，到杭州的浙江省人事厅报到，开始走进了社会。

"风物是秦余"：武义20年奠定了乡村研究的田野基础

在期待与焦虑中，我住在杭州城站红楼等分配，一直到8月，才被浙江省人事厅分配到了武义县。之后很快被县人事局分配到了浙江省武义第一中学，当上了中学教师。1982年12月调入武义第三中学任教，这一干就是20年。

武义，是一个被唐朝诗人孟浩然称为"风物是秦余"的古县城。在武义的20年中，我经历了从武汉到三江，从三江再到武义，从汉族到侗族再到汉族，这个视野与场景的转换深深地震撼了我，使我对"民族"概念的认知有了感性的提升。原来，汉族的风俗也同样丰富多彩！而且不同地域的汉族之间有如此巨大的文化差异，使我心中那颗研究汉民族的种子开始萌芽，从1979年开始了《汉民族发展史》的研究和撰写。

在武义的20年，除了撰写《汉民族发展史》之外，还做了两件有关学术的事。

一是参加了《浙江风俗简志》的编写，同时完成了《武义风俗志》的编写，参与主编了《金华地区风俗志》，考察了当时金华地区12个县的风俗，大大丰富了我对汉族民间风俗的人类学认知。

二是撰写了十几篇有关汉民族研究的论文，入职广西民族学院（2006年5月18日改为广西民族大学）后汇编成《汉民族历史和文化新探》并于1985年12月出版。此书获得学术界的高度评价，孟宪范在《中国社会科学》1989年第2期上发表《中国民族学十年发展述评》说："徐杰舜的《汉民族历史和文化新探》的发表，标志着这一课题在中国的真正开始"。费孝通先生在撰写《中华民族多元一体格局》时，向我索要了此书，并列入参考文献之中。

在武义的20年，无意中我做了一次长达20年的深度田野考察，使我深刻地认识了"乡土中国"，为20年后重返武义，完成《新乡土中国》的考察和研究打下了坚实的田野基础。

南往北复：24年扎根南宁，成了"民族大学的一条龙"

1985年3月28日，我被广西民族学院引进到民族研究所，历任民族理论研究室主任、研究所副所长；1994年8月调入学报编辑部任主任，并先后任学报副主编、常务副主编、执行主编；1996年9月，被云南大学聘为兼职教授和硕士研究生导师，并首次招收硕士研究生；1996年12月在广西民族学院兼任汉民族研究中心主任。

从1999年到2015年16年的时间，我在广西民族大学培养了40多名硕士。与此同时，2005年到2010年我曾任中央民族大学人类学博士生导师、《中央民族大学学报》执行主编，先后招过三届博士研究生，培养了6名博士；2005年到2012年任中南民族大学特聘教授、民族学学科带头人、人类学研究所所长和台湾研究所所长，带过3个硕士生。这段时期，我从南到北，又从北到南，在三所民族大学穿梭上课，被同仁们戏称为"民族大学的一条龙"，这是我一生中最忙碌、最充实的日子。

在广西民族大学的24年，我从进入人类学开始，为推进人类学学科的发展，做了三件重要的事。

（1）把广西民族大学学报办成了中国人类学的重要期刊

主持《广西民族学院学报》的工作，是我在广西民族大学工作的重要工作之一。

1994年8月，我开始主持《广西民族学院学报》工作，这时正是我进入人类学的重要时期，且深感人类学长期被边缘化，决心以"人类学研究"为特色栏目来推动人类学在中国的发展。1994年10月我收到乔健的《中国人类学的困境与前景》一文。几经周折，1995年1月我果决地在改版后的第1期发表了乔先生的大作，从而卷起了人类学民族学界的"千堆雪"，成了1995年中国的学术热点之一。

经过十年的努力，终于成就了广西民族大学学报："人类学研究"栏目于2004年成为首批入选教育部哲学社会科学名栏，2006年《广西民族学院学报》成为第二批入选教育部哲学社会科学的名刊，更改为《广西民族大学学报》。

从此，《广西民族大学学报》成了学科重建以来，中国人类学研究展示、推广、宣传的重要园地，成了中国人类学者们心中的"圣地"。

（2）开展了中国人类学家系列访谈

开展人类学家访谈，是我在广西民族大学工作的又一重要工作。

通过办学报在与学者们的交流中，我触摸到他们鲜活的思想和灵动的表达，一直思考着如何呈现这些人类学者的风姿、风貌和风采。学历史出身的我，还想到如何留下他们对人类学发展坎坷历史的记忆。我接受了周大鸣教授的建议：进行学者访谈。

从1999年4月开始，到之后的近20年，我先后访谈了包括李亦园、乔健、庄孔韶、王明珂、朱炳祥、金力、周大鸣、彭兆荣、王铭铭、范可、李辉、徐新建、赵旭东等近百位人类学者，并汇编出版了《人类学世纪坦言》《人类学世纪真言》《人类学世纪欢言》三本访谈录。2009年又总汇成三卷本的《中国人类学家访谈录》出版。

访谈录长时期的连续发表，深得好评。李亦园先生称之为"非正式口述史"，并总结其意义在于：

- 记录了中国学者进入人类学领域的转折过程与经历；
- 显示了学者们对人类学学科建设的思考与努力；
- 说明了中国人类学理论与方法的探索与建构的情况；
- 展示出人类学本土化与中国化的努力；
- 阐述了人类学者与其他学科互动的态势；

- 显示了中国人类学在不同地区发展的趋向；
- 凸显了中国人类学者与国际人类学交流的状况；
- 展现了中国人类学者对全体人类未来的关怀。

访谈录的采访和发表是一个过程，它有效地从民间层面整合并团结了人类学的队伍，近20年的历程足足哺育一代，甚至二代人类学者。不少硕士和博士都说是看着人类学访谈录长大的。我很赞同西南民族大学王璐教授在《人类学高级论坛二十年总结报告》中所说：人类学访谈录为中国人类学建立了"学者民族志档案"；通过访谈个体的学术生命史，"照见了中国人类学复兴以来的跨学科发展"，呈现出"整个人类学的生命史"。访谈录能有这样的历史作用，现在回首，心中甚为欣慰。

1993在香港中文大学

1993年12月12日与彭兆荣、徐新建在曼谷雅克先生家中

1993年12月到泰国清迈出席第二届国际瑶族协会后到曼谷拜访泰国朱拉隆功大学文学院院长提拉潘教授

1997年7月在厦门首届文学人类学研讨会上与李亦园先生交流

1996 年在桂林国际瑶族学术年会上与法国人类学家雅克

1998 年 8 月访问华盛顿大学

1998 年出席在美国威廉斯堡举行的第 14 届人类学民族学世界大会的中国代表

1999 年 3 月到香港中文大学访问时与乔健先生和张小军博士

1999 年与陈志明教授、王斯福教授在南宁出席人类学本土化国际学术研讨会

1999 年 11 月与周大鸣教授出席南宁国际民歌艺术节学术研讨会

（3）创建了人类学高级论坛

我从1995年到2000年参加了北京大学“社会文化人类学高级研讨班”的学习，从旁听到学员再到讲员，逐渐感到需要有一个更开放的形式来推动人类学的发展。2001年9月，我应台北“中研院”民族学研究所所长黄应贵先生的邀请，去作学术访问。借此机会，我把设立“人类学高级论坛”的想法在港台学者中广泛征求了意见，得到了黄应贵、李亦园、庄英章、童春发、乔健、林冠群、廖幼华、陈祥水等学者的热烈响应和支持。在途经香港时，又得到香港中文大学人类学系主任陈志明教授的积极响应。回南宁后，根据黄应贵等先生的意见，起草了“关于设立‘人类学高级论坛’的倡议书”，得到了学者们的普遍回应，以及中国社科院民族学人类学研究所、北京大学社会学人类学研究所、中国人民大学社会学理论与方法研究中心、中央民族大学民族学社会学学院、中山大学人类学系、厦门大学人类学研究所、四川大学文学人类学研究所、香港中文大学人类学系、香港科技大学华南研究中心、澳门大学人文及社会科学学院、台湾花莲东华大学原住民民族学院、台湾宜兰佛光大学人类学系等22个单位的积极响应，费孝通先生、李亦园先生欣然同意出任人类学高级论坛的顾问。

1995年与宋蜀华先生在北京大学出席首届社会文化人类学高级研讨班

这一切都给予了我极大的鼓舞。2002年5月12日首届人类学高级论坛在广西民族学院召开。费孝通与李亦园两位先生发来贺信，

1998年6月参加第三届社会文化人类学高级讲习班与费孝通先生交流

2001年10月访问台北“中研院”民族学研究所所长黄应贵研究员

给人类学高级论坛以有力的支持。我作为人类学高级论坛的创建秘书长，在赤手空拳的条件下，得道多助，竟得到了学术界的鼎力支持，建立了学术委员会，每年一个主题一届年会，奇迹般地开了20年，且每年年会的论文集都能按时出版。2004年在银川年会上发表了《生态宣言：走向生态文明》，2008年在贵阳年会上讨论了“中华民族认同”的问题，2020年在成都年会上提出了“数智文明”的概念，凡此等等，都有力地表达了人类学在中国的存在，为中国人类学发展史书写了有血、有肉、有灵魂的一页，确实难能可贵。

学科情怀：退休12年，仍为人类学学科鼓与呼

我于2009年3月退休，时年已66岁。

退休后，人类学的使命感和学科情怀，使我退而不休，依托人类学高级论坛仍力所能及地工作着。12年来，主要做了四件与人类学相关的事。

（1）创建了旅游高峰论坛，为人类学与旅游牵手搭桥

人类学高级论坛在发展中曾经孕育了2个子论坛，即原生态民族文化论坛（2010年6月）和旅游高峰论坛（2010年10月）。

在人类学与各人文社会科学纷纷牵手融合的时候，我特别关注了旅游人类学的兴起，担任了旅游高峰论坛的秘书长。从2010年在桂林金钟山召开了首届年会后，每两年开一次，先后在成都信息工程大学、乐山师范学院、武汉中南民族大学、广州中山大学召开过五届年会，并出版了5本会议论文集，很好地促进了人类学与旅游的融合和发展。

（2）组织了海峡两岸人类学青年论坛，为两岸青年人类学家的交流搭台

加强海峡两岸人类学的交流，是人类学高级论坛追求的目标。除了每届年会举行时常有的海峡两岸圆桌论坛以外，我一直努力寻求到台湾举办一届年会。2012年10月在重庆文理学院举行第12届人类学高级论坛年会时，台湾新竹清华大学的林淑蓉教授出席了会议。经商讨林教授欣然同意由她负责，2013年6月在新竹清华大学举办一届以青年人类学者为主的“海峡两岸人类学论坛”，主题为“全球化、文化多样性与地方社会”。

在林教授的努力下，论坛如期举行。人类学高级论坛秘书处组织了以彭兆荣教授为团长，周大鸣、徐新建、赵旭东、朱炳祥、关凯等知名人类学家，以及一批优秀青年学者的学术代表团赴台参会；台湾方面的人类学家几乎全部出席，包括乔健、林淑蓉、黄应贵、陈中民、王明珂、孟祥翰、潘英海、张中复、简美玲等教授，以及一批台湾优秀的青年教师与学者，可谓盛况空前。郑向春

2002 年 5 月主持首届人类学高级论坛年会

2005 年 4 月与李亦园和乔健先生出席台湾东华大学主办的“族群与社会”学术研讨会

2005 年台湾东华大学乔先生 70 大寿

2009 年与朱炳祥和徐新建在内蒙古上京出席第八届人类学高级论坛

2009 年 7 月在昆明与杜维明一起出席第十六届国际人类学民族学联合会

2010 年与徐新建、周大鸣、乔健、彭兆荣出席人类学高级论坛第四次学术委员会会议

2012 年第 11 届人类学高级论坛部分学者在塔里木

发表在《百色学院学报》(2013年第6期)上的《首届海峡两岸人类学论坛综述》记述说:“此次论坛,两岸学者就相关议题进行比较与对话,展开了一次交流自由、论争广泛且极具针对性的研讨,促成了台湾与大陆学者之间兼具现实与学理意义的对话与学习。同时,论坛也为来自国内外不同地区的年轻学者发声、表达与表现提供了一个可贵的展示平台,促进了年轻学者的成长。”特别值得一提的是,一向以主题鲜明、会风务实、民间开放和学术平等著称的人类学高级论坛,此次以“海峡两岸人类学论坛”为题,在林淑蓉、王明珂和潘英海的主持下,以跳棋式的清华-中兴-暨南三段议程,开创了一个崭新的会议模式,载入了人类学学科史。

作为回应,2014年7月12—14日,由人类学高级论坛青年学术委员会与贵州省社会科学院、贵州省人类学会合作,在贵阳召开了“‘多彩贵州’原生态文化国际论坛暨2014首届海峡两岸人类学青年论坛”。与会者近百人,其中台湾暨南大学容邵武,台湾东华大学李宜泽、杨政贤和高雄师范大学洪馨兰等10位台湾青年人类学者出席了会议。

海峡两岸青年学者通过对话与交流,就民族文化、人类学的理论与方法等问题畅所欲言,对原生态文化及保护进行了深入探究,通过并发布了《原生态文化贵阳宣言》,呼吁“人类只有一个地球,但人类有丰富多彩的文化。保护和发展人类与自然平衡的原生态文化,如同保护地球一样,世界才能多姿多彩,人类才能永续发展!”

海峡两岸人类学的这种搭台互动,不仅显示了学术的共同性,更彰显了两岸同胞血浓于水、中华民族一家亲的情谊。

2014首届海峡两岸人类学青年论坛

2015年3月14日人类学与中国研究学术研讨会与会学者合影（南宁·广西民族大学）

（3）策划了人类学学科建设贺州座谈会，为人类学的学科地位鼓与呼

人类学在中国处于二级学科的边缘地位。这种既悖于常理，亦有违公平的定位，一直是中国人类学者心中的不解之结。为了将人类学提升为一级学科，我坚持着鼓与呼，一直到2016年3月26日我与中国人类学民族学研究会、人类学高级论坛秘书处以及广西贺州学院南岭走廊族群文化研究基地联合发起，在贺州学院举行了“人类学学科建设座谈会”，共同讨论人类学成为一级学科的必要性与可能性，以及未来的趋势。出席会议的有26位全国高校人类学学科点的主要负责人，与会代表一致认为人类学是一门国际性的学科，是“强国之学、强校之学和强人之学”。解决当今世界的问题，需要人类学者这种全人类的眼光去探索新知，我们希望借助这门学科为国家发展和社会进步，为中华民族的伟大复兴，贡献更多的智慧与力量。所以我们想落实费孝通先生提出的人类学、民族学、社会学“三科并列、互相交叉、各得其所、努力发展”的建议，将人类学作为一级学科进行建设。为此，座谈会发表了《纪要》，会后，不仅在《文汇报·文汇学人》发表了十位学者讨论人类学应为一级学科的文章，还出版了《人类学之梦》的论文集，进一步从历史的发展、现实的需要，说明人类学在中国的过去、现状及未来走向，从而论证了人类学升为一级学科的重要性、必要性和可行性，努力为人类学的学科地位鼓与呼。

（4）策划出版了《新生代人类学家之路》，为中国人类学的薪火相传铺了路

人类学的永续发展靠的是人才的薪火相传。改革开放以来的40年，虽然

2016 年 3 月中国人类学学科建设座谈会合影

2018 年 5 月和王明珂教授在武汉参加
“李亦园先生学术思想研讨会”

2019 年 3 月和庄英章教授在中山大学参加
“乔健先生学术思想研讨会”

2021 年 7 月 23 日参加“广西巩固拓展全国民族团结进步示范区成果创建铸牢中华民族共同体意识示范区”论坛

学科地位比较尴尬，但在民族学与社会学的夹缝中仍然得到了长足的发展，不仅涌现出了一大批学术成果，还形成了一支老、中、青三代相结合的学术队伍，人类学中国化的道路上，已开始形成了具有中国特色、中国风格、中国气派的学术性格，在铸牢中华民族共同体意识和建构人类命运共同体中将发挥出越来越多、越来越大的作用。更重要的是他们承前启后，继往开来，培养了一大批博士，为中国人类学新生代的崛起打下了基础。

历史的发展是永恒的，学科的传承也是必然的。中国人类学经历了百年的发展，完成了老、中、新三代的传承。正如《庄子·养生主》所云：“指穷于为薪，火传也，不知其尽也。”为了记录这个薪火相传的历史时刻，我提请人类学高级论坛学术委员会讨论，决定编辑出版《新生代人类学家之路》。

经过近一年的努力，在学苑出版社的支持下，两卷本的《新生代人类学家之路》出版。该书的出版一是象征了中国人类学经历了百年的发展，完成了老、中、新三代的传承；二是尽管人类学如今依然“位卑言轻”，但人类学自身的魅力使人类学者队伍仍然呈欣欣向荣之势；三是见证了人类学者“互为镜像”的自我反思，形成了“自我民族志”的范式，建构了又一本人类学者的“民族志档案”，为中国人类学的薪火相传铺了路，既向学术界展示了中国人类学不负新时代的使命，又向世界展示了中国人类学家队伍的力量。

学术之树常青：我的学问之道

成果是学者学术生命的表达。我做学问60年来，学术之树常青的秘诀是在学术实践中形成了自己的学问之道，就是“深挖井”“滚雪球”和“建油田”。

深挖了一口“井”

我在《科研方略论》中曾说：“深挖井：在一个方向上走向学术前沿”。这也被人们戏称为“一口井主义”。

1985年出版的《汉民族历史和文化新探》开汉民族研究之滥觞，并成为费孝通先生研究中华民族多元一体格局理论的参考文献。[①]由于这本书的特殊地位，孟宪范在1989年《中国社会科学》第2期上发表《中国民族学十年发展述评》说：“徐杰舜的《汉民族历史和文化新探》的发表，标志着这一课题在中国的真正开始”。

1992年，历经12年的坎坷，《汉民族发展史》终于出版，《人民日报·海外版》(1995年8月9日)、香港《大公报》(1995年8月12日)分别以《中国第一部汉族史问世》和《首部汉族史问世》为题发表了中新社记者的电讯稿，向世人宣布“由广西民族学院徐杰舜教授编著的《汉民族发展史》已于最近出版发行，从而结束了世界上最大民族没有民族史的状况”。

1987年至1994年《汉族民间风俗》[②]出版，民俗学泰斗钟敬文先生在95岁高龄为丛书写了书评，称赞“《丛书》中每册的辐射面各不相同，面内的风俗形态大多皆描摹得清晰而又生动，八块辐射面拼合为一，便构成为绚丽多彩的诱人的汉族风俗全景”，“汉族广大，风俗博而杂，用文字的排列组合将大江南北的风土人情跃然纸上，实属不易”。

2004年《汉族风俗史》(5卷本)，历经20年后出版，建构了解读汉族民族历史的另一把钥匙。

1999年《雪球：汉民族的人类学分析》的出版，是我遵照费孝通先生“重视和加强对汉民族的人类学研究”的指教，运用人类学的理论和方法研究汉民族的成功之作。该书提出的“雪球理论”，在2008年斯坦福举行的“汉民族研究反思国际学术研讨会”上受到了国际同行的关注。

2019年《汉民族史记》(九卷本)出版，是我从事汉民族研究50年的积累，20年的团队研究，5年的撰写，深挖“一口井”60年，向世人捧出“一眼千年”的集大成之作。

滚动了一个“雪球”

做人做事切忌一棵树上吊死，做学问亦然。

我做学问虽然坚持深挖井，但并没有成为井底之蛙，而是视深挖井为基础工程，与此同时，从1988年开始，我顺势滚动了学术“雪球”。

1988年，我因教学之需，出版了《中国民族史新编》，提出了按民族发展规律研究民族历史的观点，开创了中国民族史编写的新结构图。

1991年，我与彭英明合作，完成了《从原始群到民族——人们共同体通论》，建构了一个全新的人们共同体发展路线图。

1992年，我因1988年参加第二届中国民族史学会年会，了解了中国民族政策史研究碎片化的现况，组建团队，用4年的时间，完成了《中国民族政策史鉴》和《中国民族政策通论》的出版，成功地建构了中国民族政策从历史到现实的历史链。

1991年至1997年，我因参加广西自治条例的修改，成为广西十人小组的成员，开始了对民族区域自治制度的研究，与学术团队先后出版了《民族自治权论》和《实施自治法研究》，从理论到实践上完成了对民族区域自治制度的一次完整研究。

2008年，经过20余年的田野考察，我和我的研究生们完成了对汉族一个古老族群——平话人比较系统的考察，出版了“平话人书系”：《平话人印象》、《平话人图像》和《平话人素描》。

2011年，学术“雪球”滚到了中国民族教育政策研究方面，与学术团队完成了《希望：中国民族教育政策研究报告》，提出了“教育是民族的希望之所在”的观点。

2012年，受桂林金钟山旅游发展有限公司的委托，进行了对漓江旅游的徒步考察，完成了《中国名片：黄金漓江》的出版，提出了“在保护中开发，在开发中保护”，提出将桂林设为旅游直辖市的建议。

2014年，与复旦大学李辉教授合作，完成了国家社科西部项目“岭南民族源流史”的研究，出版了《岭南民族源流史》，实现了民族史学与分子人类学从理论到方法论的牵手，实现了社会科学与自然科学的互动与跨越，获得中国出版政府奖图书提名奖的《岭南民族源流史》，成了近30多年来在民族源流史上具有创新和开拓意义的标杆之作。

凡此等等学术“雪球”的滚动，不一一列举了。

建设了一批“油田”

60年来，在“深挖井”“滚雪球”做学问的过程中，我还建设了三个学术“油田”。

（1）中国民族团结研究“油田”

2000—2003年，我的学术团队，历经两年完成了对5个自治区、2个多民族省（青海和云南）、1个民族较少省（湖南）、2个大城市（上海和深圳）的田野考察，取得了丰富的田野资料，使我全面、系统地认识了中国民族团结的现状，于2003年出版了《中国民族团结考察报告》。

2003—2007年，在《中国民族团结考察报告》的基础上，我撰写的《磐石：中国民族团结研究报告》一书，完成了对中国民族团结的理论提升，明确提出了中国民族团结发展的战略，关键是增强中华民族的凝聚力，核心是强化中华民族的民族意识，基石是确认中华民族的“国族”地位，提出了“磐石”理论。

为了进一步验证中国民族团结的“磐石”理论，我带领以我的研究生为主体的学术团队，继续做了约10年的田野考察，写了一系列民族志报告。

一是《葵花：一个民族自治县的人类学研究》。《葵花》研究的是龙胜各族自治县，其5个民族，就像葵花一样紧密地团结在一起。

二是《大象：中国民族团结南宁经验研究》。《大象》研究的是广西壮族自治区首府南宁的民族团结工作像“大象”一样是一个整体。

三是《磐石荔波：中国民族团结县域样本研究》。《磐石荔波》研究的是贵州的荔波县，表达的是荔波民族团结经验的“天地人和”，荔波民族团结县域样本的最高境界，是用中华民族共同体意识凝聚了人心。

在这个过程中，2014年到2015年左右，还有三个关于民族团结的民族志报告，即关于大化瑶族自治县民族团结经验研究的民族志报告：《阳光：中国民族团结大化经验研究》；南宁六县民族团结经验研究的民族志报告：《榕树：中国民族团结南宁县域经验研究》；南宁六区民族团结经验研究的民族志报告：《石榴：中国民族团结南宁城区经验研究》，虽然分别在2015年前后完成，但因种种原因而没有出版。

这样，历经20年，在“磐石”理论的关照下，从《中国民族团结考察报告》到《磐石荔波》的一系列民族志报告，不仅从中收获了以阳光、葵花、石榴和榕树4种中国农耕文化的象征物，来呈现民族团结的理论意境，而且在理论和实践上回答了中国民族团结何以“安如磐石”的问题。人类学的田野考察方法的运用，也助力和成就了我建构中国民族团结研究的“油田”。

（2）乡村社会研究“油田”

一百多年前，当人类学传入中国之时，谁也没有想到，一个专门研究原始简单社会的“摇篮”学科，在其发展史上出人意料地逐渐在中国拐了弯，转而关注和研究拥有上下五千年历史的中国乡村社会。我50多年“走在乡间的小路上”，建设了一个乡村社会研究的学术“油田”。

1989—1990年，我与学术团队在广西贺县南乡（现为贺州市八步区）进行社会文化变迁的田野考察，完成了田野报告——《南乡春色：一个壮族乡社会文化的变迁》，感受到了中国乡村的巨大进步。

1987 年考察广西龙州金龙乡壮族布傣人

1987年考察金秀茶山瑶

1998年在贺州八步考察客家仙婆

2006年7月，受武义县委的委托，对武义新农村建设的田野考察，是我在武义生活了20年，又离开了20年后的一次回访。武义20年的巨变激发了我20年田野体验的喷发，我与学术团队写出了《新乡土中国：新农村建设武义模式研究》。认识到乡村振兴既要关注协调人与自然的平衡，又要沟通人与人的和谐，呈现一个与费先生笔下“乡土中国”不同的“新乡土中国”：乡村振兴的关键是县域整合；乡村振兴要走城乡融合的道路；中国乡村现代化的一片欣欣向荣。

2008年到2018年，我与学术团队对桂林永福乡村振兴进行了8个月的田野调查，10年“牵肠挂肚”式的深度观察和积极参与，目睹了福村从传统向现代的转向，撰写和出版了福村三部曲：《福村迟来的转身——一个山村在景区开发中现代转型纪实之一》(2010)、《福村艰难的迈步——一个山村在景区开发中现代转型纪实之二》(2010)、《福村幸福的生活——一个山村在景区开发中现代转型纪实之三》(2018)，使我对乡村振兴、乡村现代化的道路有了更深切的感受：村落是中国乡村现代化的细胞；农民是中国乡村现代化的主体；政府是中国乡村现代化的主导；“就地发展”是中国乡村现代化的主道。这是我对中国乡村在改革开放40年中走向现代化道路的一个概括和提炼，也是在人类学田野调查基础上一次理性认识的升华。

2005 年 1 月考察中山市孙中山故居

2006 年考察云南蒙自

2008 年 7 月在西江千户苗寨考察

2008 年在甘肃阿克塞哈萨克自治县牧民草场考察

2009 年在桂林永福县南登屯与村民讨论新农村建设规划

2010 年 6 月在西江进行第二次考察参加长桌宴

2013 年在广西黄姚古镇考察

2013 年考察岳麓书院

(3)中华民族研究“油田”

当21世纪20年代中华民族研究在中国风起之时，我从1998年接受统战部的委托，做“世纪之交中国民族政策调整的思考”课题开始研究中华民族，已20余年了，中华民族研究的学术“油田”基本上建成了。

2003年，在中央民族大学举行的第二届人类学高级论坛上，我提交了一篇《对中华民族国族地位确立的思考》[③]的论文，提出只有一个共同的身份——中华民族才能代表中国。正是一石激起千层浪，也跨出了我研究中华民族的重要一步。

2008年，我对中华民族研究的理论思考:《从多元走向一体: 中华民族论》一书，从文献与理论、结构与过程、互动与轨迹、冲突与整合、文化基因、边疆与中心、草原与农业和汉族案例8个方面，论述了中华民族从多元走向一体的过程，提出了中华民族研究的“过程论”。

2009年，在中华民族认同问题的争论比较大的情况下，我在贵阳举行人类学高级论坛年会时，进行了一次学术调查，组织了一次圆桌讨论:“中华民族认同问题”，这个讨论全文收入了《中华民族认同与认同中华民族》[④]一书。通过这次调查，升华了我对中华民族的认识，使我进一步明白了中华民族认同的必然性、艰巨性和复杂性。

2006—2014年，我感到对中华民族的研究除了理论探讨以外，真正的认同要从历史开始，所以我开始尝试着从通俗读物入手，试图体现中华民族从多元走向一体的过程。在福建教育出版社张惠芳的策划下，以团队合作的形式，写了6卷本的《中华民族史记》，用图片、讲故事的形式，讲述了中华民族怎么从多元走向一体。

2012—2014年，在对中华民族认同做了田野调查的基础上，我对这个重大的实际问题进行了理论的思考，写成了《中华民族认同论》[⑤]一书。该书作为国家出版基金项目，2014年由宁夏人民出版社出版。从整体到区域，从族群多元到走向一体，从文化创造到认同文化三个维度，阐述了中华民族认同的学理问题，从理论上升华了中华民族认同的研究。

2018年初，我受云南大学政治学系的邀请，给研究生开设了“中华民族史纲”的课程，十天讲了十讲。就是在这次讲授中，我在提出中华民族从多元走向一体“过程论”的基础上，又提出了中华民族从多元走向一体“链性论”。

历史是一条大河弯弯曲曲，川流不息，而理论的创新不是一蹴而就的，需要长期的学术积累，建构中华民族史新思路，就有一个从结构论到过程论，再到链性论价值的释放与转换的渐进过程。

链性是历史链条论的简称。梁启超早就讲过："历史好像一条长链、环环相接，继续不断"[⑥]。链性论，正是整合中华民族史的密码，链性论对整合中华民族史是一个有价值的创新，这不仅为我撰写《中华民族史纲》提供了理论框架的支撑，也呈现了我对中华民族研究的升华。

从2003年出版《中华民族国族地位确立的思考》开始，到2021年《中华民族史纲》[⑦]，整整18年，才完成了对中华民族研究"油田"的建设。

人类学：我学问之道的学术灯塔

大海航行靠灯塔的指引，我做学问靠人类学理论的指导，而我走进人类学是偶然中的必然。

风从香港来。

1986年，时任广西民族学院民族研究所所长的张有隽教授，到香港中文大学参加首届国际瑶族协会回来，带来了一股人类学的南风。一是带回了一套三册美国人类学家基辛的《文化人类学》；二是带来了时任香港中文大学人类学系主任乔健教授的建议：学习人类学，融入国际学术体系，不然中国学术界将无法与国际学术界对话。

乔健的建议醍醐灌顶，带给我的是学术震撼；基辛的书则使我脑洞大开，打开了人类学的大门。1986年从香港吹来的人类学之风，从此把我带上了人类学之路。

《人类学教程》的编写（2005年）

从1986年到2005年，我阅读了30本国内外学者编写的有关人类学的教材，深感艰涩深枯。当我要给学生开设"人类学概论"课程时，感觉现有的教材都不甚合适，于是决定自己动手编写一本通畅简明，适合本科生以及初学者用的教材——《人类学教程》。这本教材从对人类本身到文化与文化现象，网罗人类进化、人种、生存策略、性爱生育、婚姻，从家庭组建到依据亲属关系结成宗族、族群与民族、社会、宗教与民间信仰，纵横文化、文化多样性及其变迁，文化适应与文化自觉，人格与文化濡化，人观、空间和时间，语言与交流等与人类学相关的各个领域。虽然名为一部教程，其实也是我近20年学习人类学理论和方法的体会，故而感同身受地旁征博引中外相关经典理论、最新研究动态与成果，努力做到严谨、全面而不繁冗，言简意赅，精当准确，既避免出现类似以往庞杂的发散性结构，也不沦入浩瀚的百科全书式论述，算是交了一份学习人类学的读书笔记，既为人类学整体建构奉献了自己的见解，又作为一次人类学宏博思想和严谨科学资料归整的实践尝试。

《田野上的教室》(2008年)

在走进人类学后，我经历了一个与学生们共同成长的过程。这一切都记录在2008年出版的《田野上的教室》一书中了。

2006年6月，广西民族大学的研究生们一手策划、筹备、组织和操办，举行了"读书、田野与学术感悟研讨会"，这次学术讨论会完全由学生按照学术会议的规范和程序组织，它让同学们真刀真枪地"实战演习"一次。在此基础上，我把十几年人类学教学法研究各阶段性的成果汇集了起来，并撰写了研究报告《田野在线：大学文科试验教学法研究》。

2008年出版时取名《田野上的教室》，是想强调人类学教学的本质核心，是从课堂走进田野，从教室走进社会。田野是另一种形式的教室，与学校里的教室相辅相成，互为补充，互相促进，从两个不同方面为人类学培养人才。

《人类学与中国传统》(2009年)

走进人类学之后，越来越感到中国传统的魅力是巨大的。中国文化一以贯之的惊人连续性，常常令人有一种跨越时空的惊奇感。所以，中国的人类学者无论以什么背景进入人类学，都离不开中国传统对其学术的影响。与此同时，学者们对于中国人类学的理论建构又一向十分关注，认为中国传统可能构成人类学的一个体系。为此，我组织了对"人类学与中国传统"的系列讨论，主要有：研究对象意义上的人类学的中国传统；研究趋向意义上的人类学的中国传统；本土文化意义上的人类学的中国传统；古代民族志意义上的人类学的中国传统，并将这些讨论汇编成《人类学与中国传统》一书，记录了我对中国传统可能构成人类学一个体系的认识和努力。

《乡村人类学》的研究(2012年)

从中央民族大学任人类学博士生导师之时，为教学之需，先后编写过《乡村人类学》《历史人类学》《族群人类学》三部教材，最后只有《乡村人类学》于2012年得以出版。

为什么《乡村人类学》能得以出版？

因为在中国走进了人类学，就走进了乡村。在中国从事人类学研究的学者，大多从乡村研究起步或起家。

我从20世纪80年代中期开始，从走近人类学到走进人类学，就进入了乡村。从此之后，30多年来，不是在乡村做田野，就是在去乡村田野的路上，身体力行地参与了中国乡村人类学学派的建设。多年的课教下来，对我也是一个学习的过程，我边教边学，读了几乎所有与中国乡村人类学有关的著作，在教乡村人类学课程的过程中，深感人类学在中国的发展离不开乡村，西方学者只要进入中国，就必然首先把目光投向乡村。

2010年是费孝通、林耀华诞辰百年，中央民族大学于6月11日举行“纪念费孝通、林耀华诞辰100周年学术研讨会”，并将“民族学中国学派理论与方法”作为会议主题。我撰写了《从费孝通·林耀华先生百年诞辰谈人类学的中国学派》并参加了研讨会。没想到此文受到许多学者的赞赏，乔健先生专门要了论文的打印稿。云南省社科联的《学术探学》杂志在当年年底的第6期，以《人类学中国乡村学派初论——从费孝通·林耀华先生百年诞辰谈起》为题公开发表。此后不久，2011年5月，《新华文摘》第9期全文转载，于是有“好事者”在QQ上评论说:“《新华文摘》全文转载《人类学中国乡村学派初论——从费孝通·林耀华先生百年诞辰谈起》一文，标志着人类学中国乡村学派横空出世”。

《新华文摘》转载此文，却使我突然感到了一种责任：要为人类学的中国乡村学派树碑立传！于是我全力投入这项工作，完成人类学中国乡村学派建构的使命！

《乡村人类学》是人类学中国乡村学派的一块纪念碑，它尽可能忠实地记录人类学中国乡村学派形成的历史过程；它是人类学中国乡村学派的一本教科书，是人类学中国乡村学派的一个田野指南，它尽可能详细地告诉有兴趣进行乡村人类学田野考察的人们，进入乡村后应具有什么样的眼光，才能洞悉乡村的方方面面，里里外外；它是人类学中国乡村学派的一个经典导读，尽可能负责地介绍了中国乡村人类学研究主要代表作的作者简介、写作背景、主要内容和相关评价；它是人类学中国乡村学派发展的垫脚石和梯子，希望能为人类学中国乡村学派的继往开来做出力所能及的贡献！

以寄语为结语

自从我走进人类学之后，虽然我的主要著作都是史学方面的研究，如《汉民族发展史》《雪球：汉民族的人类学分析》《岭南民族源流史》《从多元走向一体：中华民族论》《汉民族史记》《中华民族认同论》《中华民族史纲》，但都用心学习和运用人类学的理论和方法，使得我往往能独辟蹊径而有所创新，能在学术的崎岖小路上不断攀登上高峰。

想当年，在大学求学时，我曾经十分醉迷王国维先生在论述做学问、成大事的三个境界"昨夜西风凋碧树，独上高楼，望断天涯路；衣带渐宽终不悔，为伊消得人憔悴；众里寻他千百度，蓦然回首，那人却在灯火阑珊处"。

那时的我"独上高楼"，立志高远要研究汉民族，"望断天涯路"。而后的五六十年，为研究汉民族和中华民族，"为伊消得人憔悴"，在撰写《汉民族史记》中两次休克，最后左眼黄斑病变而失明，但"衣带渐宽终不悔"。今朝"众里寻他千百度，蓦然回首"，看到手中九卷本的《汉民族史记》和六卷本的《中华民族史记》等著作，以及即将出版的《中华民族史纲》校稿，正是"那人却在灯火阑珊处"！人生虽然苦短，但在我近八十年的人生中有幸经历了王国维先生所论述的三个境界，实乃终身无憾、无悔、无怨也！

"为伊消得人憔悴""衣带渐宽终不悔"，这是立志做学问之人的必经之路，以此与大家共勉。

唐宋八大家中，韩愈和苏轼都认为摩羯座命苦[⑧]。我虽也是摩羯座，但正因为人生苦短，做学问攀登艰难，所以我特别珍惜、珍爱上海人类学会授予我的"人类学终身成就奖"！

注释

① 费孝通.中华民族多元一体格局[M].北京：中央民族学院出版社，1989：36.

② 1998年，丛书精选汇成《汉族民间风俗》一册，由中央民族大学出版社出版；2022年云南美术出版社重版，改名为《汉族风俗志》(两卷本)。

③ 杨圣敏主编.民族学人类学的中国经验：人类学高级论坛2003卷[M].哈尔滨：黑龙江人民出版社，2005：129.

④ 详见吴晓萍，徐杰舜主编.中华民族认同与认同中华民族[M].哈尔滨：黑龙江人民出版社，2009：151-184.

⑤ 徐杰舜，刘冰清，罗树杰.中华民族认同论[M].银川：宁夏人民出版社，2014：1-43.

⑥ 梁启超.中国历史研究法[M].北京：中国人民大学出版社，2011：155.

⑦《中华民族史纲》即将由学苑出版社出版。

⑧ 参阅笑川：苏轼：我命苦，因为我是摩羯座[N].光明日报，2020-01-07.

主要论著

一、独著

[1] 徐杰舜.汉民族历史和文化新探[M].南宁：广西人民出版社,1985.
[2] 徐杰舜.中国民族史新编[M].南宁：广西教育出版社,1988.
[3] 徐杰舜.汉族民间经济风俗[M].南宁：广西教育出版社,1990.
[4] 徐杰舜.汉民族发展史[M].成都：四川民族出版社,1992.
[5] 徐杰舜.磐石：中国民族团结研究报告[M].南宁：广西人民出版社,2007.
[6] 徐杰舜.从多元走向一体——中华民族论[M].桂林：广西师范大学出版社,2008.
[7] 徐杰舜.福村迟来的转身[M].哈尔滨：黑龙江人民出版社,2010.
[8] 徐杰舜.汉民族发展史[M].武汉：武汉大学出版社,2012.

二、合著

[1] 徐杰舜,彭英明.民族新论[M].南宁：广西人民出版社,1987.
[2] 徐杰舜,徐桂兰.中国奇风异俗[M].南宁：广西人民出版社,1989.
[3] 徐杰舜,罗树杰,刘小春.南乡春色——一个壮族乡社会文化的变迁[M].南宁：广西人民出版社,1990.
[4] 彭英明,徐杰舜.从原始群到民族——人们共同体通论[M].南宁：广西人民出版社,1991.
[5] 徐杰舜,覃乃昌.民族自治权论[M].南宁：广西教育出版社,1991.
[6] 徐杰舜,杨秀楠,徐桂兰.程阳桥风俗[M].南宁：广西民族出版社,1992.
[7] 徐杰舜,陈顺宣.中国的风俗[M].北京：人民出版社,1992.
[8] 胡敏,万建中,吴崇新,等.汉族民间信仰风俗[M].南宁：广西教育出版社,1994.
[9] 徐杰舜,罗树杰.马克思主义民族观导论[M].北京：民族出版社,1997.
[10] 徐杰舜,周耀明.汉族风俗文化史纲[M].南宁：广西民族出版社,2001.
[11] 徐杰舜,等.新乡土中国——新农村建设武义模式研究[M].北京：中国经济出版社,2006.
[12] 徐杰舜,王瑞莲,李然,等.京族简史[M].北京：民族出版社,2008.
[13] 黄兰红,徐杰舜,蒋中意.新九龙山村的幸福生活[M].哈尔滨：黑龙江人民出版社,2009.
[14] 荣仕星,徐杰舜,等.中国民族地区公共政策研究[M].北京：人民出版社,2009.
[15] 徐杰舜,何月华.城中小村的“前世今生”[M].北京：民族出版社,2010.
[16] 徐杰舜,罗树杰,许立坤.中国民族政策简史[M].银川：宁夏人民出版社,2011.
[17] 徐杰舜,荣仕星,吴政富.希望：中国民族教育政策研究报告[M].哈尔滨：黑龙江人民出版社,2011.
[18] 覃锐钧,徐杰舜,黄兰红,等.接触与变迁——广西金秀花蓝瑶人类学考察[M].北京：民族出版社,2011.
[19] 徐杰舜,徐桂兰.中国汉族[M].银川：宁夏人民出版社,2012.
[20] 徐杰舜,刘冰清.乡村人类学[M].银川：宁夏人民出版社,2012.

[21] 徐杰舜，赵杨，丁苏安．风景郭洞独好：一个古村落生态文明的人类学考察[M]．哈尔滨：黑龙江人民出版社，2012.
[22] 徐杰舜，吕志辉，王清荣，等．中国名片：黄金漓江[M]．哈尔滨：黑龙江人民出版社，2012.
[23] 罗彩娟，徐杰舜，罗树杰．中国西南边疆治理模式研究[M]．哈尔滨：黑龙江人民出版社，2012.
[24] 徐杰舜，刘冰清，罗树杰．中华民族认同论[M]．银川：宁夏人民出版社，2014.
[25] 徐杰舜，孙亚楠，刘少莹．大象：中国民族团结南宁经验研究[M]．北京：民族出版社，2017.
[26] 徐杰舜，韦小鹏，孙亚楠．磐石荔波：中国民族团结县域样本研究[M]．哈尔滨：黑龙江人民出版社，2020.

三、主编

（一）著作主编

[1] 徐杰舜．民族理论政策简明教程[M]．南宁：广西教育出版社，1988.
[2] 袁少芬，徐杰舜．汉民族研究·第一辑[M]．南宁：广西人民出版社，1989.
[3] 张有隽，徐杰舜．民族与民族观[M]·第一辑．南宁：广西教育出版社，1991.
[4] 张有隽，徐杰舜．中国民族政策通论[M]．南宁：广西教育出版社，1992.
[5] 徐杰舜，韦日科．中国民族政策史鉴[M]．南宁：广西人民出版社，1992.
[6] 徐杰舜，吴淑兴．实施自治法研究[M]．南宁：广西民族出版社，1997.
[7] 荣仕星，徐杰舜．人类学本土化在中国[M]．南宁：广西民族出版社，1998.
[8] 徐杰舜．汉族民间风俗[M]．北京：中央民族大学出版社，1998.
[9] 徐杰舜．雪球——汉民族的人类学分析[M]．上海：上海人民出版社，1999.
[10] 徐杰舜．本土化：人类学的大趋势[M]．南宁：广西民族出版社，2001.
[11] 容本镇，徐杰舜（任副主编）．悄然崛起的相思湖作家群[M]．南宁：广西民族出版社，2002.
[12] 徐杰舜．中国民族团结考察报告[M]．北京：民族出版社，2003.
[13] 徐杰舜．汉族风俗史（1—5卷）[M]．上海：学林出版社，2004.
[14] 徐杰舜．人类学教程[M]．上海：上海文艺出版社，2005.
[15] 罗树杰，徐杰舜．民族理论与民族政策教程[M]．北京：民族出版社，2005.
[16] 徐杰舜，徐桂兰．平话人书系：徐杰舜，杨清媚，等．平话人印象/徐杰舜，覃锐钧，等．平话人图像/徐杰舜，林敏霞，梁冬平，等[M]．平话人素描/哈尔滨：黑龙江人民出版社，2008.
[17] 荣仕星，徐杰舜．人类学世纪真言[M]．北京：中央民族大学出版社，2009.
[18] 胡春惠，徐杰舜．少数民族——中国的一个政治元素[M]．香港：香港珠海书院亚洲研究中心，2009.
[19] 徐杰舜，许立坤．人类学与中国传统[M]．北京：民族出版社，2009.
[20] 徐杰舜．小荷尖尖[M]．哈尔滨：黑龙江人民出版社，2009.
[21] 许宪隆，石玉刚，徐杰舜，等．中国少数民族[M]．北京：民族出版社，2009.

[22] 何龙群，秦红增，徐杰舜．名刊建设与主编自觉[M]．哈尔滨：黑龙江人民出版社，2011.
[23] 徐杰舜．中国汉族通史·第一卷[M]．银川：宁夏人民出版社，2012.
[24] 徐杰舜．中国汉族通史·第二卷[M]．银川：宁夏人民出版社，2012.
[25] 徐杰舜，关凯，李晓明．中国社会的文化转型：人类学高级论坛十年论文精选[M]．北京：民族出版社出版，2012.
[26] 徐杰舜．中华民族史记（六卷本）[M]．福州：福建教育出版社，2014.
[27] 徐杰舜．中国人类学家访谈录（3 卷本）[M]．昆明：云南人民出版社，2019.
[28] 田敏，徐杰舜．李亦园与中国人类学[M]．上海：上海文艺出版社，2019.
[29] 徐杰舜．汉民族史记（九卷本）[M]．北京：中国社会科学出版社，2019.
[30] 徐杰舜，韦小鹏．新生代人类学家之路[M]．北京：学苑出版社，2021.

（二）丛书/文库主编

1. 徐杰舜主编：汉族民间风俗丛书

[1] 徐杰舜．汉族民间经济风俗[M]．南宁：广西教育出版社，1990.
[2] 徐桂兰．汉族红白喜事风俗[M]．南宁：广西教育出版社，1990.
[3] 胡敏．汉族四时八节风俗[M]．南宁：广西教育出版社，1990.
[4] 陈顺宣．汉族生养益寿风俗[M]．南宁：广西教育出版社，1990.
[5] 莫高，吴华．汉族衣食住行风俗[M]．南宁：广西教育出版社，1994.
[6] 周耀明，吴晓华．汉族民间游乐风俗[M]．南宁：广西教育出版社，1994.
[7] 周耀明．汉族民间交际风俗[M]．南宁：广西教育出版社，1994.
[8] 胡敏，万建中，徐杰舜，等．汉族民间信仰风俗[M]．南宁：广西教育出版社，1994.

2. 徐杰舜，周大鸣主编：人类学文库

[1] 何毛堂，李玉田，李全伟．黑衣壮的人类学考察[M]．南宁：广西民族出版社，1999.
[2] 吴和培，罗志发，黄家信．族群岛：浪平高山汉探秘[M]．南宁：广西民族出版社，1999.
[3] 李远龙．认同与互动：防城港的族群关系[M]．南宁：广西民族出版社，1999.
[4] 容观瓊．人类学方法论[M]．南宁：广西民族出版社，1999.
[5] 徐杰舜，徐桂兰，罗树杰，等．从磨合到整合——贺州族群关系研究[M]．南宁：广西民族出版社，2001.
[6] 徐杰舜．本土化：人类学的大趋势[M]．南宁：广西民族出版社，2001.
[7] 陈益源．民间文化图像——台湾民间文学论文集[M]．南宁：广西民族出版社，2001.
[8] 周大鸣．中国的族群与族群关系[M]．南宁：广西民族出版社，2002.
[9] 徐桂兰．中国育俗的文化叠合[M]．南宁：广西民族出版社，2002.
[10] 黄世杰．蛊毒：财富与权力的幻觉[M]．南宁：广西民族出版社，2004.

3. 徐杰舜主编：人类学高级论坛文库

第一辑

[1] 林美容．妈祖信仰与汉人社会[M]．哈尔滨：黑龙江人民出版社，2003.
[2] 周大鸣，等．当代华南的宗族与社会[M]．哈尔滨：黑龙江人民出版社，2003.
[3] 徐杰舜．人类学的世纪坦言[M]．哈尔滨：黑龙江人民出版社，2004.

[4] 莫蓉，徐杰舜.互动中的磨合与认同——广西民族团结模式研究[M].哈尔滨：黑龙江人民出版社，2004.
[5] 徐杰舜.金羊毛的寻找者——世纪之交的中国民俗学家[M].哈尔滨：黑龙江人民出版社，2005.
[6] 李富强.让传统告诉未来：关于民族传统与发展的人类学研究[M].哈尔滨：黑龙江人民出版社，2006.
[7] 徐杰舜.族群与族群文化[M].哈尔滨：黑龙江人民出版社，2006.
[8] 罗树杰.无声的危机[M].哈尔滨：黑龙江人民出版社，2006.
[9] 罗志发.壮族的性别平等[M].哈尔滨：黑龙江人民出版社，2007.
[10] 陈其斌，冼奕，等.人类学的中国大师[M].哈尔滨：黑龙江人民出版社，2008.
[11] 徐杰舜，等.田野上的教室[M].哈尔滨：黑龙江人民出版社，2008.
第二辑
[1] 秦璞，徐桂兰.河疍与海疍珠疍[M].哈尔滨：黑龙江人民出版社，2009.
[2] 蒋中意，徐榕，徐杰舜.办报人心史：《金华日报》的媒体人类学考察[M].哈尔滨：黑龙江人民出版社，2009.
[3] 冯雪红，徐杰舜，郭鸣.走进乡村人类学书林[M].哈尔滨：黑龙江人民出版社，2009.
[4] 谢林轩.越南人类学田野笔记[M].哈尔滨：黑龙江人民出版社，2013.
[5] 罗彩娟，徐杰舜，罗树杰.中国西南边疆治理模式研究[M].哈尔滨：黑龙江人民出版社，2014.
4. 徐杰舜主编：人类学高级论坛·年会论文集
[1] 徐杰舜，周建新.人类学与当代中国：人类学高级论坛2002卷[C].哈尔滨：黑龙江人民出版社，2003.
[2] 杨圣敏.民族学人类学的中国经验：人类学高级论坛2003卷[C].哈尔滨：黑龙江人民出版社，2005.
[3] 孙振玉.人类生存与生态环境：人类学高级论坛2004卷[C].哈尔滨：黑龙江人民出版社，2005.
[4] 徐杰舜，许宪隆.人类学与乡土中国：人类学高级论坛2005卷[C].哈尔滨：黑龙江人民出版社，2006.
[5] 罗康隆，徐杰舜.人类学与当代生活：人类学高级论坛2006卷[C].哈尔滨：黑龙江人民出版社，2007.
[6] 罗布江村，徐杰舜.人类学的中国话语：人类学高级论坛2007卷[C].哈尔滨：黑龙江人民出版社，2008.
[7] 吴晓萍，徐杰舜.中华民族认同与认同中华民族：人类学高级论坛2008卷[C].哈尔滨：黑龙江人民出版社，2009.
[8] 齐木德道尔吉，徐杰舜.游牧文化与农耕文化：人类学高级论坛2009卷[C].哈尔滨：黑龙江人民出版社，2010.
[9] 曾羽，徐杰舜.走进原生态文化：人类学高级论坛2010卷[C].哈尔滨：黑龙江人民出版社，2011.
[10] 罗勇，徐杰舜.族群迁徙与文化认同：人类学高级论坛2011卷[C].哈尔滨：黑龙江

人民出版社，2012.
［11］安晓平，徐杰舜．社会转型与文化转型：人类学高级论坛2012卷［C］．哈尔滨：黑龙江人民出版社，2013.
［12］谭宏，徐杰舜．人类学与江河文明：人类学高级论坛2013卷［C］．哈尔滨：黑龙江人民出版社，2014.
［13］行龙，徐杰舜．人类学与黄土文明：人类学高级论坛2014卷［C］．哈尔滨：黑龙江人民出版社，2015.
［14］陈刚，徐杰舜．人类学与山地文明：人类学高级论坛2015卷［C］．哈尔滨：黑龙江人民出版社，2016.
［15］田阡，徐杰舜．人类学与流域文明：人类学高级论坛2016卷［C］．哈尔滨：黑龙江人民出版社，2017.
［16］骆桂花，徐杰舜．道路与族群：人类学高级论坛2017卷［C］．北京：民族出版社，2018.

5. 徐杰舜主编：人类学·千手观音书系

［1］徐杰舜．一方水土养一方人［M］．哈尔滨：黑龙江人民出版社，2004.
［2］徐杰舜．走在乡间的小路上［M］．哈尔滨：黑龙江人民出版社，2005.
［3］徐杰舜，秦红增．人命关天［M］．哈尔滨：黑龙江人民出版社，2010.

6. 徐杰舜，吕志辉主编：旅游高峰论坛·年会论文集

［1］刘冰清，徐杰舜，吕志辉．旅游与景观：旅游高峰论坛2010年卷［C］．哈尔滨：黑龙江人民出版社，2011.
［2］黄萍，徐杰舜．好客中国：旅游高峰论坛2012年卷［C］．哈尔滨：黑龙江人民出版社，2013.
［3］向玉成，徐杰舜，邱云志．遗产旅游与文化中国：旅游高峰论坛2014年卷［C］．哈尔滨：黑龙江人民出版社，2015.
［4］田敏，徐杰舜．民族旅游与文化中国：旅游高峰论坛2016卷［C］．哈尔滨：黑龙江人民出版社，2017.
［5］孙九霞，韦小鹏，徐杰舜．人类学与乡村旅游：旅游高峰论坛2018年卷［C］．哈尔滨：黑龙江人民出版社，2019.

7. 吕志辉，徐杰舜主编：旅游高峰论坛文库

［1］徐杰舜，吕志辉，王清荣．中国名片：黄金漓江［M］．哈尔滨：黑龙江人民出版社，2012.
［2］徐杰舜．福村迟来的转身：一个山村在景区开发中现代转型纪实之一［M］．哈尔滨：黑龙江人民出版社，2010.
［3］丘文荣，徐杰舜．福村艰难的迈步：一个山村在景区开发中现代转型纪实之二［M］．哈尔滨：黑龙江人民出版社，2010.
［4］徐杰舜，赵杨，丁苏安．风景郭洞独好：一个古村落生态文明的人类学考察［M］．哈尔滨：黑龙江人民出版社，2012.
［5］徐杰舜，等．福村幸福的生活：一个山村在景区开发中现代转型纪实之三［M］．哈尔滨：黑龙江人民出版社，2018.

8. 徐杰舜，彭兆荣，徐新建主编：人类学高级论坛·中国人类学家口述史文库 李菲访谈记录. 乔健口述史[M]. 昆明：云南人民出版社，2014.

四、《新华文摘》转载徐杰舜教授论文汇总

(一)全文转载

[1] 徐杰舜. 文化发现与发现文化.《新华文摘》2012年第7期(摘自《学术探索》2012年第1期)

[2] 徐杰舜. 人类学中国乡村学派初论.《新华文摘》2011年第9期(摘自《学术探索》2010年第6期)

[3] 徐杰舜，韦小鹏. "中华民族多元一体格局"理论研究述评.《新华文摘》2008年第14期(摘自《民族研究》2008年第2期)

[4] 徐杰舜.《雪球——汉民族的人类学分析》题识.《新华文摘》2000年第10期(摘自《雪球——汉民族的人类学分析》上海人民出版社1999年8月出版)

[5] 徐杰舜. 中国人类学的现状及未来走向.《新华文摘》1998年第1期(摘自《广西民族学院学报》1997年第4期)

[6] 徐杰舜. 试论汉字对汉民族的内聚作用.《新华文摘》1998年第1期(摘自《浙江社会科学》1996年第3期)

(二)论点摘编

[1] 徐杰舜. 创建人类学中国学派的现实意义.《新华文摘》论点摘编《新华文摘》2011年第6期(摘自《学术探索》2010年第6期)

[2] 罗树杰，徐杰舜. 世纪之交中国民族政策调整的思考.《新华文摘》论点摘编《新华文摘》1999年第9期(摘自《广西民族学院学报》1999年第2期)

[3] 赵世怀，徐杰舜，欧以克. 新时期如何办好民族学院.《新华文摘》论点摘编《新华文摘》1998年第9期(摘自《广西民族学院学报》1998年第3期)

[4] 徐杰舜. 岭南文化解剖散论.《新华文摘》论点摘编1996年第7期(摘自《广西民族学院学报》1996年第2期)

[5] 徐杰舜. 汉民族研究的学术意义.《新华文摘》论点摘编1986年第9期(摘自《广西民族学院学报》1986年第2期)

[6] 徐杰舜，彭英明. "互为外国"论不能成立.《新华文摘》论点摘编1986年第9期(摘自《中央民族学院学报》1986年第2期)

代表作

论族群与民族*

徐杰舜
（广西民族大学）

摘要：族群概念的引入和使用为中国人类学和民族学的研究开辟了一个新天地。族群概念的界定是多义的，有的强调族群的内涵，有的强调族群的边界，有的是两者兼而有之，但简明准确的界定可以概括为“族群是对某些社会文化要素认同，而自觉为我的一种社会实体”。族群与民族的区别是：（1）从性质上看，族群强调的是文化性，而民族强调的是政治性；（2）从社会效果上看，族群显现的是学术性，而民族显现的是法律性；（3）从使用范围上看，族群概念的使用十分宽泛，而民族概念的使用则比较狭小。族群与民族的联系是：族群可能是一个民族，也可能不是一个民族；而民族不仅可以称为族群，还可以包括若干不同的族群。

关键词：族群　民族　民族理论

族群关系是当代国际人类学研究的前沿热点，族群概念的提出对人类学、民族学界研究人们共同体，是一个重大的发展。

传统的民族学研究人们共同体，是将其分为原始群、氏族、胞族、部落、部落联盟、部族和民族等层面。新中国成立后，由于受苏联民族学的影响，中国民族学界长期以来对民族共同体的研究，不仅研究方法落后、单一，而且理论陈旧，使得中国民族学界无论在学术水平上，还是在学术理论上，都落后于国际学术界，许多学术问题不能得到深入的研究。

---

* 原载于《民族研究》2002年第1期。

20世纪90年代以来，族群的概念被介绍到了中国，在中国人类学和民族学界产生了强烈的反响。虽然有的学者不主张使用族群概念，但是多数学者认为族群概念的使用有利于人类学和民族学研究的深入。从一定意义上说，族群概念的使用为中国人类学和民族学的研究开辟了一个新天地。

一、关于族群概念

族群（ethnic group）概念是西方人类学研究社会实体的一种范畴分类法。从语源学的角度看，ethnic源于希腊语，是经拉丁语进入英语系统的形容词。最初使用ethnic是在15世纪晚期，在英语世界里用以指称非犹太教和基督教徒的各种族成员，是野蛮人和异教徒的代名词，①显现出区分我群（in-group）与他群（out-group）的含义。所以牟小磊在他的硕士论文《"中国少数民族"的族性过程与研究策略》中指出："从这个意义上来说，ethnic具有指涉非西方宗教世界的人群成员的概念传统。"此后，在西方学术研究中出现了ethnic group一词，由于group包含有着共同利益以及一定连带感的人们的意义，所以ethnic group可以视为ethnic的复指名词形式。对应于英语中nation一词具有国家、国民、民族的多义现象，在已有的描述非西方人群的学术文献中，ethnic group具有比nation较为下位的意义。对此，美国人类学家郝瑞（Stevan Harrel）先生1996年9月在厦门大学人类学研究所作的《民族、族群和族性》的报告中曾明确地指出：要具体解释族性，应先区分民族（nation）与族群（ethnic group）。英文nation是指有state（国家）或government（政府）的一个族群，含有国家和民族两层意思。而族群本身并不一定含有state（国家）或government（政府）的意义，它只是有意识、有认同的群体中的一种。②可见在西方话语中，ethnic不指称具有明显政治优势地位的群体，这是一方面。另一方面，ethnic在希腊语词源中的名词形式是ethnos。③费孝通先生在《简述我的民族研究经历和思考》中曾说："ethnos是一个形成民族的过程，一个个的民族只是这个历史过程在一定时间空间的场合里呈现的一种人们共同体。"（《北京大学学报》1997年第2期）从这个意义上来看，ethnic group这一概念范畴与nation相比，指涉外延较宽，更具动态性和灵活性。

但是，正如人们所知，族群的概念与文化的定义一样，在西方学术界是多义的。早在1965年，日本学者涩谷和匡就将族群界定为"由于具有实际或虚构的共同祖先，因而自认为是同族并被他人认为是同族的一群人"。④这种意义上的族群具有共同的文化传统、相同的语言和相似的生活方式。其后，1969年，挪威人类学家弗雷德里克·巴斯（亦译为巴特或巴尔特）在著名的《族群与边界》一书的序言中说："族群这个名称在人类学著作中一般理解为用以指（这样）一个群体：① 生物上具有极强的自我延续性；② 分享基本的文化价值，实现文化形式上的统一；③ 形成交流和互动的领域；④ 具有自我认同和他人认同的成员资格，以形成一种与其他具有同一秩序的类型不同的类型。"[5]此后，学者们从不同的视角，对族群作出各自的界定，主要有以下一些：

——族群意指同一社会中共享文化的一群人，尤其是共享同一语言，并且文化和语言能够没有什么变化地代代传承下去。[6]

——族群是一个有一定规模的群体，意识到自己或被意识到其与周围不同，"我们不像他们，他们不像我们"，并具有一定的特征以与其他族群相区别。这些特征有共同的地理来源，迁移情况，种族，语言或方言，宗教信仰，超越亲属、邻里和社区界限的联系，共有的传统、价值和象征，文字、民间创作和音乐，饮食习惯，居住和职业模式，对群体内外的不同感觉。[7]

——族群是指一个较大的文化和社会体系中具有自身文化特质的一种群体；其中最显著的就是这一群体的宗教的、语言的特征，以及其成员或祖先所具有的体质的，民族的，地理的起源。[8]

——族群就是一种社会群体，其成员宣称具有共同世系或在继嗣方面相近，并宣称具有历史上或现实的共同文化。[9]

——族群是指"能自我区分或是能被与其共处或互动的其他群体区分出来的一群人，区分的标准是语言的，种族的，文化的……"[10]

——族群是一种有着共享文化的某些观念的一种分类或群体，这些观念的一个或更多的方面原生性地（primordially）构成群体成员资格的许可证。[11]

——族群这个概念习惯上是指享有同一种文化，讲同一种语言，从属于同一个社会的人。[12]

——族群这个群体包括两个特点：一是族群成员认为拥有共同祖先和共同文化，这种认同可以是客观实在的，也可以是虚拟的（artificial）；二是群体用共同祖先、共同文化来有意识地与其他群体相区别，形成内部的统一和外部的差异。[13]

在关于族群的定义中，较有代表性的是科威特人类学家穆罕默德·哈达德的界定。他说：族群是指在社会上具有独特的因素，因文化和血统而形成不同意识的群体。可以说，它是因体质或文化上的特点而与社会上其他群体区别开来的人们共同体。他认为可识别性（identifiability）、权力差别（differential power）及群体意识（group awareness）是族群的三个基本特点。[14]这是从广义上给族群下的定义。但是学术界比较常用的是马克斯·韦伯（Max Weber）的定义："某种群体由于体质类型、文化的相似，或者由于迁移中的共同记忆，面对他们共同的世系抱有一种主观的信念，这种信念对于非亲属社区关系的延续相当重要，这个群体就被称为族群。"[15]

除上述西方学者对族群概念所作的界定以外，中国学者对族群的概念也作出了自己的界定。孙九霞主张在较广的范围内使用族群定义，即可以等同于民族一词，也可以指民族的下位集团"民系"，还可以在超出民族的外延上使用，并给族群下定义说："在较大的社会文化体系中，由于客观上具有共同的渊源和文化，因此主观上自我认同并被其他群体所区分的一群人，即称为族群。其中共同的渊源是指世系、血统、体质的相似；共同的文化指相似的语言、宗教、习俗等。这两方面都是客观的标准，族外人对他们的区分，一般是通过这些标准确定的。主观上的自我认同意识即对我群和他群的认知，大多是集体无意识的，但有时也借助于某些客观标准加以强化和延续。"[16]

上述种种有关族群的概念从不同的背景出发，或强调群体内部的共同特征，即族群的内涵；或强调群体的排他性和归属性，即族群的边界；或既强调群体的内涵，又强调群体的边界。由此反映了族群概念的多义性。

笔者认为，如果对族群概念作一个更简明准确的界定，可以这样概括，即所谓族群，是对某些社会文化要素认同而自觉为我的一种社会实体。这个概念有三层含义：一是对某些社会文化要素的认同；二是要对它"自觉为我"；三是一个社会实体。

族群的概念必须包含对某些社会文化要素的认同。在诸多学者关于族群的概念中，有的主张以文化、语言、社会为族群认同的要素；有的主张以文化、社会为族群认同的要素；有的主张以宗教、语言、民族为族群认同的要素；有的主张以信仰、价值、习惯、风俗、历史经验为族群认同的要素。其实族群是一个更为灵活、操作性更强的概念，而文化又是一个动态的多变的东西，因此大可不必对社会文化要素的认同作机械的规定，而要从实际出发。

族群的概念必须包含对它"自觉为我"，因为这种对它"自觉为我"，就是族群的自我意识，自我意识既是个哲学概念，也是个心理学概念，是主体在对象性关系中对自身及其与对象世界的关系的意识。[17]族群的自我意识具有认同性、相对性、内聚性、自主性、稳定性，是族群形成的灵魂所在。

族群概念还必须包含这样一点，即族群是一个社会实体，这是因为，族群作为一种人们共同体是一种社会存在，具有社会属性，只有在一定的社会条件下才能形成和发展。如汉族中的客家人，就是在中国历史上出现动乱之时，由中原汉族南迁而逐渐形成的。在客家人迁入华南时，华南的平原沃土早已被当地族群及早期移民所占有，他们只好居山开垦，加上自认为祖先是中原望族，有着一种文化上的优越感，而少与华南当地族群交往，从而过着自给自足的、封闭式的家族生活，并形成了“诗礼传家”、“书香门第”的家风，以及勤劳、勇敢、豪爽、深沉的性格。如果离开了这些社会背景和条件，客家人也就不成其为客家人了。

二、族群与民族的关系

关于民族概念问题，从民族译名到对斯大林民族定义[18]的争论，在中国由来已久，且众说纷纭。笔者在《从原始群到民族——人们共同体通论》第六章《民族共同体》中曾对“民族”一词的译名，民族概念在东西方的历史发展，以及对斯大林民族定义的质疑和争论作了详细的论述。笔者认为，在对斯大林民族定义的争论中所提出的种种修改方案，“并没有对斯大林的民族定义有实质性的理论突破，无论补充一点或减去一点，稍加分析推敲，就可见其与斯大林的定义大同小异，有的在表述上还没有斯大林的简练和明确”。[19]因此，从宏观上看，笔者认为斯大林的民族定义是具有学术性、科学性和普通性的。问题在于，我们在考察民族时应该从实际出发，具体情况具体分析。

那么，族群与民族之间又有什么区别和联系呢？

民族与族群虽然都是历史上形成的人们共同体，但两者的区别十分明显。

从性质上看，族群强调的是文化性，而民族强调的是政治性。

族群这个人们共同体的根本属性在于它的文化性，无论哪一位学者的定义都认同这一点。正因为如此，笔者才认为族群是一个对某些文化要素的认同而自觉为我的一种社会实体，以此突出它的文化性特征。

民族这个人们共同体虽然也具有文化性，但这不是它最重要的基本特征，一些民族虽然没有鲜明的文化特征，但仍然认同为一个民族，因为他们仍具有民族的自我意识。民族强调的是它的政治性。这是因为：

第一，在从部落发展成民族的历史过程中，国家是孕育民族的母腹。恩格斯说过："建立国家的最初企图，就在于破坏氏族的联系，其办法就是把每一氏族的成员分为特权者和非特权者，把非特权者又按他们的职业分为两个阶级，从而使之互相对立起来。"[20]所以国家的产生，对民族形成的最大作用，就是氏族制度的彻底瓦解。"氏族组织不知不觉地变成了地区组织，因而才能和国家相适应。"[21]关于这一点，恩格斯又说："以血族团体为基础的旧社会，由于新形成的社会各阶级的冲突而被炸毁；组成国家的新社会取而代之，而国家的基层单位已经不是血族团体，而是地区团体了。"[22]

民族与氏族、部落等人们共同体最本质的区别在于，前者以地缘关系为基础，后者以血缘关系为纽带，所以，国家的产生也就表现为血缘关系向地缘关系转化的完成。对于民族形成更重要的是，最初的国家无不用战争来扩大自己的地域范围，以站稳脚跟并充实自己的力量。于是在原始社会末期，随着国家的产生和战争的频繁，杂居现象也愈加明显，无形之中国家成了形成民族共同地域的纽带。

一般情况下，最初的国家都要使用行政的手段统一语言、文字，加强人们的经济联系。有的学者正确地认为："国家是民族共同体形成的工具，只有借助于国家的推动，并在国家的强力作用下，把不同的人们共同体聚集在一起，利用国家的力量对它们加以融合，一个稳定的民族共同体才有可能形成。"[23]

第二，民族与国家政体有着密不可分的关系。对此，周星在《民族政治学》中有很好的论述。他认为：任何国家政治体系的发育与存续，都必须以既定的民族、民族社会或多民族社会政治生活的存在为背景。在国家的制约下，当民族规模与国家政治体系相吻合时，该民族的政治体系便采取了国家的形态，以其民族构成的单一性而成为民族国家。而民族规模与国家政治体系不相吻合的情形有两种状态：一为国家以多民族社会为背景，因民族构成复杂而为多民族国家；二是一个民族分为若干个民族社会，形成若干个民族国家政治体系。多民族国家的政体形式既有单一制，又有联邦制和邦联制等，情况也不尽相同。但是，不管怎样，多民族的构成，总会以各种不同的方式或渠道对国家的政体结构产生影响，或者导致联邦与邦联，或者采取单一体制之内的民族区域自治。

由此可见，在民族国家的条件下，民族与国家的界限及利害多是相互一致的，民族的愿望可以直接成为国家政策的基础。国家的政治社会化与一体化努力，久而久之，便可以改变民族性格，而民族性格又会在相当程度上影响到国家政治制度。在多民族国家中，国家的宪法也可能对国内各民族的利益做出保障。

总之，国家政治体系（无论单一民族国家还是多民族国家）需要以民族、民族社会或多民族社会为依据。民族的某种程度与性质的统一，常常都以国家为归宿，并构成国家存续的重要条件。这种民族的统一，也可在若干不同的层面上实现，从而在国家政治体系内部构成某种自治或特区政治体系的基础。所以，民族构成是国家分类乃至于国家政体分类的尺度之一。单一民族国家与多民族国家，单一制的多民族国家和联盟制的多民族国家等，正是这样的分类。㉔

正因为民族从形成到发展都与国家有着密不可分的关系，所以周星先生还明确地提出了“民族的政治属性”这一命题。他说：对于民族政治学的民族观而言，民族除了它在文化、地理、语言和心理等方面的特性之外，还有着十分突出的政治属性。这种政治属性是如此重要，以至于民族共同体与民族社会的其他任何特性，在特定的条件下，都可能受到民族政治属性一定的影响，并且，常常也作为民族政治属性的某些资源和表现形式。民族政治属性的命题，是民族政治学得以成立并展开研究的根本基点，这不仅意味着承认民族现象与政治现象之间固有的千丝万缕的内在联系，而且意味着把政治属性视为民族共同体及其范畴的本质内涵之一。㉕

在中国现实的社会政治生活中，族群的文化性特质与民族的政治性特质之间的反差更是凸显无遗。作为族群，无论是汉族的客家人、广府人、闽南人、平话人、东北人、陕北人、昆明人等，还是瑶族的盘瑶、山子瑶、花蓝瑶、茶山瑶、过山瑶、布努瑶等，抑或是彝族的黑彝、撒尼、阿细、红彝等，还是苗族的花苗、红苗、白苗、青苗，都以文化为边界，既不享有政治权利，也不谋求政治权利。而民族则不然，作为一个民族，在中国必须得到国务院的承认。而一个共同体一旦被确认为一个民族，就享有国家赋予民族的一切政治权利，哪怕这个民族只有几千人，在全国人民代表大会中也必须有其代表，并享有建立民族自治地方的政治权利。

从社会效果上看，族群显现的是学术性，而民族显现的是法律性。

族群概念的引入，尤其是我们运用族群概念来研究民族内部的支系或民系，如近几年对客家人的研究，所显现出来的社会效果都是学术性的，既无经济利益的驱动，也无政治权利的追求。所以，不具任何政治色彩的族群概念的使用，从学术上为我们更深入、更细致地研究人们共同体提供了一个很好的"武器"或"工具"，或者说是给人们提供了一个不会引起争议的话语。

如前所述，民族的政治属性决定了民族问题在涉及经济利益和政治权利时，必须依靠法律给予保障和进行裁决。所以，为了解决中国的民族问题，国家制定了《民族区域自治法》来保障少数民族的经济利益和政治权利，从而在社会效果上显现出来的是法律性。可以说，民族及民族问题从古到今一直是一个敏感的政治因素。

从使用范围上看，族群概念的使用十分宽泛，而民族概念的使用则比较狭小。

在传统的民族学理论中，原始群、氏族、胞族、部落、部落联盟、部族和民族都有严格的界定。斯大林所定义的民族是前资本主义民族和资本主义以后的民族，而不包括氏族、部落等前民族共同体在内的所谓"广义的民族"，这一民族概念的使用范围比较狭小，故人们称之为"狭义的民族"。正因为民族概念有不同的学术含义，所以在使用中，由于理解的不同和环境的差异，往往引起误解和争论，中国学术界20世纪60年代初就民族译名而引起的争论就是一例。而当中国学术界对民族概念争论不休之时，美国人类学界主要关注的则是族群及族群关系。

由于族群强调的是文化性，它的形成和发展与政治无关，所以它的使用范围宽泛，外延可大可小。一方面，它可以泛指从古到今的一切人们共同体：若从历史上看，可以指原始族群、古代族群、现代族群；若从结构上看，可以指大的人们共同体集团，如中国古代的百越集团、苗蛮集团、戎狄集团等，以及当代的中华民族，也可以指一个具体的民族共同体，如汉族、壮族、蒙古族、维吾尔族、藏族等。另一方面，它还可以指民族内部的一个支系或民系，如汉族的客家人、广府人、东北人、陕北人等。周大鸣就说过："族群可以是一个民族亦可是一个民族中的次级群体，如汉族中的客家人、闽南人、广府人等；而民族一词无法包含这些内容。"[26]总之，族群的使用没有什么限制，虽然看起来似乎无所不指，无所不包，但在实际应用中，操作方便，一目了然，绝不会像民族概念那样容易产生歧义，引起不必要的概念争论。

族群概念的使用虽然有灵活方便的特点，但它也不能完全替代民族。民族共同体是历史上形成的，具有政治性质的，被赋予法律地位的一种人们共同体。如前所述，其与国家有密切的关系，故而用“族群”取代“民族”不可取，也无法实施。尤其在我们中国，每一个民族的地位都是经过国务院批准后确认的，并且有宪法和民族区域自治法给予法律保障，人们对此不能视而不见。我在1998年8月初访问美国华盛顿大学人类学系时，曾与郝瑞教授讨论这个问题，请美国学者在研究中国族群问题时，要充分考虑民族在中国的法律地位。

明确了族群与民族的区别，两者之间的联系也就容易理解了。简言之，一个族群可能是一个民族，也可能不是一个民族；而民族不仅可以称之为族群，还可以包含若干不同的族群。

注释

① 参见Robert H. Winthrop, Dictionary of Concepts in Cultural Anthropology, Greenwood Press, 1991: 94–95。转引自牟小磊硕士论文：《“中国少数民族”的族性过程与研究策略》(1997年打印本)。

② 参见美郝瑞：《民族、族群和族性》,《中国人类学会通讯》第196期。

③ 参见Robert H. Winthrop. Dictionary of Concepts in Cultural Anthropology, 1991: 94–95。转引自牟小磊硕士论文：《“中国少数民族”的族性过程与研究策略》(1997年打印本)。

④ 转引自M·G·史密斯：《美国的民族集团和民族性》,《民族译丛》1983年第6期。

⑤ Barth, F., Ethnic Groups And Boundaries. Waveland Press, Inc., 1996.转引自［挪威］弗雷德里克·巴斯著，高崇译、周大鸣校、李远龙复校：《族群与边界(序言)》,《广西民族学院学报》1999年第1期。

⑥ 参见Barfield, Thomas, ed., The Dictionary of Anthropology, Blackwell Publishers 1997。转引自周大鸣：《论族群与族群关系》,《广西民族学院学报》2001年第2期。

⑦ 参见Themstrom, Stephan, ed., Harvard Encyclopedia of American Ethnic Groups, 1980。转引自周大鸣：《论族群与族群关系》,《广西民族学院学报》2001年第2期。

⑧ 参见Nathan Glazer & Daniel P. Moynihan, Ethnicity Theory and Experience, Harvard University Press 1975。转引自周大鸣：《族群与文化论——都市人类学研究(上)》,《广西民族学院学报》1997年第2期。

⑨ 参见Stephen Cornell. The variable ties that bind: Content and circumstance in ethnic processes［J］. Ethnic and Racial Studies, 2010, 19(2): 265–289。

⑩ Seymour-Smith, Charloue Dictionary of Anthropology, Macmillan, 1986: 95.转引自孙九霞：《试论族群与族群认同》《中山大学学报》1998年第2期。

⑪ 参见Negata Judith. In Defense of Ethnic Boundaries: The Changing Myths and Charters of Malay Identity, in Ethnic Change, Charles F. Keyes, ed., Sharp Publishers Inc., 1987: 37。转引自牟小磊硕士论文:《"中国少数民族"的族性过程与研究策略》。

⑫ 参见Charles F. Keyes: Ethnic Adaptation and Identity: The Karen on the Thai Frontier With Burma, 1979。转引自乔健:《族群关系与文化咨询》,周星、王铭铭主编:《社会文化人类学讲演集》下,天津人民出版社1997年版,第483页。

⑬ 参见[美]郝瑞:《民族、族群和族性》,《中国人类学通讯》第196期。

⑭ 参见[科]穆罕默德·哈达德著、晓兵摘译:《科威特市的民族群体和民族等级结构》,《民族译丛》1992年第5期。

⑮ Marx Weber, The Ethnic Group, In Parsons and Shils Etal(eds), Theories of Society, Vol, 1, The Fiee Press, 1961: 306.转引自孙九霞:《试论族群与族群认同》,《中山大学学报》1998年第2期。

⑯ 孙九霞:《试论族群与族群认同》,《中山大学学报》1998年第2期。

⑰ 参见彭英明、徐杰舜:《从原始群到民族——人们共同体通论》,广西人民出版社1991年版,第272页。

⑱ 斯大林关于民族的定义是:"民族是人们在历史上形成的有共同语言、共同地域、共同经济生活以及表现于共同民族文化特点上的共同心理素质这四个基本特征的稳定的共同体。"见《斯大林全集》第11卷,人民出版社1955年版,第286页。

⑲ 彭英明,徐杰舜:《从原始群到民族——人们共同体通论》,第189页。

⑳ 恩格斯:《家庭、私有制和国家的起源》,《马克思恩格斯选集》第4卷,人民出版社1972年版,第107页。

㉑ 恩格斯:《家庭、私有制和国家的起源》,《马克思恩格斯选集》第4卷,第148页。

㉒ 恩格斯:《家庭、私有制和国家的起源》,《马克思恩格斯选集》第4卷,第2页。

㉓ 张敦安:《国家在民族形成中的作用探究》,《民族学研究》第8辑,民族出版社1987年版,第151页。

㉔ 参见周星:《民族政治学》,中国社会科学出版社1993年版,第88—92页。

㉕ 参见周星:《民族政治学》,第31页。

㉖ 周大鸣:《现代都市人类学》,中山大学出版社1997年版,第139页。

再论族群与民族*

徐杰舜
（中南民族大学人类学研究所）

摘要:“民族”一词在近代中国开始广泛使用之时，对其语义有种种解读。斯大林的民族定义虽然一统中国“民族”一词语义40年，但在中西语义的对接中不断地受到批判，从挑战发展到解构。20世纪90年代之后“族群”一词传入并强烈冲击着中国学术界和中国社会。“民族”一词的中西语义对接的争论，将从“族群”一词语义的争论中延续下去。

关键词: 族群　民族　冲击波　语义

大约6年前，笔者在当时的背景和条件下，初步研究了族群的概念后，在《民族研究》2002年第1期上发表了《论族群与民族》一文，给族群概念作了一个简明的界定，即“所谓族群，是对某些社会文化要素认同而自觉为我的一种社会实体”。强调了族群的文化性、民族的政治性。但今天面对争论和纷扰，觉得从历史文化语义学的角度还有几句话要说，故成此文，求教于方家。

一、近现代“民族”语义历史文化解读

“民族”一词在近现代中国有种种语义解读，大致分为以下三个时期。

1. 鸦片战争之后学者的解读。鸦片战争之后，西方学说涌入，受西方影响，国人对民族语义的解读也呈纷纭之彩。蔡元培认为：“凡种族之别，一曰血液，二曰风习。”[1]柳亚子认为：“凡是血裔、风俗、言语同的，是同民族；血裔、风俗、言语不同的，就是不同民族。”[2]汪精卫认为：“民族云者，人种学上

---

* 原载于《西北第二民族学院学报（哲学社会科学版）》2008年第2期。

用语也，其定义甚繁，今举所信者，曰：民族者同气类之继续的人类团体也。兹所云气类，其条件有六；一同血系（此最要件，然因移往婚姻，略减其例），二同语言文字，三同住所（自然而之地域），四同习惯，五同宗教（近世宗教信仰自由，略减其例），六同精神体质。此六者皆民族之要素也。”[3]梁启超则于1903年在翻译、介绍欧洲法学家伯伦奇里时以民族语义作评论，指出伯伦奇里论述了民族具有地域、血统、肢体形状、语言、文字、宗教、风俗和生计8个特征。

2. 孙中山的解读。1924年，孙中山在系统阐述三民主义的内涵时，对民族的语义作了解读，他说："英文中民族的名词是'哪逊'，'哪逊' 这一个词有两种解释：一是民族，一是国家。" 他认为自然力造成民族，霸道力造成国家；民族具有血统、生活、语言、宗教、风俗习惯等特征[4]。孙中山对民族语义的解读影响极大，被人们称为民族五要素说。在对民族的解读中，孙中山还受到西方民族——国家（nation-state）理论的影响，把对民族的解读与国家政治问题联系在一起。他把 "nation" 译为 "国族"，并明确指出，"民族主义就是国族主义，中国人最崇拜的是家族主义和宗族主义，所以中国只有家族主义，没有国族主义……中国人的团结力，只能及于宗族而止，还没有扩张到国族"[5]。在孙中山的解读中，并没有区分民族、种族、国族等词的语义和用法，他把民族当作种族下面的分支。"人类的分别，第一级是人种，有白色、黑色、红色、黄色、棕色五种之分。更由种细分，便有许多族。"[5]在造成民族的原因之中，他认为首要的是 "血统"。此观点极具代表性，当时的学术著作大都把民族作为种族下的分支加以论述，"ethnology"（民族学）也长期被译为 "人种学"。

3. 20世纪50—80年代由西方话语向苏联话语的转向。中华人民共和国成立后，由于受国际政治背景的影响，中国对 "民族" 一词的解读由西方话语向苏联话语转向，其标志就是斯大林的民族定义占据了民族语义的主导地位。其实早在20世纪30年代末，斯大林的民族定义已在中国共产党党内传播。1938年时任中共中央宣传部副部长的杨松在延安所作的 "关于民族殖民地问题" 系列讲座第一讲 "论民族" 中，基本上是按照斯大林的民族定义展开论述的。王明在1938年中共六中全会上的发言中也介绍了斯大林的民族定义[6]。此外，在当时的学术界，斯大林的民族定义也开始传播。周传斌撰文指出，1929年郭真所著《现代民族问题》中对斯大林民族定义的直接引用和翻译可能是斯大林民族定义最早的汉文翻译[7]，笔者很认同这一观点。中华人民共和国成立后，对民族一词的语义解读一边倒地按照苏联的话语展开，在斯大林民族定义的主导下，“民族主义”被界定为“资产阶级的民族观”，而“无产阶级的民族观”则是“国际主义”，西方式的“民族—国家”也被苏联式的“社会主义民族大家庭”所替换；对民族的特征则重语言、地域等客观特征，视血统、宗教与民族的本质特征无关。在这种语义解读的观照下，“民族”逐渐演变为专指与“少数民族”有关的事项，如民族问题、民族地区、民族事务委员会、民族研究、民族教育等。

二、对斯大林民族定义神话的解构

斯大林的民族定义虽然定型化了，一统中国“民族”一词语义40年，但在中西语义的对接中不断地受到批判，从挑战发展到了解构。

1. 范文澜对斯大林民族定义的挑战。1953年苏联学者格·叶菲莫夫在《论中国民族的形成》中以列宁、斯大林关于民族问题的论述为理论依据，认为汉族是19世纪下半期随着外国资本主义的入侵和本国资本主义的发展而形成的，此前的其他人们共同体都是“部族”，中国除汉族外再无“民族”，都是“部族”，就是汉族也是19世纪以后才形成的。无论从语义，还是从事实出发，中国学者对此都很难接受。尽管斯大林的民族定义正处在一边倒的风头上，范文澜还是以极大的理论勇气和巧妙的策略，于1954年发表《自秦汉以来中国成为统一国家的原因》提出汉族形成于秦汉的观点并认为汉族是一个独特的民族[8]，实际上是公开地对斯大林的民族定义作了挑战。

2. 牙含章、林耀华对斯大林民族定义的挑战。由汉民族形成问题的讨论所引起的关于民族形成的讨论，引起了“民族”一词中西语义的对接，先是牙含章，继而是林耀华从民族译名问题切入，向斯大林的民族定义提出了挑战。20世纪60年代牙含章任中国社科院民族研究所所长，为了弄清民族译名问题，他组织有关人员查阅了马克思、恩格斯、列宁、斯大林的原著，确认“民族”与“部族”之争主要是翻译问题引起的，马克思、恩格斯论述了由部落发展成民族和国家的一般规律，斯大林论述的则是“现代民族”即“资本主义民族”[9]。查清了这个情况后，1962年春天，由中国科学院哲学社会科学部和中共中央马恩列斯著作编译局联合召开座谈会专门讨论了经典著作中“民族”一词的译名统一问题，建议不再使用“部族”一词。这次讨论的结果发表在1962年6月14日的《人民日报》上。与此同时，牙含章发表了《关于“民族”

一词的使用和翻译情况》《关于民族的起源与形成问题》两篇文章，对此作了深入的论述。牙含章的挑战，使人们对斯大林民族定义从神圣性和刻板性中解放了出来。紧接着林耀华从中西语义对接的技术层面，对斯大林的民族定义进行了挑战。1963年林耀华在《历史研究》上发表《关于"民族"一词的使用和译名的问题》的长文，详细讨论了马克思、恩格斯、列宁、斯大林原著中使用的民族类词汇，并将俄文"民族"一词的词义归纳为四类:(1) 用以指最一般意义的人们共同体;(2) 指阶级社会产生以后的各个时代的共同体;(3) 指的是资本主义上升以后的现代民族;(4) 指与相对的前资本主义民族，以及没经过资本主义而直接过渡到社会主义阶段的民族[10]。林耀华的挑战，使得人们从技术层面上弄清了斯大林民族定义的刻板性，从而为人们进一步质疑斯大林的民族定义，并进而提出修订或否定斯大林的民族定义作了铺垫。

3. 对斯大林的民族定义的修订。经过范文澜、牙含章、林耀华等学者的挑战，到了20世纪80年代，斯大林民族定义一统民族语义天下的状况被打破了，不少学者对斯大林民族定义提出了修订的意见。在1986年4月23日至25日召开的"民族理论专题学术讨论会"上主张修改和另拟民族定义的学者提出了以下四个方案供研究，探讨:(1) 民族是人们于社会历史发展的各阶段，在语言、地域、文化诸方面共同因素的基础上形成的相对稳定的共同体;(2) 民族是在一定的地域和经济联系的基础上历史地形成的，具有统一的语言特点、文化特点、心理特点和自我意识的人群;(3) 民族是由若干基本要素或共同因素构成，其本质特征和运动规律寓于要素的构成以及诸要素的相互联系、相互制约和相互作用的基础上历史地形成的人们共同体;(4) 民族是人们在历史上由共同地域、共同语言、共同经济生活等要素所组成的共同文化传统和民族自我意识的比较稳定的共同体[11]。接着，曾强在《民族研究》1987年第1期发表《略论人们共同体的共性与个性》指出:"民族是历史上形成的人们共同体，由若干个基本要素或共同因素构成，由本质特征和运动规律寓于要素的构成以及诸要素间的相互联系、相互制约和相互作用的方式之中。"贺国安在《民族研究》1988年第5期发表《关于人们共同体和民族共同体的思考》指出:"民族是一些具有自我意识的语言文化共同体。"更有甚者，有的学者提出否定或抛弃斯大林的民族定义，有的更是按照自己的见解重新定义了民族。

4. 对斯大林民族定义的解构。"民族"一词在中国的定型化，建构了以斯大林民族定义为核心的一整套关于民族符号意义的系统结构。随着中国学者学术素养的提升，从对斯大林民族定义的修订，发展到了对民族定义的解构。首先对斯大林民族定义进行解构反思的是王联，他认为汉文“民族”的层次性对应着民族主义的不同表现层面，民族定义的争论实质上“都是出于定义者要维护他本人所属的那个民族的利益”[12]。接着马戎就民族概念本身的歧义性进行了探讨。他分析了英文文献中与“民族”相关而又常用的三个词：“ethnicity”、“race”和“nationality”，列举了14种具有代表性的观点，认为“要在‘民族’定义及其内涵方面形成共识，达成完全统一的认识，是非常不容易的”[13]。潘蛟则借用福柯的“权力”（power）观来解构民族概念，认为“有关民族概念的争论并不是单纯的认知问题，而是十分复杂的权力问题”，民族“是人们借以构建现代政治的核心概念，它的定义连带着对于‘他们’和‘我们’的界定，对于忠诚的预期，对于记忆的导引，对于历史的裁剪，对于屈辱和尊严的分配，对于自决、自治权利的承认和拒认，对于政治单元，国家的建构和解构”。这样，关于民族概念的论争，其实质不在于“民族到底是什么”，而在于“人们为何如此关心民族是什么”[14]。

三、族群：对接中的冲击波

20世纪90年代，“族群”一词的传入，犹如一股巨大的冲击波冲击着中国学术界，冲击着中国社会。族群（ethnic group）概念是西方人类学研究社会实体的一种范畴分类法。从语源学的角度看，ethnic源于希腊语，是经拉丁语进入英语系统的形容词。最初使用ethnic是在15世纪晚期，在英语世界里用以指称非基督教徒的各种族成员，是野蛮人和异教徒的代名词。在西方学术研究中，由于group包含着共同利益以及一定连带感的人们的意义，所以ethnic group可以视为ethnic的复指名词形式。对应于英语中nation一词具有“国家”、“国民”、“民族”的多义现象，在已有的描述非西方人群的学术文献中，ethnic group具有比nation较为下位的意义。对此，美国人类学家赫瑞（Steven Harrel）先生1996年9月在厦门大学人类学研究所作的《民族、族群和族性》的报告中曾明确地指出：要具有解释族性，应先区分民族（nation）与族群（ethnic group），英文nation是指有state（国家）或government（政府）的一个族群，含有国家和民族两层意思。而族群本身并不一定含有state（国家）或government（政府）的意义，它只是有意识、有认同的群体中的一种。在西方话语中，ethnic不指称具有明显政治优势地位的群体，这是一方面。另一方面，ethnic在希腊词源中的名词形式是ethnos。费孝通先生在《简述我的民族研究经历和思考》中曾说，ethnos是一个形成民族的过程，一个个的民族只是这个历史过程在一定时间空间的场合里呈现的一种人们共同体。从这个意义上来看，ethnic group这一概念范畴与nation相比，指涉外延较宽，更具动态性和灵活性。

但是，正如人们所知，族群的概念与文化的定义一样，在西方学术界也是多义的。早在1965年，日本学者涩谷和匡就将族群界定为"由于具有实际或虚构的共同祖先，因而自认为是同族并被他人认为是同族的一群人"[15]，这种意义上的族群具有共同的文化传统、相同的语言和相似的生活方式。1969年，挪威学者弗里德里克·巴斯认为："族群这个名称在人类学著作中一般理解为用以指（这样）一个群体：(1) 生物上具有极强的自我延续性；(2) 分享基本的文化价值，实现文化形式上的统一；(3) 形成交流和互动的领域；(4) 具有自我认同和他人认同的成员资格，以形成一种与其他具有同一秩序的类型不同的类型。"[16]此后，学者们从不同的视角，对族群作出界定，不胜枚举。

中国学者对族群的语义也作出了自己的界定。孙九霞认为，族群是在较大的社会文化体系中，由于客观上具有共同的渊源和文化，因此主观上自我认同并被其他群体所区分的一群人。其中共同的渊源是指世系、血统、体质的相似；共同的文化是指相似的语言、宗教、习俗等。并主张在较广的范围内使用族群定义，既可以等同于民族一词，也可指民族的下位集团"民系"，还可以在超出民族的外延上使用[17]。徐杰舜对族群的概括是：所谓族群，是对某些社会文化要素认同而自觉为我的一种社会实体。这个概念包括三种含义：一是对某些社会文化要素的认同；二是要对它"自觉为我"，三是一个社会实体[18]。纳日碧力戈认为，族群兼具"种族"、"语言"、"文化"含义，本质上是家族结构的象征性扩展，它继承了家族象征体系的核心部分，以默认或者隐喻的方式在族群乃至国家的层面上演练原本属于家族范围的象征仪式，并且通过构造各种有象征意义的设施加以巩固[19]。

就是这样一个"族群"的引入，一下子在中国学术界炸了锅，赞成者有之，主张取代"民族"，广泛应用族群，马戎为此写了一本洋洋50万字的专论族群的《民族社会学》。另外，我国一些重要的政府机构的名称也开始采用ethnic group来代替nationality这一固有表达法，如"国家民族事务委员会"中的“民族事务”就已从自20世纪50年代沿用至90年代末的Nationality Affairs改变为Ethnic Affairs，还有一些相关的行政机构和学术机构也正在酝酿更名事宜。否定者有之，他们反对使用“族群”概念或拒绝认为ethnic group指的就是“民族”。阮西湖就认为：“作为单词，group一词就是指民族”[20]。朱伦亦同此，他认为汉语中的民族与英文中的ethnic group并非对等的概念，不能将其等同起来，“族群”更多的是文化人类学意义上的概念。主张对中国各民族应坚持以nationalities而不应以ethnic group来界定，并对民族事务委员会的英名改译感到不安[21]。折衷者有之，他们承认“族群”概念的特定学术价值，但反对“泛族群化”，反对“复杂主义”，“族群”定义是以文化为基础的，“民族”实际上是一个十分中国化的概念，其内涵及外延在政治上、学术上及民间对话中，都已约定俗成，对于“民族”概念而言，文化不是其唯一的基准指标。因此，两个概念实际上是居于不同的层次。故而用“族群”取代“民族”或者认为两个概念可以兼用的观点是不恰当的，但可以相互补充，构成合理有效的研究网络[22]。反思者有之，他们认为“族群”概念的背后是弥漫在全球范围的西方话语霸权，是一场从一开始就注定不平等的全球对话。一些学者对当今西方国家借助其强大的政治、经济实力在全球范围内推广其文化价值观的战略行为进行了思考，认为“族群”概念及其相关的学术之争也是这样一个过程的体现。

笔者认为还是马戎谈得较合理。马戎的观点虽然得到不少学者的共鸣，却得不到有关政府部门人士的认同。可见“族群”一词虽强调的是人类共同体的文化性，但一旦与政治有了联系，发生了影响，也就因其本身的“政治化”而复杂化了。这表明“民族”一词的中西语义对接的争论，将从“族群”一词语义的争论中延续下去。

参考文献

[1] 李毅夫.汉文“民族”一词的出现及其初期使用情况(1903年)[A].蔡元培选集[M].北京：中华书局，1959.
[2] 柳亚子.民权主义、民族主义[J].复报，1907(9).
[3] 精卫(汪兆铭).民族的国民[J].民报，1905(1～2).
[4] 孙中山.民族主义[A].孙中山选集[M].北京：人民出版社，1981.
[5] 孙中山.民族主义[A].民权与国族——孙中山文选[M].上海：上海远东出版社，1994.

[6] 王明. 目前抗战形势与如何坚持战斗争取最后胜利[A].1938年10月20日在中共六届六中全会上的发言提纲[Z]. 北京：中央档案馆.
[7] 周传斌. 论中国特色的民族概念[J]. 广西民族研究，2003(4)：19–30.
[8] 范文澜. 自秦汉以来中国成为统一国家的原因[J]. 历史研究，1954(3).
[9] 牙含章，孙青. 建国以来民族理论战线的一场论战——从汉民族形成问题谈起[J]. 民族研究，1979(2)：3–8.
[10] 林耀华. 关于“民族”一词的使用和译名的问题[J]. 历史研究，1963(2)：171–190.
[11] 纪闻. 民族和民族问题的争鸣与新探索[J]. 民族研究，1986(4)：2–8.
[12] 王联. 关于民族和民族主义的理论[J]. 世界民族，1999(1)：2–12.
[13] 马戎. 关于“民族”定义[J]. 云南民族学院学报，2000(1)：5–13.
[14] 周旭芳. “1998年‘民族’概念暨相关理论问题专题讨论会”综述[J]. 世界民族，1999(1)：78–81.
[15] M·G·史密斯，何宁. 美国的民族集团和民族性——哈佛的观点[J]. 民族译丛，1983(6)：4–19.
[16] 弗里德里克·巴斯，高崇，周大鸣，等. 族群与边界[J]. 广西民族学院学报（哲学社会科学版），1999(1)：21–32.
[17] 孙九霞. 试论族群与族群认同[J]. 中山大学学报（社会科学版），1998(2)：24–31.
[18] 徐杰舜. 论族群与民族[J]. 民族研究，2002(1)：12–18+106.
[19] 纳日碧力戈. 民族与民族概念再辨正[J]. 民族研究，1995(3)：9–16.
[20] 阮西湖. 关于术语“族群”[J]. 世界民族，1998(2)：81.
[21] 朱伦. 浅议当代资本主义多民族国家的民族政治建设[J]. 世界民族，1996(2)：8–17.
[22] 乔玉光. 民族与族群：不等位的判断价值[J]. 中央民族大学学报，2003(4)：36–38.

第六章

▼

沧海横流南岛情

2022年人类学终身成就奖获得者
——邓晓华

人类学生涯回顾

"嫋嫋兮秋风,洞庭波兮木叶下"。

时光冉冉,岁月如梭,不知不觉中我也到了人生的"秋叶"时节。回顾来路,我常在学术的孤独与愉悦中前行。聊以自慰的是,我的研究兴趣总使我在学术的前沿中探索。数十年间,身处这个风云际会的大时代,难免有未尽人意处,不过总能"无欲自然心似水",超脱坦然地面对各种境遇。于学术,于人生,确是"我见青山多妩媚,料青山见我应如是"。

回首学术之路,能有今日之成果,我特别感激和怀念那些对我有过帮助的老师和朋友们!

我的童年

我出生于1957年2月,父母是新中国基层公社干部,虽不能说生活富裕,但比起那个时代我的许多同学和亲戚的孩子来说,已经是衣食无忧了。我的整个青少年阶段都在家乡度过,先后在闽西的连城县文亨中心小学、连城一中求学,罗坊知青场插队。故乡的文化对我早期人生影响甚大,特别是建筑、景观、仪式和方言等文化形态,是一个古老文化最重要的元素。年少时,跟随父母,先后住过文陂和罗坊的"老公社",即客家的著名建筑"九厅十八井"。(虽然它的主人是大地主,不过我一向以为,地主也是中国文化的重要代表,从古代孔、孟、李、杜,到近代闻一多、朱自清等。孔孟为代表的礼教为乱世所不容,但为盛世所推崇。)在各地生活使我会讲多种方言土语。

记得儿时的文陂四周千年古樟、古松林立,小桥流水、古塔古寺交相辉映。如古人云"所谓故国者,有乔木之谓也",即一个有着悠久文化历史的地方,必定有高大的树木如樟树,这就是我对早期故乡的记忆。记得我小学基本没读书,看大人武斗、跟小伙伴爬树、砍柴、摸鱼、到附近部队机场看电影……那就是每天的作业。记得父辈们常说的就是打土匪。1950年国民党派了一个少将特务，竟然在闽西拉起万人队伍与新政权为敌，许多农民被胁迫为匪，最后的结果是一个9万人的小县被镇压了数以千计的土匪。客家人的反叛与革命特点一样突出。儿时的教育就是亲近自然与民间政治。我的初中、高中在连城一中完成，感谢我的老师傅干春先生、饶慧群先生、陈焕南先生、陈元琪先生对我的教诲。当时的野营拉练训练了我的生存能力和集体思想；我家楼下的废品收购站里收购的书籍培养了我的阅读习惯；同学吴新成的传抄名人语录激发了我的力争上进的精神；而参与政治议题写大字报则训练了我的写作能力。

知青生活

1976年在龙岩市连城县罗坊公社的知青生活

1974年6月，我下乡当知青到了罗坊公社，就几乎完全没有读书了。因为父母原因也没挨饿过。两年多的知青生活，我担任过手扶拖拉机手、武装基干民兵排长、司务长。生活虽艰苦，但心情愉悦充实，比较深刻地体验了中国农民的艰辛。受时代狂潮的影响，对过去的乡绅、新社会的敌人“地富反坏右”有过过激的行为，虽未造成后果，但于心戚戚焉久矣。其时追求上进是我一直坚持的信念，我们与时代同呼吸、共命运。连城是我的故乡，对于故乡的情感，虽然不似鲁迅笔下对他自己故乡情感的描述“冷、冷、冷”，然而每次回乡，面对着这个建于南宋、盛于明清的客家古城的凌乱街市，难免悲情伤怀，常思故乡文化“风流已被雨打风吹去”，感叹地方政治之纷扰与现代工业化之草率对一个古老文化的伤害之甚。

初识语言学

1976年10月，我被招工到龙岩招待所当服务员，也曾接待服务了一些名人，如齐燕铭、江华、项南、李敏等人。1977年我参加高考，当时正在龙岩招待所喂猪，以候补生资格被扩招就读于龙岩师范（龙岩学院前身）大专班。仅仅读了1年零3个月，承蒙母校厚爱留校任教。我无任何关系，留校是因为校长赵澄清先生坚持以分数高低排名录用，不讲情面。我有一门课“现代文学”竟然考了班上最高分。后来我留校整理学生档案，发现我高考分数高于许多同学，才知道当年高考并未完全按分数录取，有工作者、多兄弟考取者、年纪大者、家庭成分不好者，都要视情让位于其他考生，或降级录取。我因已有工作，故降取。77级大学生堪称世界教育史上的奇迹，这批人普遍学养基础不高，却成为40年来推动中国进步的中流砥柱。我们同学经历“文革”苦难岁月，虽过去读书不多，但大多较为聪慧，许多人后来成了龙岩老区各界的领导人物，这也印证了西方主流大学校长对大学教育目标的共识——大学的根本就是启蒙和启智，而非仅限于传授具体的知识。当时龙岩师专聚集了一批极其优秀的教师，如郭启熹老师、王展彩老师、郭义山老师、简启梅老师、雷丰畴老师、赖丹老师、吴瑞裘老师、刘宗涛老师、黄今许老师、沈天民老师，他们基本上都是“文革”前福建各个大学的老师，因为“文革”，下放返乡任教。他们是那个年代的学术精英，都是闽西大地哺育出来的优秀儿女。校长是新四军的老革命赵澄清先生，我们有幸接受他们的教诲，拳拳感恩之情难以言表！在龙岩师专任教期间，我下乡时的公社书记，时任龙岩团地委书记李天喜同志曾推荐我任团地委副书记，后来曾任福建省人大常委会副主任袁锦贵同志多次来师专商调，均被校长李逢蕊拒绝。李校长爱才出名，曾是上海地下党大学生领袖，也是上海博物馆馆长马承源的同学，多年后他见到我还自诩他当年的正确决定。

1981年暑假，我参加由中国语言学会在武夷山下的建瓯古城主办的暑期汉语方言研究班。当时给我们上课的先生有：李如龙、黄典诚、许宝华、黄家教、吴宗济、王福堂等著名语言学家。虽仅月余时间，这个宁静的闽北古城的学习生活给我留下了深刻印象，此时为初识语言学。该班结束时，许宝华先生写信推荐我参加当年9月由先师严学宭主办的华中工学院（华中科技大学前身）音韵学高级研究班，我至今很感激许先生的荐举。彼时严老是华中工学院顾问、中南民院副院长、中国音韵学会会长，以他老人家良好的人脉和优势资源，音韵班云集了国内高校一批优秀的古汉语骨干教师，聘请的授课先生都是国内最杰出的语言学家。其中，严老讲语言学方法论，孙宏开老师讲国际音标和羌语记音，赵诚老师讲甲骨文形音义以及商代音系，李新魁老师讲等韵学，瞿霭堂老师讲汉藏语比较，授课设计完全体现了严老的语言学理念，即注重从古代文献、汉语方言、少数民族语言三方面结合研究中国语言学，这种思想至今仍是很先进的。音韵班每天上下午上课，晚上自习，历时三个月，我们接受了传统语言学、语言调查、历史比较法的严格训练，该班为后来的学术界培养了一批优秀的语言学家。

考入华中工学院开启学术生涯

1984年我考上了严老的研究生，攻读汉语史。其间印象较深的是，1985年6月美国华盛顿大学罗杰瑞先生应邀从北大到华工讲学，专门给我讲了几次有关古闽语构拟的课，由此我对西方比较语言学有了更深刻认识。我后来出版《人类文化语言学》分别请严老、罗杰瑞先生、李新魁先生作序，反映了我这一阶段的学术训练背景。我的硕士论文是《闽西客话的音韵研究》，这篇硕士论文后来分成几篇论文发表，《中国语言学年鉴》几次给予好评，还得了省社科奖。等到20年后，我在恩师尉迟治平的帮助和指导下，重返母校获得博士学位。

1985年与恩师严学宭教授及师母合影

严学宭，字子君，我国著名的语言学家，明朝严嵩之后，其父是民初江西“状元”。严老是我国现代语言学开创人罗常培先生1935年在北大时的研究生，创建了中山大学、华中科技大学语言学科，是中南民大的创院院长、中国音韵学会创会会长。我整个青少年时期教育缺失，没读过什么书，多亏遇上严老指导，才有幸走上学术之路，他教会了我如何选题、组织材料写论文，如何创新。我深深地怀念他！

在厦门大学的三十年岁月

求学和治学是需要经济基础的，如达尔文的贵族身份帮助他完成进化论研究。早年严老曾跟我谈起，他1935年读北大研究生时，助学金可以租一座四合院、养一家妻小外，还有余钱旅游。我家境还好，大学及研究生有足够助学金，无需养家。到厦门大学开始几年很辛苦，月薪100多元，养家不够。1990年前后在市外资委工作的高中同学吴新成介绍我到台资企业兼职，每月有大几百元额外收入，总算渡过难关了。这里要特别感谢我的妻子李小芬对我的支持和帮助！她除了担任厦门大学外文学院繁重的教学任务外，还要操持家务，很不容易。1991年我参加了台湾“中研院”李亦园院士与厦门大学的合作项目，有一些美元支持，就此退出兼职回归学术了。可是，我的一些很有才华的好友出去浪迹江湖后，未及时回归，叹息终老，抱恨悠悠，真是人生难测。

1987年6月，我通过同学安钢联系到厦门大学人类学系工作，当时系主任陈国强先生欣然接受。陈先生故去多年，我至今怀念他。我开始给考古学专业和人类学专业学生讲授“汉学导读”和“语言人类学”等课程，并担任1988级人类学专业班主任，带学生参加考古学及人类学野外实习，开始接触和学习人类学和考古学知识。记得当时我跟吴诗池、吴春明老师一起带人类学、考古学学生到江西广丰蛇山头一处新石器遗址搞发掘。江西考古所徐长青教授是系友，他负责帮助，他后来做了海昏侯遗址发掘总领队。野外很辛苦，冬天也在河里洗澡，女生在上游洗，男生在下游洗。我带了这届人类学学生跟郭志超、董建辉老师一起到南靖塔下村做田野，最担心的就是怕女生出事。我看不惯村书记的作风，志超兄还开导我做人要学习神仙、老虎、狗。这个时期是厦门大学人类学最鼎盛的时候，人类学系包括了人类学传统4个分支，即考古学、体质人类学、文化人类学/民族学、语言学的完整建制，陈国强老师、

2000年在“21世纪人类生存与发展”国际学术会议上作报告，后排为费老和陈国强教授

蒋炳钊老师、吴绵吉老师、叶文程老师都是当时国内顶级的著名学者，培养了一批杰出的人才，如果当时发展下去，厦门大学人类学可能成为亚洲乃至世界的重要人类学基地之一。可惜历史没有假如，当时的厦门大学领导撤系肢解了人类学专业，错过了最重要的发展时期。接下来知名学者退休，团队弱化，又错过了教育部人文社科基地平台的申报。这种局面一直到费孝通先生的到来才被打破。2000年我联系到杰出系友北大王铭铭教授，请他出面邀请人大常委会副委员长费孝通先生来厦门参加中国人类学学会与厦门市政府合办的“21世纪人类生存与发展”国际会议。费老专程来厦门参加会议，并全程参加了5天会议。一批国内外著名学者与会。费老与省、市领导和厦门大学领导共商人类学发展。为表示对费老的感谢，我请著名篆刻家人类博物馆王守桢先生刻了一枚印章送给费老，费老很高兴。至此，学校开始重新审视人类学的学科发展问题。

我在人类学系的学术收获：认识到考古学跟语言学都具有共同的理论和方法论基础，即重视区域视野、类型学比较、层次分析以及域外文化因素移植。通过人类学与民族学的田野调查认识到整体性的重要意义与语言学本体结构分析的不足。按照王士元先生说法，语言是个复杂系统，其变化与人类生理、认知、文化因素、社会网络关系极其密切。语言史的研究必须结合族群史，以汉语史为中心的历史观必须重新审视。

我有10余年时间负责厦门大学人类博物馆工作。其间主持了2次厦门大学校庆人类博物馆大修及布展，虽然责任重大，但由于工作关系，我对博物馆以及文化遗产专业有了切实了解和认知。同时，我长期担任国家一级学会中国人类学学会法人代表及秘书长，常常组织协调各种人类学学术活动，这些工作经历让我具备相关学科的知识背景，从多学科视野来审视语言学研究比起仅仅单一学科的角度来说，应该更具备学术优势和特点。至今为止，我在厦门大学语言学、人类学与民族学、考古学招收博士生，也曾是这几个博士点的学术带头人。这种长期以来的多学科的学术积累和熏陶，给我带来许多学术乐趣，我也特别愿意给年轻朋友介绍多学科研究的优点和乐趣。

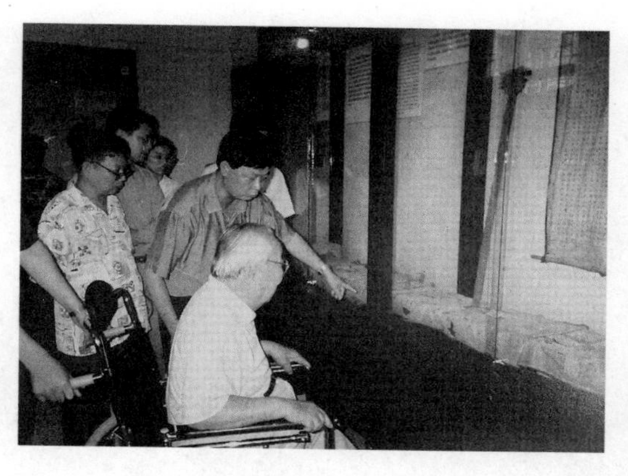

2000年费老视察厦门大学人类学博物馆

与国际学术界交流是学术创新的动力，记得先师严老多次训导，学术是在前人的肩膀上前进的，不是独自的拓荒。我1990年代先后参加美国斯坦福大学Arthur P. Wolf教授，台北“中研院”李亦园院士、庄英章教授和厦门大学合作的“台湾与福建社会文化比较研究”项目；与日本东京外国语大学亚非语言文化研究所三尾裕子教授合作“福建民间信仰研究”项目，调查闽南“王爷”信仰；与东京都立大学著名宗教人类学家佐佐木宏干教授合作调查闽南“萨满”信仰；应台湾姓氏渊源研究会会长林瑶祺医师邀请赴台湾调查客家文化2个月，我很感谢林先生的资助，让我有机会体验台湾社会的生活。2000—2001年经好友日本文教大学潘宏立教授介绍，我应日本国立民族学博物馆田村克己教授邀请，担任民博客座教授一年。2012年时任国立民族学博物馆副馆长的田村克己教授再次邀请我任民博客座教授半年。日本国立民族学博物馆是国际上最重要的人类学研究基地之一，其研究实力堪称亚洲第一。我在民博更新了知识，度过了难忘的美好时光，我很感谢田村老师、韩敏老师、宏立兄的热情帮助。2001—2003年我应王士元老师邀请到他在香港城市大学的语言工程实验室工作，并任城大语言学研究所客座研究员，参加由王士元院士、王明珂院士主持的“中研院”主题研究计划“中国民族的起源”。在王士元老师的指导下，我开始学习如何使用现代科技手段，处理语言数据，并尝试从遗传学、考古学、语言学这三个窗口研究语言学，由此产出了一系列的研究成果。香港的生活愉快而难忘！感谢王士元院士、郑锦全院士、邹嘉彦院士和同学潘海华教授给予的许多关照和帮助。

探索语言人类学新境界

我的研究特点是长期使用语言学、人类学与考古学等多学科结合的方

1997 年在台湾地区进行语言调查（左 1 为本人）

1997 年开展客家田野调查

2000 年在日本国立民族学博物馆任客座教授时与田村克己教授夫妇及老友日本文教大学潘宏立教授合影（左 1 为本人）

2000 年在日本国立民族学博物馆作学术报告

法，研究成果受到国内外同行的高度重视和赞扬。承担多项国家社科重大、重点和一般项目，相关研究成果获得国家以及省部级学术奖励，其中关于使用自然科学方法研究中国语言及方言分类以及人类学的研究成果达到了一流研究水平，而关于客家话的多元结构来源，以及客畲族群性差异的语言人类学分析的研究结论，具有独创性意义。

在语言与族群的关系分类研究、南岛语族的起源及形成研究、南方土著语言的底层研究、语言与族群分类的计量模型研究、华南族群的语言及方言的分区研究具有创新性意义，取得了具有特色的研究成果。其特色如下：

（1）把分子生物学中的系统发生方法应用于处理语言数据。这种方法论上的创新在于采取了定量的方法研究语言的分类，国际权威刊物Journal of Chinese Linguistics（美国加州大学伯克利分校出版、SSCI\A&HSCI收录）2010年第2期发表著名语言学家澳大利亚墨尔本大学Yongxian Luo教授对我的著作《中国的语言及方言的分类》的长篇英文书评，给予高度评价。认为“几十年来，中国大陆的语言学研究一直失于理论和分析的精确性。自20世纪80年代开始，新的语言数据的流入，使学界意识到应该重新评估传统的中国语言分类理论。这部书的出版给中国语言学研究领域带来了崭新的理念和方法，必将被看作中国语言学的一个里程碑，将改变中国语言学的面貌。该书从跨学科的视角呈现中国的语言和方言分类的创新研究。该书在处理中国语言和方言分类的未决问题方面作了全新的尝试，是作者近年研究的集大成。作者重新审视了传统上被视作汉藏语言的汉语、藏缅语、苗瑶语、卡岱语以及其他一些语言间的发生学关系，最值得瞩目的是，该研究把用于分子生物学的系统发生方法应用于处理语言数据。这种方法论上的创新在于采取了定量的方法研究语言的分类，尤其是把Felsenstein于1990年 和Cavalli-Sforza等于1994年开发的距离-特征法与标准的词源统计法相结合。迄今为止，只有少数学者把这样的方法用于自然语言（Gray 2000, Gray&Atkinson 2003）。”

“该书是中国语言学历史上具有里程碑意义的一部著作，不仅进一步增进了对东亚和东南亚语言环境的理解，对汉藏语系当前和未来的历史比较研究和类型学研究也具有深远的影响，对于研究亚洲和东南亚的人类学、民族学、历史学和语言学的学者来说，也是不可或缺的主要信息资料。”

（2）在南岛语族的起源及形成研究领域，受到学术界好评。例如考古学家焦天龙认为：中国语言学家最近的研究使南岛语族起源及扩散这一问题有了重大突破。邓晓华等语言学者的研究表明，在当今的闽南方言中，存在着相当多的南岛语系词汇，并进而推论南岛语是福建史前及上古时代先民的语言；邓晓华、王士元进一步挑战西方学者关于南岛语系单向由台湾向太平洋地区扩散的说法，认为南岛语在东南沿海形成后，至少有两个扩散方向：其一是东南沿海经云南和东南亚岛屿，然后再到台湾；其二是由东南沿海直接传到台湾。目前分布在中国西南地区的壮侗语言与南岛语的关系远比与汉藏语系的关系更密切，应该是由南岛语分化出来的。而邓晓华关于闽方言中存在“南岛语系底层”的观点，是历史语言学关于南岛语系发源的重要发现（焦天龙，2010）。

（3）注重汉语方言与南方少数民族语言的比较研究，主张多元互动发展的民族语言进化观点，与传统的“汉族中心地位”的观点不同，强调“非汉语”的异文化比较。这方面的研究成果还体现在长期以来与国际著名语言学家王士元院士合作的“中国南方濒危语言研究”、“中国濒危语言预测模型研究”等多项研究计划及其学术成果。民族语言以及濒危语言的数理比较研究在理论和方法上做出了贡献，具有学科前沿性特点。丰富和完善了语言与民族（族群）分类的不同知识体系，更正了原有的族群分类认识，其结论已为民族学界重视，其理论和方法对我国族群理论的建立和发展具有重要的推动作用（王明珂，2013）。

（4）首次从语言学角度，系统论证客家话的形成年代（1988，1991），为客家族群的建构提供参照；首次系统地从语言学角度，论证畲族的起源与形成以及与客家及苗瑶、壮侗之间的关系（1998，1999）。提出与传统学术界认为客家和畲族形成是单线演化结果所不同的观点，认为客家和畲族的形成是多元互动演化的结果。

（5）注重语言人类学理论与方法的研究和学科的发展，建构具有多学科视野的语言人类学理论体系（1993）。以大量研究实践，强调以系统的历史地理及文化演化为参照点的语言人类学方法，对传统语言学、民族学的单一历史类型的比较，具有方法论的突破意义。从传统来看，语言学与人类学关系密不可分，美国人类学大师博厄斯研究印第安人文化，其《美洲印第安人语言手册》是语言民族志与描写语言学的开山之作，创建了语言人类学。2009 年受老友美国加州大学洛杉矶分校东亚系著名语言学家陶红印教授邀请，我在该校访问时，有意调研语言人类学与语言学的关系，发现美国语言人类学的刊物内容与其他语言学专业刊物几无差异。日本国立民族学博物馆研究结构沿袭美国人类学四大分科传统，即语言、考古、生物和文化人类学，该馆聚集了杰出的语言学家。总的来说，语言人类学的研究方法及研究目标重视人类学研究取向。

2003 年与日本小熊诚教授一行在南靖客家土楼做田野调查

2004 年在上海博物馆参会

2004 年与吴新智院士合影

2004 年在宁夏进行田野调查

2004 年在宁夏考察贺兰山岩画

2005 年在广东潮州凤凰山调查畲族

2008 年在英国访学

2009 年在美国加州大学洛杉矶分校访学

2012 年与孙宏开先生参会合照

2012 年第二次任日本国立民族学博物馆客座教授与田村克己教授合影

2012 年在西藏田野调查

2013 年与日本人类学家一起（左起韩敏、田村克己、横山广子、野林厚志）

2015年与王士元院士、沈钟伟教授在厦门调研考察

2018年在德国访学

代表性论著

［1］邓晓华.语言、族群与演化：语言人类学的传统与超越［M］.北京：商务印书馆，2019.

［2］邓晓华，王士元.中国的语言及方言的分类［M］.北京：中华书局，2009.

［3］邓晓华.人类文化语言学［M］.厦门：厦门大学出版社，1993.

［4］杨文姣，邓晓华，王传超.多学科视域下的侗台语和南岛语亲缘关系探究［J］.广西民族大学学报（哲学社会科学版），2022，4：30–41.

［5］范志泉，邓晓华，王传超.语言与基因：论南岛语族的起源与扩散［J］.学术月刊，2018，10：175–184.（人大复印资料2019年第2期全文转载）

［6］邓晓华，童芳华.类标记的跨语言对比［J］.山西大学学报，2018，3：60–68.

［7］邓晓华，杨晓霞，高天俊.试论语言演化网络——以藏缅语为例［J］.语言研究，2015，3：12–19.

［8］邓晓华，高天俊.演化语言学的理论、方法与实践［J］.山西大学学报，2014，37(2)：72–75.

［9］邓晓华，高天俊.语言研究新视野：演化语言学［J］.厦门大学学报（哲学社会科学版），2014，2：28–39.

［10］Deng X H, Deng X L. Cognitive models obtained by studying body-part names of Hakka and She［J］. Journal of Chinese Linguistics, 2013, 41(2): 359–391.

［11］邓晓华，邓晓玲.论壮侗语和南岛语的发生学关系［J］.语言研究，2011，4：34–41.

［12］邓晓华.对中国南部地区的濒危语言的消亡时间、速度和方向的模仿预测［M］//郑培凯，鄢秀.文化认同与语言焦虑.桂林：广西师范大学出版社，2009.

［13］邓晓华，王士元.壮侗语族语言的数理分类及其时间深度［J］.中国语文，2007，6：536–548.

[14] 邓晓华.闽客方言一些核心词的“本字”的来源[J].语言研究,2006,26(1): 85–89.
[15] 邓晓华.论客家话的来源[J].云南民族大学学报,2006,23(4): 143–146.
[16] 邓晓华.论客家话[J]//台湾“中研院”语言学研究所.语言暨语言学(专刊之三),2005.
[17] 邓晓华.中国境内的语言关系与语言分类[J].国家社科基金项目成果选介汇编(第一辑).北京：全国哲学社会科学工作办公室,2005.
[18] Deng X H, Yang Y. Lineage society and folk religion—An anthropological study of the folk religion at Three Hakka Villages[M]//三 尾 裕 子.The revitalization of folk culture in the coastal areas of East Asia: An anthropological study. 東京：株式会社風響社,2005.
[19] 邓晓华,王士元.古闽客方言的来源及其历史层次问题[J].古汉语研究2003,3: 8–12.
[20] 邓晓华,王士元.藏缅语族语言的数理分类及其分析[J].民族语文,2003,4: 8–18.
[21] 王士元,邓晓华.苗瑶语族语言亲缘关系的计量研究[J].中国语文,2003,3: 253–263.
[22] 邓晓华.从林惠祥越语“马来”说假设谈南方汉语的发生[M]//汪毅夫，郭志超.记念林惠祥文集.厦门：厦门大学出版社,2001：122–136.
[23] 邓晓华.古百越语言在今闽客族群中的遗存//蒋炳钊.百越民族研究.长春：吉林人民出版社,2001：156–186.
[24] 邓晓华.关于南方汉语非“汉”说//陈支平.林惠祥教授诞辰一百周年纪念论文集.厦门：厦门大学出版社,2001：70–87.
[25] 邓晓华.福建族群关系的比较研究[M]//周大鸣.中国的族群与族群关系.南宁：广西民族出版社,2001：192–211.
[26] 邓晓华.试论南中国汉人及汉语的来源[J].日本国立民族学博物馆调查报告,2001.
[27] 邓晓华.试论古南方汉语的形成[J].古汉语研究，2000，3：1–10.(人大复印资料2001年第1期全文转载)
[28] 邓晓华.客家话与畲语及苗瑶语、壮侗语的关系[J].民族语文，1999，3：42–50.(人大复印资料1999年第6期全文转载)
[29] 邓晓华.考古学文化区系理论与古南方汉语假说[J].东南考古，1999，1：26–36.
[30] 邓晓华.福建境内的闽、客族群及畲族的语言文化关系比较[J].日本国立民族学博物馆研究报告，1999.
[31] 邓晓华.论闽客族群方言文化研究中的几个问题//庄英章.华南农村社会文化研究论文集.台北：台湾“中研院”民族学研究所，1998.
[32] 邓晓华.客家话与赣语及闽语的比较[J].语文研究，1998，3：48–52.
[33] 邓晓华.论客家方言的断代证据及其相关音韵特征[J].厦门大学学报，1997，4：101–106.
[34] 邓晓华.客家形成时代的语言学证据[J].客家学研究，1997：118–122.
[35] 邓晓华.闽南文化中的古南岛语文化底层的证据[J].台湾源流，1996，3：33–38.
[36] 邓晓华.客家方言的词汇特点[J].语言研究，1996，2：88–95.
[37] 邓晓华.南方汉语中的古南岛语成分[J].民族语文，1994，3：36–41.(人大复印资料1994年第6期全文转载)

[38] 邓晓华.闽客若干文化特征的比较研究[J]//庄英章，潘英海.台湾与福建社会文化研究论文集(二).台北：台湾中研院民族学研究所，1995：265-285.
[39] 邓晓华.古音构拟与方言特别语音现象研究[J].语文研究，1993，4：66-78.(人大复印资料1994年第1期全文转载)
[40] 邓晓华.从语言推论壮侗语族与南岛语系的史前文化关系[J].语言研究，1992，1：110-122.(全文收入《中国语言人类学百年文选》，北京：知识产权出版社，2008)
[41] 邓晓华.客家方言与宋代音韵[J].汉语言学国际学术研讨会论文集·语言研究增刊.武汉：华中理工大学语言研究所，1991：74-79.
[42] 邓晓华.闽西客话韵母的音韵特点及其演变[J].语言研究，1988，1：75-98.
[43] 邓晓华.闽语时代层次的语音证据[J].华中工学院研究生学报，1986，2；又刊《语言研究》1998年增刊.

主持主要的项目

[1] 国家社科基金重大项目2项(多学科视角下的南岛语族的起源与形成，在研，2021—2025；中华南方民族的起源及形成，2005—2009)
[2] 国家社科基金重点项目(台湾原住民问题研究，2011—2018)
[3] 国家社科基金一般项目(汉语方言与语言接触理论的研究，1998—2001；汉藏语系谱系分类及其时间深度的研究，2010—2013)
[4] 教育部项目4项(基于汉藏语同源词数据库的语言数理分类及语言年代学研究；汉语方言研究与语言演变理论的建构；人类学数字博物馆；大学数字博物馆技术升级和应用示范)
[5] 中国科协重点项目(土楼客家人的生存技术专题虚拟博物馆，2009)
[6] 福建省社科基金重大项目(世界文化遗产客家土楼的保护、开发和利用研究，2011)
[7] 福建省软科学重点项目(海峡西岸河口区域城乡开发与海洋永续利用的模式研究，2010—2013)

以第一作者获得的主要获奖成果

先后获教育部高等学校科学研究(人文社会科学)优秀成果奖
国家民委优秀社会科学成果奖
福建省政府优秀社科成果奖(其中二等奖2项，三等奖3项)
联合国教科文组织2007世界信息峰会大奖
福建省优秀博士论文奖(指导老师，2018语言学)

主要海外研究经历

日本国立民族学博物馆客座教授(2000.9—2001.8；2012.12—2013.6)
香港城市大学语言学研究所客座研究员(2001.9—2003.9)
美国加州大学洛杉矶分校亚洲系访问教授(2009.1—2009.3)

代表作

壮侗语族语言的数理分类及其时间深度*

邓晓华[1]　王士元[2]
（1. 厦门大学；2. 香港城市大学）

摘要： 本文运用词源统计分析法，对壮侗语族语言做出数理分类以及亲缘关系程度的描述，并通过树枝长短来表示距离关系，显示壮侗语族语言的类簇和分级层次。同时计算出壮侗语族诸语言的时间深度，并分析其形成过程。以新的资料和方法，质疑传统的语言分类理论和方法，并试图提出新的东亚语言区域形成的理论解释。

关键词： 壮侗语族　数理分类　树形图　亲缘关系　词源统计分析法

一、壮侗语族语言的传统分类

（一）诸家的分类

1937年，李方桂（Li, 1973）分侗台语族（壮侗语族Kam-Tai①）为两大语支，即台语支和侗水语支。

20世纪50年代，罗常培、傅懋勣（1954）分中国境内的壮侗语族为3个语支②。即壮傣语支、侗水语支、黎语支。罗、傅的分类近似李。

《中国语言地图集》（1988）（以下简称地图）则沿袭罗、傅的分类，分壮侗语族14种语言为3个语支。

壮傣语支：壮语，布依语，傣语，临高话。

侗水语支：侗语，水语，仫佬语，毛南语，佯僙语，莫语，拉珈语。

黎语支：黎语和村话。

此外，仫佬语是否作为语支未定。

---

* 原载于《中国语文》2007年第6期（总第321期）。

梁敏、张均如(1996)的分类与《地图》相似,只是将仫佬语支作为独立语支单立。

本尼迪克特(Benedict, 1990)有关壮侗语的分类跟别家不同处是:分别将仫佬、黎语、临高(Be)在三个不同的层次上独立,他运用的是树图的每个分叉点上的二分法,不同于李方桂等人的三分或四分法。本尼迪克特同时认为他的"澳台语系"跟"南亚语系"有一定的底层上的联系。

壮侗语专家W. J. Gedney(1993)将壮侗语族三分,即泰、侗水、黎和临高。其特点是将黎语、临高(Be)单立为一支。

(二)分类的标准

历史语言学认为,语言的分类主要依据语言的相似性特征。语言的相似性特征可以有四种解释:① 语言平行演变的结果,由于语言发展的普遍性特征导致不同语言的相似性特征;② 由于语言演变的偶然性造成语言之间的相似性特点;③ 语言共同来源的保留;④ 语言之间相互借用。

李方桂(1977)采用的是音韵学的标准,根据声母、韵母、声调在壮侗语中的不同反映模式来做壮侗的分类,语音的演变有严格的对应规律可循。他认为采取词汇的标准很危险,因为有大量的文化词不容易排除,而文化词是借用的结果。

但是,据最近罗永现(Luo, 1997)对壮侗语族分类的研究,表明采用语音的分类标准与采用词汇的分类标准的结果是不同的。马提索夫(Matisoff, 1985)认为语音的变化模式反映东南亚语言区域较晚期的面貌,例如声调的产生只在公元1500年代(中国的元、明时期),东南亚语言受到北方汉语的扩散,导致"北方化"的结果。他认同本尼迪克特依据基本词汇作为分类的标准的方法。

《地图》划分语支的依据主要是同源词的比率。各语支内部同源词约有45% ~ 75%,壮傣与侗水语的同源词约为25% ~ 45%,壮傣语、侗水语与黎语的同源词约为22% ~ 27%。《地图》所依据的同源词的比率跟经典的语言分类的标准有一定的距离;传统的分类理论认为语言分类的标准有沟通度和基本词汇相似率两种。一些语言学家假定80%的基本词汇相似率,作为语言与方言的切分点。大于80%的基本词汇相似率的是方言,而小于80%的基本词汇相似率的则是不同的语言。但是,这也是个很任意的标准。

造成同源词比率统计差别的主要原因是用作统计的同源词的总数不同,

例如梁敏、张均如（1996）用斯瓦迪什（Swadesh，1952，1955）提出200词作算术统计，就跟《地图》的算术统计结果不同。用作统计的词目越多，同源词比率就会越小。因为这存在词频和词的统计“权重”的问题。出现频率较多的词与出现频率较少的词放在一起统计，与完全统计出现频率高的词，其结果会不同。此外，统计的词目数量越大，就越难排除语言之间相互借用的成分。因为我们的目的是研究语言的发生学分类，并计算出语言的进化树分枝的时间深度，所以，应当尽量分清语言之间的同源和借用的关系。

（三）各种分类的主要分歧

关于壮侗语族的系属问题，李方桂（1976）一直坚持将壮侗语族与苗瑶语族、藏缅语族一起组成汉藏语系。而本尼迪克特（1975，1990）则坚持壮侗语族与汉藏语系分离，两者之间没有发生学关系，只存在接触关系。壮侗语族与苗瑶语族和南岛语族组成澳泰语系。国内学者大多数支持李方桂的观点。但自沙加尔（Sagart，1993）发现南岛语跟古汉语有近60条的同源词，提出南岛-汉同源体系，近年来国内部分学者开始重视本尼迪克特和沙加尔提出的证据，在李方桂的基础上，又重新建立了一个更大范围的语系，即所谓的“华澳语系”。这些观点的核心仍然是强调汉-台同源。例如曾晓渝（2003）从借词的声调对应支持邢公畹的汉台同源说。关于壮侗语族的内部分类，主要是临高话（Be）的分类不同。本尼迪克特、Gedney（1993）都将黎语和临高话（Be）单立出来，但《地图》和梁、张则将临高话（Be）归为壮傣语支，认为与壮语关系最近。桥本万太郎（Hashimoto，1980）认为临高话（Be）是“混合语”，临高话（Be）的语音特征受到汉语特别是闽语的影响，而基本词则与泰系语言关系很深。法国的萨维那（Savina，1965）认为它是黎语的一支。德国人类学家史图博（Stubel，H.）认为临高话（Be或OngBe）“可能是黎语和泰汉语的混合语”[③]。

二、壮侗语族语言的计量分类

（一）分类的标准和方法：100词及数量方法

我们主要依据斯瓦迪什的100词表做同源词的数理分析，同时参考选择雅洪托夫（Yakhontov）的35词表[④]。具体统计方法说明：斯瓦迪什的基本词汇表已成功适用于世界上的多种语言［例如“罗赛塔计划”（Rosetta Project）］。但是各语言不见得会有完全相对应的词汇语义范畴；极可能有找不到对应词汇，或对应词汇的意义有相当距离（王士元，1994）。运用于壮侗语族语言需经过一定的修订，例如壮侗语专家倪大白在确认笔者提供的同源词表时认为：斯瓦迪什的基本词汇表第9词“地”并不很适合于壮侗语言比较，所以我们换用词义确切的“水田”。类似的例子还有第66词“站”换用雅洪托夫35词表中的“盐”，第97词“全部”换用雅洪托夫35词表中的“风”。雅洪托夫35词表虽然在历史比较语言学有较大影响，但在我们前期的汉藏语关系计量研究中，发现此词表并不很适合，分析其原因在于词目太少；认为斯瓦迪什词表最具有词义稳定性，用于统计的词目太少，误差必大，难以反映语言间关系的信息，但统计的词目太多，则难以排除语言间的借用。如果我们不能很好地排除借词，则画出来的树图会与事实相去甚远。

其次，词目与义项往往纠葛不清，同一个词目下，不同的语言会有不同义项的对应形式。我们采用的是“词根词源统计法”，采用较严格的语义对当原则。例如：“灰（草木灰）、叶（树叶）、根（树根）、虱（衣虱）、角（牛角）、乳（乳房）、肉（肌肉）、皮（皮肤）、名（名字）。”

我们具体分析了壮侗语族的12个语言，最后画出壮侗语族语言树形图。树形图包含两个重要的信息：① 语言集团的呈阶级式的聚合分类；② 树枝的长度可以反映语言从祖语分离的时间距离以及各语言间亲缘程度。

生物学家发明的一些研究生物种系发生分类的程序，对语言学家很有用。因为语言学与生物学有相似的生物遗传基因系统，生物学的分类与语言学分类很类似，而科学研究的一个重要特点，就是可用公式来反复验证和测量研究对象。生物学家为生物种系发生分类设计出很好的计算程序。最有影响的是，1967年由Fitch和Margoliash发明的以及1987年由Saitou和Nei发明的程序，1990年由Felsenstein将此两种合成为称PHYLIP的软件。有关的具体运算过程说明，请参看有关文章⑤。

历史语言学认为同源词的证据对于重建语言史，比起语音等其他语言特征来说，更为重要⑥。同源词的意义并不仅仅是原始祖语的“保留”（retention），它同时还具有“创新”或“突变”的意义，“创新”是语言再分类的极重要标志，它有一个重要特点：同源词的“创新”是一种单向的演变，不可逆向变化，例如，“腹”的原始词义，在语言中可能演变为“肚（du）子”、“肠子”、“胃”等词义，但是，“肚子”、“肠子”、“胃”等词义不可能再演变回去“腹”的原始词义。我们则可根据词义的再分化或“突变”的程度，对语言重组做次生的分类。例如，我们可根据各语言对核心词义变化的特征“创新”的“共享”的差异，利用计算机生物学程序PENNY重新对中国的语言及方言做出发生学的分类。“创新”既可反映语言的同源关系，也可反映语言的接触关系。而许多学者却往往忽略了“创新”对重构的重要意义。

同源词数据显示：同一个大簇（语支）的同源词语音形式较接近，而又以同一个小簇的语音形式更为接近。

（二）数理分类的主要步骤

1. 相似矩阵（similarity matrix）

首先优选出同源词[⑦]，编制同源词表，然后计算出每对语言的同源百分比。

表1同源词数据说明：南岛与汉藏不是单一的语言，我们采用的是其早期形式与壮侗语早期形式进行比较来判断它们之间的同源关系，进而统计同源数目。这种方法广用于历史语言学。在具体计算时，先确立南岛与汉藏分别作为一个已知的较为疏远或最早分离出来的语言来作为树图的参照系数（outgroup），完全不会影响所比较的壮侗语12支语言的分类计算。我们采用经南岛语专家，例如：Dempwolff（1934，1937，1938），Dyen（1971），Dahl（1973，1976），Blust（1980，1989，1996）认可的原始南岛语的同源词构拟形式跟李方桂（1976）、梁敏和张均如（1996）、吴安其（2002）等构拟的原始壮侗语形式比较，以便发现和解释它们之间的同源关系。原始汉藏语的同源词构拟形式采用马提索夫（2003）、吴安其（2002）等的研究成果。[⑧]

表1

	壮	布依	临高	傣西	傣德	侗	仫佬	水	毛南	黎	泰	老挝	南岛	汉藏
壮														
布依	92													
临高	66	61												
傣西	77	73	54											

（续 表）

	壮	布依	临高	傣西	傣德	侗	仫佬	水	毛南	黎	泰	老挝	南岛	汉藏
傣德	72	69	51	89										
侗	63	63	51	54	49									
仫佬	62	57	46	47	44	77								
水	67	64	51	52	49	82	81							
毛南	59	57	41	49	44	75	79	76						
黎	47	41	43	47	43	36	35	35	37					
泰	79	73	53	85	82	53	50	51	48	42				
老挝	78	71	52	90	86	50	47	49	46	40	92			
南岛	33	31	29	31	31	32	28	31	27	24	33	32		
汉藏	15	15	12	15	13	12	12	12	12	12	15	14	8	

2. 距离矩阵（distance matrix）

由于数理树形图是通过分枝的长度来反映语言间的距离的，所以我们必须把上面的相似矩阵转换为距离矩阵。

3. 从无根树到有根树

从距离矩阵转换成无根树有许多种方法，我们采用的计算程序则是最有影响的，1967年由Fitch和Margoliash发明的以及1987年由Saitou和Nei发明的程序。虽然有的学者批评Saitou和Nei提出的毗邻连接法（neighbor joining）无法完全排除语言间的借用成分，但目前为止，学术界仍然公认毗邻连接法较为科学，可信度较高。

图示说明：树图上的数字表示距离的长短，树枝的距离只计算每一树枝的端口到根部的横向距离，以及各个树枝横向距离的相加，而不管纵向的关

表2

	壮	布依	临高	傣西	傣德	侗	仫佬	水	毛南	黎	泰	老挝	南岛	汉藏
壮	0.00	3.62	18.05	11.35	14.27	20.07	20.76	17.39	22.91	32.79	10.24	10.79	48.15	82.39
布依	3.62	0.00	21.47	13.67	16.12	20.07	24.41	19.38	24.41	38.72	13.67	14.87	50.86	82.39
临高	18.05	21.47	0.00	26.76	29.24	29.24	33.72	29.24	38.72	36.65	27.57	28.40	53.76	92.08
傣西	11.35	13.67	26.76	0.00	5.06	26.76	32.79	28.40	30.98	32.79	7.06	4.58	50.86	82.39
傣德	14.27	16.12	29.24	5.06	0.00	30.98	35.65	30.98	35.65	36.65	8.62	6.55	50.86	88.61
侗	20.07	20.07	29.24	26.76	30.98	0.00	11.35	8.62	12.49	44.37	27.57	30.10	49.49	92.08
仫佬	20.76	24.41	33.72	32.79	35.65	11.35	0.00	9.15	10.24	45.59	30.10	32.79	55.28	92.08
水	17.39	19.38	29.24	28.40	30.98	8.62	9.15	0.00	11.92	45.59	29.24	30.98	50.86	92.08
毛南	22.91	24.41	38.72	30.98	35.65	12.49	10.24	11.92	0.00	43.18	31.88	33.72	56.86	92.08
黎	32.79	38.72	36.65	32.79	36.65	44.37	45.59	45.59	43.18	0.00	37.68	39.79	61.98	92.08
泰	10.24	13.67	27.57	7.06	8.62	27.57	30.10	29.24	31.88	37.68	0.00	3.62	48.15	82.39
老挝	10.79	14.87	28.40	4.58	6.55	30.10	32.79	30.98	33.72	39.79	3.62	0.00	49.49	85.39
南岛	48.15	50.86	53.76	50.86	50.86	49.49	55.28	50.86	56.86	61.98	48.15	49.49	0.00	109.69
汉藏	82.39	82.39	92.08	82.39	88.61	92.08	92.08	92.08	92.08	92.08	82.39	85.39	109.7	0.00

系；属于同簇内的各语言比簇外的各语言关系更密切。

树图一显示壮侗语族12支语言，可分为4个较大的聚类，即黎、临高、壮傣（壮、布依、傣西、傣德、泰、老挝）、侗水（侗、仫佬、毛南、水）。其分级和层次可假设为：由于黎、临高的树根分离点的数字为零，所以可以认为树图的第一层为三分，即黎、临高和壮侗等；第二层次为壮傣与侗水；第三层次为壮、布依与傣西、傣德和泰、老挝组成一个簇类，而侗与水（仫佬、毛南）组成一个簇类；第四层次为傣西、傣德和泰、老挝组成一个小簇，显示平行关系。水与仫佬、毛南组成一个簇类：侗水的远近关系依次为侗、水、仫佬、毛南。树图的数字代表树枝的长短距离，反映语言之间的亲缘关系的远近，树枝长的表示两种语言亲缘距离远，树枝短的则表示两种语言亲缘距离近。而同一个小簇类里的语言关系则比外簇类的语言关系近。树图一显示在壮侗语族12支语言中，各个树枝的分离点到树根的距离长度能够反映语言分化时间的先后和早晚，即树根与各层次的分离点的远近与分离时间一致，可表现从母语分离的时间深度。同时，其分类的结果以及各语言之间的亲缘关系的远近的描述则应是可信的。

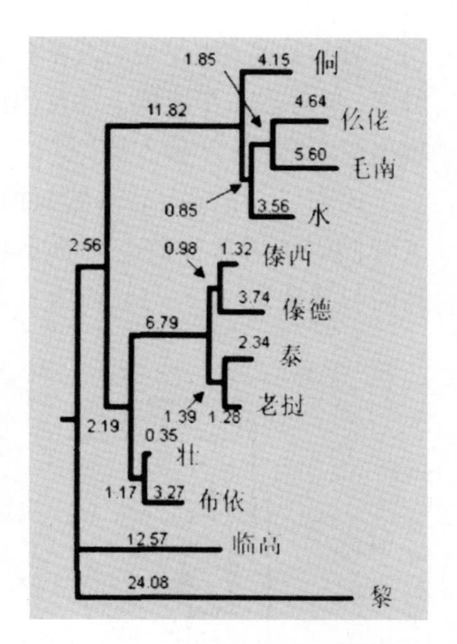

树图一　壮侗语族语言数理树形图，12种语言，采用毗邻连接法，黎语作为参照系数

从树图结构来看，如果各个语言的发展是均衡的，那么，代表各个语言的各个分枝的末端应显示出对齐均等；事实上，由于存在语言接触，必然导致大量语言成分的相互借用，因此语言变化速度的均衡发展是不可能的；树图充分反映出语言的发展速度是不均衡的。从树图发现，壮侗大簇中的壮-布依小簇分离点距离树根最近，这种结果可从两方面获得解释。

① 保守原则：壮-布依小簇语言比起其他各簇语言较多地保留了母语的成分，所以距离树根较近。

② 接触原则：这暗示中心语言与周边语言的关系，如果侗语保留母语的一个同源词，但壮语未保留，后来壮语向侗语借过来，则壮语会离根部近。这反映了语言间借用的方向。

黎与临高相比，黎的线条长，而临高的线条较短，这暗示临高与壮语簇接触多，黎与临高在地理上体现出南、北区别，北部更靠近大陆壮侗地区。这体现出树图结构的“俭省原则”。

树图二：增加南岛语的数据，变成13支语言，发现树图一计算出来的12支壮侗语族语言各个小簇的分类格局仍然未变。只是，南岛语独立为一支，临高与黎合为一簇，与壮傣为一大簇。总分为三大簇：南岛、壮傣黎、侗水。

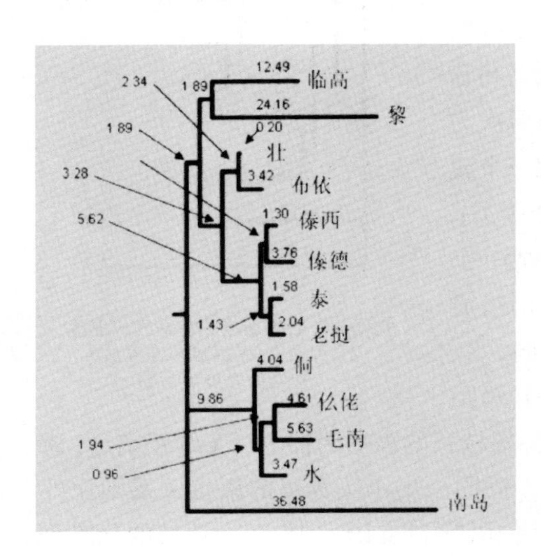

树图二　13种语言，采用毗邻连接法，南岛语作为参照系数

树图说明：

植根的位置：经验的做法是取树图中距离最长的两支语言的中点做根。但可根据实际更合理的情况植根点，只要在树图整体结构不改变的情况下，计算时，用来给树加根的最好办法，就是先确立一个已知的较为疏远或最早分离出来的语言来作为树图的参照系数，而树根一定是在与其他语言的线条之间的两分的位置上。例如我们把黎、南岛和汉藏作为参照系数的语言。必须说明的是，树根和分离点的概念所代表的意义不同，树根代表假设中的祖语，而分离点则代表各语言分枝。

树形图说明：我们分别使用了毗邻连接法和由Fitch、Margoliash发明的计算方法，结果都一样。各分级层次的类簇相同。增加南岛语和汉藏语数据，由12支语言变成13或14支语言（树图三）后，分类的情况仍然相同。这就说明分类结果是可信的。

从几个不同的树图比较来看，我们有个重要的发现：随着语言数的增加和减少的数据的变化，树图的反映很敏感，都会体现在树图的不同变化上；而最重要的变化是临高这支语言在图中出现的位置，即摆放的位置，这就反映了临高这支语言在壮侗语言中的特殊地位，而传统的分类正是认为临高系属未定。数理分析的结果跟传统的定性分析相合。

（三）结论

在壮侗语族12支语言中，可以分为4大聚类：黎、临高、壮傣、侗水。

如果增加南岛语，并以南岛语为参照系数，可分为4大聚类：南岛、黎语与临高、侗水、壮傣。如果抽去黎的数据，则临高归入壮傣大簇，显示临高分别跟黎与壮语最为接近。

壮侗语族的阶级分层为：如果以黎为参照系数，则分三大聚类：临高、壮傣、侗水。如果以汉藏为参照系数，则第一层次为南岛与黎、临高、壮傣、侗水；第二层次为黎与临高、壮傣、侗水；第三层次为临高与壮傣、侗水；第四层次为壮傣和侗水。

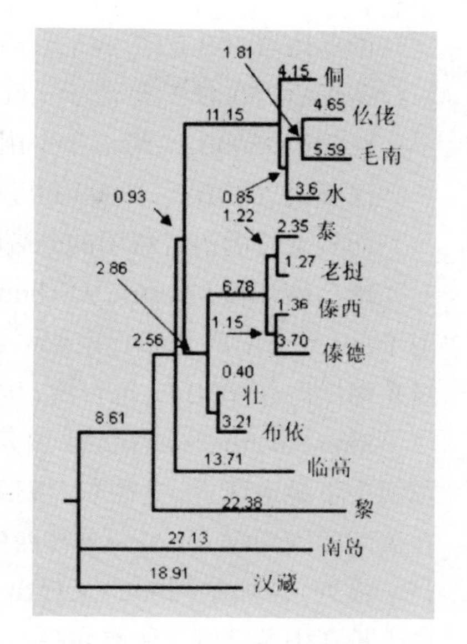

树图三 14种语言，采用毗邻连接法，汉藏语作为参照系数

（四）计量分类跟传统分类的比较

最重要的差别：

① 传统分类只重视语言差异的定性分类，而无法做语言间亲缘关系距离程度的量的分析。而计量分类可做亲缘关系程度的描述，并通过树枝长短来表示距离关系。

② 虽然两者使用的材料和方法不同，但分类的结果却大致一样。

③ 计量分类可以显示语言的类簇和分级层次。

④ 传统分类没有细分出壮侗语族各语言的较小的聚类和关系程度，《地图》和梁敏、张均如都认为临高话只跟壮语关系最近，而与黎关系最远，因系不同语支之间的差别。而我们的研究表明：各种的树图结构显示，临高分别跟黎和壮的亲缘关系最近。

我们认为语言学的数理分类确认临高分别与黎语和壮语的亲缘关系最为接近，这一结论是较科学的，其可信度较高。

三、壮侗语族与南岛语族的分离时代以及壮侗语族内部各语言的分离时代

已有的相关语族的语言年代学的结论：

南亚语6000B.P.，形成于中国西南地区。

南岛语6000B.P.,形成于台湾。

苗瑶语2500B.P.,形成于长江下游地区。

壮侗语2500B.P.,形成于中国东南部。

汉藏语7000B.P.—6000B.P.,形成于黄河中上游地区。

而较上位的语言集团的分裂时间深度则更长。

南岛南亚说（Austric）（Schmidt, Reid, Blust）: Austric 9000B.P.,形成于中南半岛（或云南西北部），南亚语族向中南半岛扩散；而南岛语族7000B.P.,则经东南沿海,6000B.P.,抵台湾,开始真正的南岛语向南太平洋岛屿扩散。

Sino-Austronesian说（沙加尔，1993，1999）: 8500B.P.—7500B.P.，形成于黄河中下游地区，一支往东，携带稻米耕作技术到达台湾的为今南岛民族，留在大陆的则为汉人；另一支往西南，成为今藏缅语族。

最近，沙加尔（2002）又提出了一个“扬子语族”（Yangzian，9000B.P.），形成于长江中游地区。主要包括南亚语和苗瑶语。

由于以上的语言年代学的年代顺序，主要是依据考古学的证据，并没有经过语言学数据的计算，而考古学的证据无法为历史上的族群断代编年，所以，各家的编年顺序有较大的分歧。

词源统计分析法可以帮助我们计算出语言分裂的时间的深度，虽然这种方法受到许多人的批评，认为语言学不同于生物学，用作测量单位的基本词汇受横向传播的干扰较大，不像生物学的基因单位那样稳定⑨。但是，任何科学的方法都有其局限性，如果这种计算方法的结果能够跟其他学科诸如考古学、民族学、人类学的研究结论一致，那么，就有相当高的可信度。

两种语言的核心词共享程度的比率不同，其分裂年代的时间深度是不一样的。我们采取一种较为合理的计算方法，即分别统计出每1 000年的保留率为75% ～ 95%的结果，采取它的平均值，其公式：$P(La, Lb)=r^{2t}$ $t=lg(P(La, Lb))/(2lgr)$

其中r表示每千年的保留率，t表示每千年分离的值。

最后认为采用每1 000年的保留比率为85%的计算方法较合理，采用每1 000年的保留比率为85%的计算结果较符合人文学科的研究结论。可比较表3：$t=lg(P(La, Lb))/(2lgr)$

以上只能计算出每对语言的分离时间，但是，我们的目的是需要在树图上反映出各个语言进化分枝的时间深度，而树图各个分离点的时间深度的表

表 3

85%保留率	壮	布依	临高	傣西	傣德	侗	仫佬	水	毛南	黎	泰	老挝	南岛	汉藏
壮	0	257	1 278	804	1 011	1 421	1 471	1 232	1 623	2 323	725	764	3 411	5 837
布依	257	0	1 521	968	1 142	1 421	1 729	1 373	1 729	2 743	968	1 054	3 603	5 837
临高	1 278	1 521	0	1 896	2 072	2 072	2 389	2 072	2 743	2 597	1 953	2 012	3 808	6 523
傣西	804	968	1 896	0	359	1 896	2 323	2 012	2 195	2 323	500	324	3 603	5 837
傣德	1 011	1 142	2 072	359	0	2 195	2 526	2 195	2 526	2 597	611	464	3 603	6 277
侗	1 421	1 421	2 072	1 896	2 195	0	804	611	885	3 143	1 953	2 133	3 506	6 523
仫佬	1 471	1 729	2 389	2 323	2 526	804	0	648	725	3 230	2 133	2 323	3 916	6 523
水	1 232	1 373	2 072	2 012	2 195	611	648	0	844	3 230	2 072	2 195	3 603	6 523
毛南	1 623	1 729	2 743	2 195	2 526	885	725	844	0	3 059	2 258	2 389	4 028	6 523
黎	2 323	2 743	2 597	2 323	2 597	3 143	3 230	3 230	3 059	0	2 669	2 819	4 391	6 523
泰	725	968	1 953	500	611	1 953	2 133	2 072	2 258	2 669	0	257	3 411	5 837
老挝	764	1 054	2 012	324	464	2 133	2 323	2 195	2 389	2 819	257	0	3 506	6 049
南岛	3 411	3 603	3 808	3 603	3 603	3 506	3 916	3 603	4 028	4 391	3 411	3 506	0	7 771
汉藏	5 837	5 837	6 523	5 837	6 277	6 523	6 523	6 523	6 523	6 523	5 837	6 049	7 771	0

示也使我们对整个语言群的进化时间有较为全面的整体理解，而不限于仅了解每对语言的时间深度。所以，我们采用一个公式来转换：

分离点时间=（最大值+最小值）/2

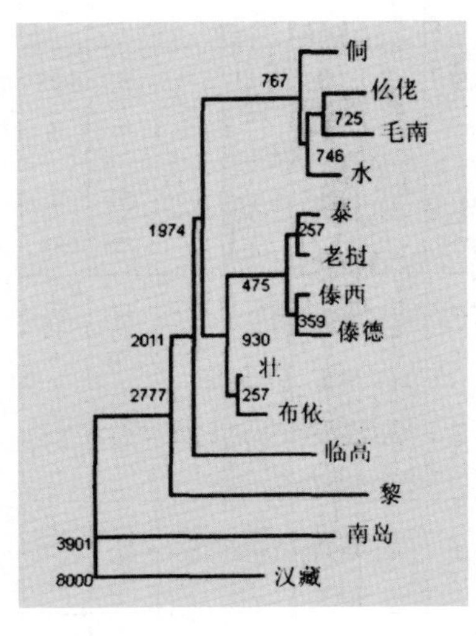

树图四

树图说明：数字为距今的相对年代。我们知道，如果语言的演变是以均衡的速度发展的话，那么，从同一个祖语分裂出来的语言的进化结果是相同的，体现在树图上，代表各个语言的分枝的末端应等齐，即各个分枝的末端距离树根的距离是等同的。但事实上语言的演化速度不可能是均衡的，所以体现在树图上各个分枝的末端并不等齐，这主要是因为：① 各个语言从祖语分裂后，有不同的保留率；② 各个语言之间存在着程度不同的借用。所以各个语言的树枝长短不同，距离根部的距离亦不相同。采用词源统计法计算语言的分离时间，只能计算出一个大致的时间范围，无法做精确的计算，所以，树图的分离点时间跟树形的不一致处，这是计算语言学家都已认识到的事实。

树图四中的年代序列与民族史的编年大体相当，即百越-南岛集团形成于约4000B.P.，其主要地域为中国东南部。秦汉帝国统治后，南岛退出中国大陆。黎族在海南原住民的基础上，约2700B.P.相对独立出来，而临高则在唐宋之前的汉人移民潮的影响下，约2000B.P.逐步独立出来；魏晋时，在百越主体上，约1900B.P.形成僚俚民族集团，最初独立出壮，约900B.P.分离，至南宋，文献正式出现壮族的“僮丁”的族称。部分壮族亦被称为“狼人”、“侬人”。而各语支内部诸民族如壮与布依，傣、泰与老挝则为明清时代分离；壮、侗、水、毛南、仫佬诸族均是从魏晋时期的“僚”族发展而来的。而水与侗、毛南、仫佬则为宋元时代分离，元明时期，侗族从一个地理行政单位的名词——峒转化为民族共同体的族称。宋代文献未见有侗的记载。文献记载茅难蛮（毛南族）与忼水蛮（水族）共同居住在一个地域，“明显是近亲的民族群体”（王文光，1999）。语言学的时间分离统计结果与民族史实相符。

必须特别说明的是，为了画树图植根的方便，我们采用了原始汉藏语的数据，因为几可确信，原始汉藏语跟我们所比较的13支语言来说，关系最为疏远，所以，可以作为植根的分离点。在比较的100词中，认为原始汉藏语有8个词跟南岛语有对应关系，即“火、角（牛角）、眼、骨、血、飞、盐、风”。还有一些借用可能性较大的核心词，例如“水田（低洼地）、弩（箭、射）、头、发、肚子、蛇、路、巫、马。”事实上，原始汉藏语与原始南岛语互相对应的这批词是很难区分同源和借用的关系的。张光直（1989）和Van Driem（2002）都认为北方强大的龙山期文化深刻地影响到东南中国的原南岛语文化，时间约在6000B.P.，如果考古学的结论属实的话，这批词中的大部分词则是东南中国“北方化”的结果——借词。由于原始汉藏语与原始南岛语之间的亲缘关系还未确定，所以，我们在计算时只是将原始汉藏语的数据作为一个参照系数来处理，这样并不影响用其他语言数据得出的结果。

四、几种语言演变理论的假设

（一）语言扩散（diffusion）的多向性

传统认为南岛语的扩散是单向的观点应重新审视，从目前的发现看，至少应有两个方向：

① 大陆东南沿海4000B.P.→云南及东南亚岛屿-台湾。例如“稻米、弩、萨满”等南岛、南亚同源词。

② 大陆东南沿海→台湾。例如“狗、脚、穿山甲”等。

这说明不同的语言形态传播的方向可以不同，即语言漂流（drift）无定式。

壮侗语跟南岛语的分离，可以从Dixon（1997）“聚变-裂变语言演变模型”（punctuated equilibrium）和Bellwood（1996）“网状结构模型”（reticulate models）的理论得到较合理的解释。秦汉帝国对中国东南百越-南岛区域的完全统治，导致大陆原南岛语的突变，打破了语言渐变的本来的平衡系统，而北方移民的大量涌入南方以及北方中原文化的优势地位的确立，加速百越-南岛语言的“汉化”速度，导致壮侗语跟南岛语的彻底分离，这就是南岛语在大陆“突然消失”的根本原因。这种现象也可以用物理学上的“复杂适应系统理论”（complex adaptive system）解释，如同水持续加温后，会突然非线性地从液体转变为气体，这种从量变到质变的非线性过程叫“相变”（phase transition）或“涌现”现象。

（二）语言同质成分具有多层次和多源性

东亚五个语言集团有一些不同层次上的同质性，其来源包括：

（1）远古人类南北蒙古人种的共同来源，如“火、盐、猪、路、骨、鸟、头”等同源词。

（2）南方蒙古种（马来人种）的同质性，如“狗、老虎、蛇、杀（死）、水田、村落（坂）、萨满（南方称‘童’，与北方的称‘巫’不同，这反映南北两个不同的文化系统）、手、五”等。

（3）新石器时代以来北来文化的传播和扩散，如“马、犬、弩、稻米、针、铁”等同源词。

Pulleyblank（1996）等认为汉语的“马”*mraʔ来源于印欧语的*marko-；比较英语mare；蒙古语morin；原始藏缅语*m-rango。

“犬”，*kwhənʔ来源于希腊语kuon。但是分子人类学的研究认为：世界范围的驯化“狗”起源于东亚。我们的最新研究则认为驯化“狗”起源于藏缅语族地区。

“车”，*kwla与北高加索语*kwolo-有同源关系。

“骨”，*kut与北高加索语*kǒća，Basque语khotx有同源关系；Na-Dene语：s-kut。

“巫”，*m(r)jaʔ跟古波斯语magu对应。中国北方萨满宗教传播当来自西北亚地区。

Starostin（1995）和Bengtson（1999）曾经提出一批重要的核心词如“血、骨、死、火、角、风”等，论证汉语跟高加索及印欧语的同源关系。

如果这种同源关系确立的话，只能有两种解释。

① 人类13万年前走出非洲，6万～4万年前到达亚洲南部及北部，这是远古人类语言底层的保留。

② 先汉民族与原藏缅民族的蒙古人种与高加索人种的早期经济文化接触的结果。我们认为远古人类共同体的语言的保留率应是很少的。事实上，同源词证据表明蒙古人种与高加索人种的接触主要为文化上的接触。

（4）不同语言集团的边缘的“文化交互作用圈”的影响。

但是应把握好语言同质性的“度”的问题。区别不同层次的“同”，对重建语言树的各个不同的阶级的原始母语至关重要；“汉–台”同源是较晚期层次上的“同”，有大量的北来移民和文化词的移借，而“Greater Austric”则包括旧石器时代以前的人种的“同”，过分强调了史前的“同”。所以，应重新评估语言树的科学价值。

（三）充分注意到不同语言集团的区域性特征的异质性，即各个语言树的阶级（hierarchical structure）“创新”（innovations）

例如在壮侗语族大区中各个较小的区域，其区域性特征明显。“侬（人）、骹、团”等反映古闽越区域特征，而不同于两广的“骆越”区域系统。

五、结论

本文运用词源统计分析法的原则，对壮侗语族语言做出数理分类以及亲缘关系程度的描述，并通过树枝长短来表示距离关系，显示壮侗语族语言的类簇和分级层次。表明计量统计的结果跟传统的定性分类大致相同，但计量的方法更为科学，分类更为合理准确，提出了跟传统分类的不同看法：即临高分别跟黎和壮的亲缘关系最近。而传统的分类都认为临高话只跟壮语关系最近，而与黎关系最远，临高与黎体现为不同语支之间的关系。我们从壮侗语言进化树形图的结构变化，提出应当重新检讨传统的西方历史语言学的谱系分类框架，即仅仅只在一个语言的平面上，人为地划分语族、语支、语言、方言；这种方法太过于简单，并已过时，体现不出语言的分层和整体结构。从树图看，语支这一层次不是固定不变的，呈动态状态，而且是多层次的，即不止一个语支，树图的每个分离点都可等同于“语支”的位置。语言进化树形图能够改进传统的分类理论，更好地反映语言的分层和整体之间的相互关系。

我们采用每千年保留率的平均值的计算方法，不仅仅计算出每对语言的分离时间，而且计算出树图各个分离点的分离时间。其结论与考古学及人类学的最新研究成果一致。我们的计算结果不同于许多语言学家已有的看法。

我们讨论了壮侗语族的形成过程及其时间深度，认为南岛语族生活在以华南为中心的广大区域，约4000B.P.开始分离，并经东南沿海或西南-中南半岛向台湾及南洋群岛扩散。最后，我们试图采用几种不同的语言进化理论来解释东亚语言区域的形成过程，以便建立一种多学科的理论框架。

注释

① 中国境内的壮侗语族语言的分布：主要分布在广西、贵州、云南、湖南、广东、海南等地区。其中临高话、村话和拉珈语集中分布在海南。必须指出的是这些语言在地理上的

表现，只是现在的共时的分布，不代表历史上的历时的状态。由于历史上不断的移民潮以及更重要的因素——"汉化"的影响，历史上"真实"的语言与民族的分布的版图发生了很大的变化。据研究，战国、秦汉时代的"闽越"为福建，"骆越"为两广，"瓯越"为江浙，这些都是属于百越即古壮侗民族居住的区域（陈国强等《百越民族史》中国社会科学出版社1988年版），而六朝、唐宋以后，由于"北方化"的结果，则多发展成为"汉化"地区。

② 壮侗语族的民族的来源：民族学比较一致的观点，壮侗语族是古代百越民族的后裔，其谱系的历史演化过程：传说时代的"蛮"、"三苗"——商周时代的"瓯"、"越沤"——春秋战国汉时代的"百越"——晋唐时代的"俚"、"僚"民族集团——宋元时代的壮、侗、水、傣、毛南诸壮侗语族民族。百越是中国东南和南部地区古代民族的名称。新石器时代晚期是奠定民族形成的时期，南中国的人群已开始具有明显的区域特点。

③ 参看Hashimoto（1980）文中的讨论。

④ 参看Starostin（1995），文中附录依据雅洪托夫的35个稳定词项的构拟形式比较统计表。

⑤ 参看Wang（1994）及邓晓华、王士元（2003a.b）文中的讨论。

⑥ 参看Greenberg（2001）的论述。

⑦ 因限于篇幅，本文的词汇附录省略。

⑧ 参 看Dempwolff（1934, 1937, 1938），Dyen（1971），Dahl（1973/76），Blust（1980, 1989, 1996），Matisoff（2003），何大安（1999），吴安其（2002），梁敏和张均如（1996），Sagart（1999, 2002）等关于同源词的讨论。

⑨ 参 看Colin Renfrew, April Mcmahon, Larry Trask (eds.) (2000) Time Depth in Historical Linguistics (2000) 的讨论。

参考文献

邓晓华，王土元.2003.苗瑶语族语言亲缘关系的计量研究——词源统计分析方法.中国语文，(3)：253-263.

邓晓华，王土元.2003.藏缅语族语言的数理分类及其形成过程的分析.民族语文，(4)：8-18.

何大安.1999.论原始南岛语同源词//石锋，潘悟云主编.中国语言学的新拓展——庆祝王士元教授65岁华诞.香港：香港城市大学出版社.

梁敏，张均如.1996.壮侗语族概论.北京：中国社会科学出版社.

罗常培，傅懋勣.1954.国内少数民族语言文字的概况.中国语文，(2)：21-26.

潘悟云.1995.对华澳语系假说的若干支持材料//Wang W S-Y. The Ancestry of the Chinese Language. Journal of Chinese Linguistics Monograph No. 8. Berkeley: University of California.

王文光.1999.中国南方民族史.北京：民族出版社.

吴安其.2002.汉藏语同源研究.北京：中央民族大学出版社.

曾晓渝.2003.论壮傣侗水语古汉语借词的调类对应.民族语文，(1)：1-11.

张光直.1989.新石器时代的台湾海峡.考古，(6)：541-550.

中国社会科学院，澳大利亚国立大学 .1988. 中国语言地图集 . 朗文出版社 .

Bellwood P. Phylogeny versus reticulation in prehistory. Antiquity, 1996, 70: 881–890.

Benedict P K. Thai, Kadai and Indonesian: A new alignment in Southeast Asia. American Anthropologist, 1942, 44: 576–601.

Benedict P K. Austro-Thai: Language and Culture, with a Glossary of Roots. New Haven: HRAF Press, 1975.

Benedict P K. Japanese-Austro-Thai. Ann Arbor: Karoma, 1990.

Bengtson J D. Wider genetic affiliations of the Chinese language. Journal of Chinese Linguistics, 1999, 27(1).

Blust R. Austronesian etymologies Ⅰ. Oceanic Linguistics, 1980, 19(1): 1–181.

Blust R. Austronesian etymologies Ⅱ. Oceanic Linguistics, 1983–84, 22(2–3): 29–149.

Blust R. Austronesian etymologies Ⅲ. Oceanic Linguistics, 1988, 25: 1–123.

Blust R. Austronesian etymologies Ⅳ. Oceanic Linguistics, 1989, 28: 111–180.

Blust R. Beyond the Austronesian homeland: The hypothesis and its implications for archaeology//Goodenough W H. Prehistoric Settlement of the Pacific. Transaction of the American Philosophical Society, 1996, 86(5): 117–137.

Dahl O C. Proto-Austronesian. Scandinavian Institute of Asian Studies Monograph Series, Vol. 15. Lund and London: Curson Press, 1973/76.

Dempwolff O. Vergleichende Lautlehre des austronesischen Worschatzes. Vol. 3. Beiheft zur Zeitschrift für Eingeborenen-Sprachen, 1934: 15; 1937: 17; 1938: 19.

Dixon R M W. The Rise and Fall of Languages. Cambridge: Cambridge University Press, 1997: 3–4.

Dyen I. The Austronesian language of Formosa//Sebeok T A. Current Trends in Linguistics. Vol. 8. The Hague: Mouton, 1971: 168–199.

Felsenstein J. PHYLIP Manual Version 3.3. University Herbarium. University of California, Berkeley, 1990.

Fitch W M, Margoliash E. Construction of phylogenetic trees. Science, 1967, 155: 279–284.

Gedney W J. William J Gedney's the Saek Language. Edited by Hudak T J. Michigan Papers on South and Southeast Asia No. 41. Ann Arbor: Center for South and Southeast Asia Studies, University of Michigan, 1993.

Greenberg J H. The methods and purposes of linguistic genetic classification. Language and Linguistics, 2001, 2(2): 1–26.

Hashimoto M. The Be Language: A Classified Lexicon of its Limkow Dialect. Institute of the Study of Language and Cultures of Africa and Asia, Tokyo, 1980.

Kitching I J, Forey P L, Humphries C J, et al. Cladistics: The Theory and Practice of Parsimony Analysis. 2nd ed. New York: Oxford University Press, 1998.

Li F K. Languages and dialects of China. Journal of Chinese Linguistics, 1973: 1–13.

Li F K. Sino-Tai. Computational Analysis of Asia and African Languages, 1976, 3: 39–48.

Li F K. A Handbook of Comparative Tai. Hawaii: The University Press of Hawaii, 1977.

Luo Y X. The Subgroup Structure of the Tai Languages: A Historical-Comparative Study. Journal of Chinese Linguistics Monograph Series, 1997.

Matisoff J A. Linguistics of the Sino-Tibetan Area: The State of the Art//Thurgood G, Matisoff J A, Bradley D. The Australian National University, 1985.

Matisoff J A. Handbook of Proto-Tibetan-Burman. Berkeley: University of California Press, 2003.

Pulleyblank E G. Early contacts between Indo-Europeans and Chinese. International Review of Chinese Linguistics, 1996, 1: 1–24.

Ruhlen M. A Guide to the World's Languages. Vol. 1: Classification. Stanford University Press, 1991.

Sagart L. Chinese and Austronesian: Evidence for a genetic relationship. Journal of Chinese Linguistics, 1993, 21(1).

Sagart L. The Roots of Old Chinese. Amsterdam: John Benjamins, 1999.

Sagart L. The Vocabulary of Cereal Cultivation and the Phylogeny of East Asian Languages. IPPA, 2002.

Saitou N, Nei M. The neighbor-joining method: A new method of reconstructing phylogenetic trees. Molecular Biology and Evolution, 1987, 4: 406–425.

Savina F M. Le Vocabularies (Presented by Haudricourt A-G). Hanoi and Paris: Ecole Française d'Extrême-Orient, 1965.

Embleton S. Lexicostatistics/Glottochronology: From Swadesh to Sankoff to Starostin to future horizons//Renfrew C, McMahon A, Trask L. Time Depth in Historical Linguistics. The McDonald Institute for Archaeological Research, Cambridge: University of Cambridge, 2000.

Starostin S. Old Chinese vocabulary: A historical perspective. Journal of Chinese Linguistics, 1995: 225–251.

Swadesh M. Lexico-statistic dating of prehistoric ethnic contacts. Proceedings of the American Philosophical Society, 1952, 96: 452–463.

Swadesh M. Time depths of American linguistic groupings. American Anthropologist, 1955, 56.

Van Driem G. Tibeto-Burman replaces Indo-Chinese in the 1990s: Review of a decade of scholarship. Lingua, 2002, 111: 79–102.

Wang W S-Y. Glottochronology, lexicostatistics, and other numerical methods// Encyclopedia of Language and Linguistics. Oxford: Pergamon Press, 1994.

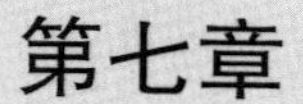

山野秘辛绘奇画

2023年人类学终身成就奖获奖者

——庄孔韶

人类学生涯回顾*

庄孔韶先生学术概述

庄孔韶教授是1978年录取的第一批硕士研究生，1984年继续追随中国著名人类学家林耀华先生，也是我国培养的第一位民族学（人类学）博士（1988）。1990—1993年在美国华盛顿大学人类学系做博士后研究和访问学者，博士后期间完成林耀华先生《金翼》的学术续本《银翅——中国的地方社会与文化变迁》（首版1996年，英文版2018年）和人类学纪录片《端午节》英文版（1992），同年入围玛格丽特·米德电影节，是中国“文革”后在国外出版的第一部反映中国文化的人类学纪录片。

1994年庄孔韶回国任中央民族大学民族学系主任兼民族学研究所所长；2003—2011年受聘在中国人民大学，创建人类学研究所，任所长和责任教授；2010年起受聘浙江大学恒逸讲座教授，并复建人类学研究所和创办《人类学研究》杂志；2018年又受聘云南大学民族学与社会学学院“魁阁”学者和文化人类学首席专家。庄教授先后出任中国民族学人类学研究会副会长，中国影视人类学学会副会长，主持第16届世界人类学民族学大会电影节；现任中国民族学人类学研究会交叉学科专业委员会主任，致力于跨学科的学术研究与实践。

1984年，庄孔韶和导师林耀华先生联名出版《父系家族公社研究》，他的硕士论文是关于世界上三个地理区域生态环境与父系家族制度的比较研究，庄孔韶在基诺族、哈尼族和景颇族居地发现了两种不同的游耕类型：前进型游耕和螺旋型游耕，为山地少数民族未来生存与生计转变研究提供了第一手的真知灼见。

---

* 本文作者为杜靖，张杰。

1989年出版中国大陆第一本《教育人类学》专著，是庄孔韶接受中国教育家庄泽宣先生比较教育的学术遗产，再交叉人类学的跨学科新作，从此推动了中国用人类学观察教育的新方向。庄孔韶回国后积极推动人类学教学改革，探索创新性课程，并由个人主编一个学科完整系列教材，包括中国大学招生与教学选用最多的人类学教科书，分别是研究生教科书《人类学通论》（2003，国家图书奖提名奖）和本科生教材《人类学概论》（2006，国家级规划精品教材），配套教辅读物《人类学经典导读》（2008，国家级规划教材）和高校双语教材"Anthropology"（2008）。庄孔韶的文字与电影合璧是其大学人类学课堂改革和教材编写的鲜明特征，十余部人类学电影进入教材和课堂，同时关于中国文化题材的人类学影片从20世纪90年代已经陆续进入中国和欧美大学课堂和图书馆系统。

在汉人社会研究领域，庄孔韶传承林耀华教授的中国宗族乡村研究方向，已形成学术团队。他的代表作《银翅》强调田野-书斋观察中的中国文化"过化"原理、"理念先在"、古今关联的反观法和文化直觉主义是其代表的本土理论。他推动学术同行对旧日著名田野点做回访式调研，出版重要作品《时空穿行——中国乡村人类学世纪回访》（2004），随后设计了中国各地14个相同与不同于宗族社会的多省多点调研，2019年推出《离别东南——一个汉人社会人类学的分解与组合研究》，凸显了中国古典文化在多样性基层社会整合中的巨大变通力量。

庄孔韶致力于人类学的学术理论与应用的转化。在医学人类学与公共卫生领域，20世纪90年代后期因特定机缘带领研究生率先在仪式戒毒和临终关怀两个方面调研，引进人类学的生物文化整体论。庄孔韶发现云南小凉山彝族用古老的仪式戒毒的成功案例，总结了以文化力量战胜人类生物性成瘾性的方法论意义上的结论，直接促成地方人民高戒毒率和卓有成效的健康教育效果。庄孔韶团队拍摄的人类学纪录片《虎日》（结合学术论文）荣获中英项目和亚洲"最佳实践"奖，庄孔韶应邀赴哈佛大学亚洲研究中心，并在第16届世界公共卫生大会作专题演讲。在国内，《虎日》在彝族地区透过电视推广，落实了"应用的影视人类学"的学术意义，扩大了成功戒毒的地域。而多名师生在仪式戒毒和临终关怀研究中渐渐能熟练运用人类学的生物-文化整体性原理，为2000年后中国方兴未艾的大规模公共卫生/医学人类学发展培养了骨干。

经过多年以人类学为先导的横向跨学科的学术实践，庄孔韶在1995年提出了“不浪费的人类学”的本土理念，实际是为了推动文化表现的多元方法综合实验。“不浪费的人类学”是指“人类学家个人或群体在一个田野调查点上将其学习、调研、阐释和理解的知识、经验、体悟以及情感用多种表征手段展示出来，著书立说以外，尚借助多种形式，如小说、随笔、散文和诗，以及影像手段，邀集地方人士的作品或口述记录，甚至编辑和同一个田野点相关的跨学科作品，以求从该族群社区获得多元信息和有益于文化理解与综观。”从四十年前至今，庄孔韶带领的学术团队延续这种做法，带出了一批撰写学术专著的人类学家（含合作与应用的人类学），又兼诗人、散文家、民族志电影人/剧作家和画家等。导源于认识论改造的思路与实验，其交叉学科实验的原理是：

第一，人类学研究的目的经常不是为了解决某个具体的问题，而更多的是处于文化认识论的过程研究之中，这正是人文学科和实证问题导向学科的重要差别。因为可清晰解说的某些“问题”答案有时是被修剪过的，它们去除了非理性、情感和不可言说的内涵。因此，庄孔韶团队长时段的写文化、拍文化、画文化和演文化的过程性合作研究就是为了寻找知识交叉产生互补的魅力，就是为了达成“触类旁通”的智慧佳境。其目标是达成人类憧憬的认识论综观，尽管绝对的综观可能永远也达不到。

第二，保持文字写作的探索，又从文字的限度中走出来。该团队把人类学的文学艺术实践和“不浪费的人类学”原初理念相结合。他们既需要科学的“事实的世界”，但不能省掉“意义的世界”。为此，探讨电影、戏剧、表演、诗学、新媒体之间的不可替代性及其整合的学术意义已开始得到重视。

第三，以往人类学的文化撰写因清晰无误的要求，忽视了隐喻与直觉，然而它却可以被引申到绘画特有的复合型思维实践之中：《虎日》里高亢的押韵诗，不只是情感与灵感突至的文学意义，而是族群认同被调动的诗性智慧动力，而电影的线性记录特点和绘画中的复合性思维在并置中各得其所。于是庄孔韶团队也体会到了跨学科、专业和方法之无限互补的时空存在，因而各自已难分伯仲主次，这也正是“不浪费的人类学”实验的原理、预想与结果。

最近五年，庄孔韶团队横向的多专业成果多多。例如代表多种写法于一炉的《银翅》英文版已经出版（2018），引出了关于写文化的国际学术对话——《对话：文化直觉、艺术实验和合作人类学》（2020），直接聚焦于文化的直觉和隐喻的表达。庄孔韶还主编了团队系列作品《人类学的诗学探讨》（2021）、《国学、足球与艺术》（2020）和《绘画人类学》（2019），以及重要的本土理论论文《文化表征的多元方法与跨学科实验》（2020），着力探讨人类学横跨学科在写文化、拍文化与画文化等的不同手法，以及完善本土理论的努力，如《本土理论如何进入绘画人类学》（2023）。这两篇文章和庄孔韶团队的十余篇论文（英文、法文）将在2024年以《不浪费的人类学》和《中国的绘画人类学》两个专刊在法国出版发行，彰显中国人类学本土理论的传播成果。

庄孔韶导演的纪录片作品《金翼山谷的冬至》（2018），先后获中国国家级奖项和两个重要国际人类学电影节奖项（参见后文近五年庄孔韶学术成果目录）。在写作、诗学和电影之后，进入庄孔韶团队最晚的绘画专业团队也已经有24年，约60人的绘画成员和人类学家一同做田野工作，到我国福建、云南以及波兰和法国，跨界完成论文和绘画作品。到2021年已经有6人的人类学绘画作品（油画）入选全国“民族百花”精品展。

庄孔韶团队成员之间相互欣赏各自的研究，已经形成多学科互助互补的学术风尚（先祖庄存与亦有不分门派的古典文人风尚）。跨学科团队沙龙最近深度交流的主题是，本土理论如何进入绘画人类学的具体层面，提倡在田野工作和创作成品的过程中，进一步追寻不可言说、难以言说和言外之意的隐喻与直觉发现，深化中国绘画人类学创作的本土擅长，因此继续引导本团队跨学科成员的诗性智慧，提升人类学学术与艺术造诣，是庄孔韶教授正在思考的问题。

庄孔韶先生小传

初识庄孔韶先生是在2002年秋季，那时他在中央民族大学的家里给我们几个学生上“汉人社会研究”专题课，有时是观看和讨论民族志电影，或让我们汇报各自研究题目和田野中的发现，庄先生总是很精准地说出存在的问题，指明方向。令人印象深刻的是，学问和人格魅力外庄先生还用美食“诱捕”学生，每次课后他都领我们下馆子，在师生笑谈间继续课堂上的讨论。大家集体的感受是，凡经先生调教的学生，一生无论顺逆，都始终把人类学作为志业，应该与他的这种带学生的方式有关。

几十年交往，庄先生时常与笔者讲述家世。我们知道，家世与思想的传承总是会适应学术的蜿蜒长河，每一时代的精英都会做出自己不同的选择与贡献。应该说，庄先生是打开20世纪80年代以来中国人类学叙事的一把钥匙，一个学理上的关键人物。只有了解了他，才能清楚中国人类学的世纪起承转合与深深扎根于中华大地上的思想脉络与新的进取面向。

出身望族，家学渊源

庄孔韶出身江南文化望族，其家世累代因学术、科举、仕宦而闻名天下。从晚明到整个清代数百年里，以迄近代，该家族的学术都对中国产生了重要影响。美国历史学家本杰明·艾尔曼称之为清代的“进士生产工厂”，即出了90名举人、29位进士，11位入职翰林院（实际上，在宗族开创初期的明代也产生了7位举人和6位进士）。

常州庄氏认15世纪后期的庄秀久为毗陵开基祖，先后出了庄起元、庄存与、庄述祖、庄绶甲等著名学者。而后庄氏又传给了刘逢禄和宋翔凤等外亲。通过逢禄之手再传魏源和龚自珍，以迄近代史上康有为和梁启超。由此造就了一个非常著名的学术流派——清代今文经学派。

庄孔韶为毗陵庄氏第22世孙，家族宗族的背景塑造了庄孔韶的学术人生，使他的思考能紧扣现代中国的命脉。其八世祖庄起元重视以血缘关系为基础的家族在维护和强化地方的道德观念和社会秩序方面所发挥的积极作用，主张心与自我修养的联系是文人“经世”“治国”的出发点。为了修炼宗族子弟的人心，他起草了家训，后世名为《鹤坡公家训》。庄孔韶回忆，小时一家人坐在一起吃饭，时常提及家世家训家风，其后续家谱一直保留至今。

出生在这样的家族，其祖先的抱负、古典精粹与学理实际上一代代潜移默化，因而庄孔韶的学问眼界不像一般大学职业者那样为稻粱谋，而是寻求一种志业。《鹤坡公家训》中说：“富贵不在爵禄，而自有至富至贵者存。”庄孔韶对祖训和中国古典思想传承的同时又在新时代里将其转化与升华，其学术视野包括中西教育思想之兼收并蓄，使人类学的教育增添了国家情怀，即寻找新教育中国化的卓有成效的路径。他认为一个地理区域在数千年传承的思想精粹自有其价值，大至儒道全观小到民俗家礼，无不是观视中国历程不可或缺的原因与出发点。

在传统中国向近现代中国转换中，庄氏后人职业选择上慢慢流入现代教育和学术领域。庄孔韶的六叔公（叔祖）庄泽宣考取了庚子赔款到美国留学，最后获得哥伦比亚大学教育学博士，归国后成为中国壬戌学制改革的发起人和起草者，并于1928年在中山大学建立了近代中国第一个教育研究所。庄泽

小学时期的庄孔韶与姐妹

庄孔韶的父亲研究生期间在辅仁大学生物学实验室

1980 年代田野调查

1980—1981 年与黄淑娉先生、程德祺和向导考察西双版纳攸乐山基诺族

1986 年与林耀华先生一同访问哈萨克毡房

20 世纪 80 年代初于中缅边境界碑

1989 年 6 月在闽江上

20 世纪 90 年代初博士后期间于华盛顿大学图书馆

宣著作甚多,尤以《民族性与教育》一书最为接近教育人类学。在中央民族大学读书期间,庄孔韶时常翻阅该书,职是之故,1988年庄孔韶以教育人类学做博士答辩,日后出版《教育人类学》。

庄孔韶的父亲庄之模是20世纪50年代中国显微生物教学片摄制的带头人,受父亲影响,庄孔韶初中就描摹《生物学通报》上的插图,喜欢摆弄照相机,至今家里还保存着小时候用过的照相机。庄先生从小喜爱足球,至今踢了60多年。他的足球研究别具一格,他从人类学的视角出发,率先提出了足球人类学的概念。

庄孔韶身材高挑,在当今中国人类学圈,堪称温文尔雅君子。不论行事、言语还是著述,堪称儒雅。在上课过程中,娓娓道来,令人如沐春风。庄氏家族教育特点还有不分门派和出身,长期熏陶的家风还显露在因材施教上,庄孔韶很少主观给学生命题,而是在考查学生知识结构与擅长的基础上进行开导。故多年来他的民主、平等、温和的人格魅力使得一代代年轻人愿意追随他做学问,其跨学科和兼收并蓄的吸引力,有力启发了年轻人做学问的多元选择,在数十年间形成了多支交叉学科的研究团队,其学术贡献的特点是潜在的和累积性的。

继承师志,思考中国

庄孔韶跟随世界著名人类学家林耀华攻读博士学位,1988年成为新中国培养的第一位民族学(人类学)博士。1944年,林耀华的代表作《金翼》在纽约出版,该书考察了闽江上游的义序和金翼黄村两个聚落的宗族组织、结构和各种活动与功能,认为宗族是地方人民赖以为生的重要文化设施。晚清以来,中国遭遇西方列强入侵,民生凋敝,社会破损,迫使中国走向新生——建设平权的现代法理社会,百余年来宗族被理解为阻碍现代中国生长的一个障碍。林耀华的《金翼》与费孝通的《江村经济》、杨懋春的《一个中国的村庄:台头》、许烺光《祖荫下》被西方学术界称为了解中国经验的四大经典著作。《银翅——中国的地方社会与文化变迁》是庄孔韶的代表作,《银翅》重访了林耀华先生描写过的金翼之家,也是中国人类学重建以来能代表整个学科水准的民族志著述之一。它最早于1997年出版,2020年美国南加州大学民族志出版社出了英文版。可以说,它是西方了解20世纪中国文化实践与变迁的一个重要窗口。这部作品自问世以来不断孵育新一代中国人类学学子,许多年轻人正是因为读了他的人类学教科书和这本深度田野著作而走上了人类学道路。

宗族是儒家意志和人类亲亲性的一个表达，一向被世界学术界理解为认知中国的关键。《金翼》重点关注的是《朱子家礼》及其附带的儒家观念如何向下播化，如何被实践和文化模塑人，如何又在20世纪50至70年代特定历史情境下被中断，继而在80年代以来改革开放的语境中再度复兴的问题。作者看到，金翼之家在传统文化理念支持下，借助种植银耳和链接外部的市场而再度振兴起来。由此感悟到，中国文化具有很强的韧性，当场景不适时它就处于隐在状态以理念方式保存下来，一旦遇到适合土壤时就再度兴盛。应该说，金翼和银翅皆是一种文化意象。在民间看来，20世纪三四十年代金翼之族兴旺发达是因为村落背后的山像一种金鸡展翼；而在庄孔韶看来，改革开放后的金翼之家是因为抓住了"闪亮的银翅（银耳）契机"而再度兴旺。"银翅"这一文化意象准确地向国际上抽绎出改革开放以来中国复苏的内生思想源泉和"察机"的文化动力。

从学术上评估，《银翅》是当代国际人类学长时段研究的重要作品，中国金翼山谷研究团队研究超过80余年。长时段的世代回访的田野调查之重要意义在于，可以寻获一次性调查提不出的问题，即经过数十年社会变故后，为什么还是金翼之家的后辈再度成功？因此长时段调查极为有利于加强对人类学田野点的跨时空的和整体性的把握，庄孔韶做了学理的重要诠释。这本书在20世纪80年代开启了中国经典田野点的回访之风，将中国问题放在千年时间轴里加以思考。在庄孔韶的带领下，20世纪90年代和21世纪第一个十年内，中国一大批新生代人类学家纷纷再访我国老一辈学人的田野研究地点，形成了中国人类学的"回访潮"（见庄孔韶主编的《时空穿行》）。这些学者主要有阮云星、潘守永、兰林友、周大鸣、孙庆忠、张华志、覃德清等人，他们均为当今中国人类学各领域的领军人物。

不仅如此，庄孔韶率先掀起回访潮之后，金翼山谷团队开始了新兴的长时段跨学科综合研究。他们的国际合作团队，法国的多贡人研究教授擅长著作、电影和诗学，而中国金翼山谷团队还增加了多种文体写作、诗歌、戏剧、新媒体、绘画人类学和策展等。他们从传统的中国宗族研究为出发点，逐渐扩展实现多学科交叉与兼收并蓄的新型综观实验，已经形成了自己的本土理论和庞大的跨学科团队成果，经国际交流，近年来尤为法国人类学界称道。

现代性意识从西方蔓延出来，几乎扩及全球每一个角落，其主要特征是追求普遍价值，由此造成了全球人类生活的单一性和均质性。几十年人类学生涯，庄先生足迹几乎遍及中国每一省份，看到了多姿多彩的民族文化，但也目睹了现代性观念影响下大量地方文化的消失。他为此组织团队，在全中国布点，对汉人社会进行了20多年的田野考察，最后出版了《离别东南——一个汉人社会人类学的分解与组合研究》。中国汉人社会除了常见的宗族社区组成以外，究竟还有何种不同的基层组织原理，正是人类学关心的。该书涉及的田野地点有山东、四川、河北等国内诸多省份。这些地点既有东部的汉人社会，也有中国的地理边缘地带的汉与非汉杂处的汉人社会，囊括村落、宗族、五服、戏剧、茶馆、宗教、轮养、情感、生态环境、水利与民族间的文化濡化与涵化等内容。团队对不同地理区域的细致研究不仅回答了汉人社会宗族构成的基本原理，也找到了宗族缺失社会的各色组织卷入的置换逻辑。这种团队性的分解与组合研究，大大有助于理解中国多样性基层社区何以在中国古今变故中因地制宜地调适环境、组织与精神生活，并得以安居乐业。

在迈向现代化的进程中，中国实践了一条独特道路，今后仍要推动中国式现代化建设。在这个过程中，文化多样性的保护与发展问题仍需要人类学的田野思考。庄孔韶及其团队的设计与实践，无疑为这项事业奠定了基础。

2000年代初，庄孔韶进入凉山彝区拍摄了人类学电影《虎日》。这部纪录片再度揭示了传统文化资源对于当下的意义：云南彝族人民借助民间结盟

20世纪90年代初博士后期间与郝瑞教授、朋友范可于西雅图

1990年在西雅图写作《银翅》时定位金翼山谷平面图

1993年回国看望林耀华先生

带硕士、博士研究生回访“金翼之家”

与法国知名人类学家范华教授和瓦努努教授

与陈宜安教授带研究生于西观藏书楼拜访已故著名汉学家施舟人先生

21世纪初与景军参加公共卫生会议

2017年6月绘画人类学跨学科团队在厦门美术馆林建寿画作《囍临门》前

2019年身穿彝族服装于小凉山

2019年4月“虎日”跨学科团队回访

2019年云南跨学科考察团摄制组于小凉山

2021年绘画人类学创作团队金翼之行

2021年绘画人类学团队专程前往
上海观看“抽象艺术先驱康定斯基展”

与黄剑波、宋雷鸣和金翼后人林仁翔于金翼村

《金翼山谷的冬至》海报

2022年访问古田县现代化银耳工厂

仪式戒毒成功。进一步讲，其“虎日”戒毒仪式人类学的研究之所以引起国际学术关注，在于他的这一仪式研究老题目中找到了方法论级别的发现，即这一民族自救的仪式行动，包含了以文化的力量战胜人类生物性的成瘾性的成果，而且他的团队利用追踪性纪录片展映实现了有效戒毒的应用性成果。该纪录片获得了2004年中英项目（HAPAC）和亚洲“最佳实践”奖，庄孔韶被邀请到哈佛大学亚洲研究中心做了题为“‘虎日’民间戒毒禁毒仪式的人类学发现与应用实践”演讲。

根植中国经验，汲取古典智慧

对于中国而言，人类学是一个舶来品。中国的从业者，尤其那些留学西方回来研究中国的学人，如果仅仅是启用西方的人类学概念和理论来套用中国，只能把中国经验流于西方理论的检验标本。一些人追求顺应西方论文系统的一个个思想浪潮，还不如老一辈留洋者拥有对中国文化的自信力。一些人并没有深谙本土儒道思想，便随意评论，想一想一个几千年的宏大思想体系延续至今，自有其道理，其今人何以汲取其精粹，完全在于学者本身，所谓兼收并蓄是以学者的文化自信力为前提的。

“文化”是人类学的核心概念，文化人类学就是因思考人类的文化而建立的一门学科。庄孔韶汲取中国传统文化智慧，从古今关联出发，创造性地赋予文化以动态与生成的意义，即“哲学家所发明、政治家所强化、教育家和乡土文人传播，并最终由农人（甚至所有中国人）所实践”，由此达成“中国何以成为中国”的学术目标,看到了中国文明的生生不息。可以说,在百年中国人类学发展史上用我们本土的文化概念来解释自我,庄孔韶独具慧心。后来,他又在“文化”基础上进一步提出“过化”概念。庄孔韶说:“中国先秦古典文论中出现的一个重要术语叫‘过化’,即指圣贤之经过某地而感化人民之谓。”通过比较可以看出,“过化”重在地域性,而“文化”(以文化之)侧重教化和风化的手段。

庄孔韶“文化”中国的思考也与时代的氛围有关。20世纪80年代正值改革开放初期,中国社会百业待兴,生机盎然,而被压抑已久的传统文化也开始复苏,这促使中国知识界从更深层的文化角度思考中国的前途与命运。但当时大部分知识分子只是在文献里面想问题,而庄孔韶却是走进了中国经验里考察。

除了“文化”“过化”概念外,庄孔韶主要看到科学主义和实证对论文撰写要求的缺陷。在他看来,人文社会学科的写作要求只达成了部分诠释,其“清晰的”结论丢掉了丰富的人文信息,如情感、诗学、隐喻与直觉。为此,庄孔韶在《银翅》里率先在逻辑实证以外,增加了多种文体和文学手法,甚至提出“文化的直觉”专论,这一点曾在西文学者写作中一直受到质疑或令学者望而却步,然而却是不可丢失的认识论组成。

庄孔韶明确指出,文化撰写补充诗学的新的民族志叙述方式导致了一种通向整体性认识论和综观的有效努力,尤其是他郑重区分了“诗学观察和科学实证属于不同的方法论”,因此从田野工作到文化表征成为一个实证与诗学的复合过程。笔者提前阅读了他在2024年法国专刊上的学术新作,就展现了以中国古典经验与批判性文论精粹重构人类学的田野-写作流程框架,提出田野工作中的互动极致与文化表征的认知极致的新模式,已经容纳了他的团队诗学与直觉实验的成果。其研究的出发点是人类学,却跨越出人类学,直接进入交叉学科,而新理论的形成主要依据中国古典文化精华思想,又不脱中西学术思路之融通。

庄孔韶及其团队近四十年一直致力于一项跨学科的实验。他们关注哲学综观与对认识论的批判性改造,却没有拘泥于空泛思辨,而是在团队的大规模交叉学科实验中扩充互动与文化表征的新理论。

在跨文化的比较中,庄孔韶近年来专攻文化表征的不可言说部分,特别在文字以外的图像表达系统探索隐喻、直觉与不可言说,这是探索一个新的表达范畴，并以团队的长年学术实验加以推动。显然不可言说和难以言说的人类知识超过了要求“清晰的”明述知识结论的理解范畴，而不可言说的表达经验刚好是中国古典文论的诗性智慧。这是庄孔韶及其团队“不浪费的人类学”理论框架的重要学理内涵。

庄孔韶创造性地移用了中国古典文论中的直觉主义作为一种田野工作方法与技巧，以区别于西方人类学一套成熟且刻板化的格式化实证调查方法。人类学的直觉主义方法拼得是学者对田野的熟悉程度，只有当学者对所研究的社会、文化和人群非常熟悉的时候才能使用。从事田野工作时，研究对象一举手、一投足，甚至一个眼神、一种声音以及发出声音的高低都能让调查者迅速而准确地捕捉到对方的心意和情感。这种无需借助询问、反复的核实和调查问卷等常规手段获取信息的办法，实际上是一种得鱼忘筌的技巧，即直接获得了鱼，而不必再寻觅捕鱼的工具——筌。直觉主义的源头可以径直追到《庄子》，不假外物而得道，正是直觉主义的精髓所在。当然，如果仔细耙梳林耀华的著述就会发现，他在《拜祖》一文中就强调古典与理念跟现实的对接。所以，也可以说，庄孔韶发现古典文献中的一些概念可用于今日之人类学，也是承继了林的思想。

强调跨学科，凸显“不浪费”

20世纪90年代中期以来，庄孔韶特别强调人类学研究的跨学科视野，提出了“不浪费的人类学”概念，其意涵是在通常的人类学民族志表现手法之外，在启用随笔、诗歌、绘画、电影等表现手段来展现中国经验并表达研究者的体验和感受。可见，体现庄孔韶的创造性精神的“不浪费的人类学”，实际不只是方法，而是来自涵摄不同方法论的研究模式。其主要内涵是，不能只会用文字呈现我们的研究对象，还应该借助其他手段，比如摄影、电影、随笔、小说、诗歌和绘画创作等形式和手段来展现在调查研究中的体验和互动感受。文字民族志外，他分别实践了影视人类学、文学与诗学人类学、绘画人类学等样式，形成了相关学术团队，并各自发掘了跨学科创意与理论归纳。

绘画人类学团队由中国艺术研究院的签约画家林建寿、上海戏剧学院的胡继宁、福建大学厦门工艺美术学院的蔡志鸿等50余人组成。这是一个以油画为主的学术群体。庄先生的系列油画作品之一《金翼山谷的婚礼·入洞房》入选中国油画院和中国少数民族美术促进会“民族百花”油画精品展。

文学与诗学人类学队伍由中央民族大学的徐鲁亚、中国社会科学院民族学与人类学研究所的周泓和方静文等人组成，此外还有国际学者，如法国远东学院研究员范华（Patrice Fava）等人。庄孔韶本人在这方面也取得了不少成就。先后出版了《情人节》《北美花间》和《独行者——人类学随想丛书》（包括《家族与人生》《文化与性灵》《自我与临摹》《表现与重构》《远山与近土》五种）等著作。在人类学电影方面，庄孔韶先后拍摄过《端午节》《长江沿岸田野纪行》《怀想——北京“新疆街”的时空变迁》《虎日》《我妻我女》《金翼山谷的冬至》等著作，其中《端午节》《金翼山谷的冬至》和《虎日》三部纪录片在国际上影响最大。《端午节》于1992年入围美国玛格丽特·米德电影节，《金翼山谷的冬至》于2018年入选英国皇家人类学会电影节、2019年入选中国台湾国际民族志电影节。它们合起来也可以叫作跨学科的影视人类学。

应该说，这些多样的形式并不鲜见，亦非庄孔韶首创。比如，鸟居龙藏在从19世纪末到20世纪初对我国东北地区、内蒙古东部、台湾和西南“苗族”的考察中就拍摄了大量照片，同时每到一地他都撰写日记。又如，1922年马凌诺斯基出版的《西太平洋的航海者》和1940年福忒斯出版的《努尔人》中也配有不少照片。西方很早就拍摄了关于因纽特人的无声民族志电影，记录了他们的捕捞与生活。我国摄影人类学家庄学本在20世纪三四十年代深入藏区拍摄了大量照片，成为中国人类学史上最杰出的摄影人类学家。新中国成立后，中国科学院民族学研究所在少数民族识别和历史调查中也拍摄了多部人类学纪录片。但是，中外人类学家没有一个像庄孔韶这样将各种形式整合成一个整体，而彼此又互文，即构成一个语义结构链而互相阐释。

跨学科的人类学是双向的探索。比如，既有人类学的诗学创作，也有关乎诗学的人类学理论思考；既有人类学立场的绘画创作，也有关于绘画的人类学学理思索。一般说来，学术发挥理性认知的长处，而艺术注重体验和情感的表达，但在民族志随笔、文学人类学、电影人类学、绘画人类学诸形式中庄孔韶能有机融合，而且在这些探索中他再次发挥了中国古典文论中有关思想的价值。他将袁氏三兄弟“公安派”的性灵、真、趣和刘勰《文心雕龙》中的“隐秀”等批判性文论警句等，纳入其人类学田野过程研究的互动极致与认知极致的学理关联定位之中，从而开启了中国学者以本土理论进入国际人类学理论之林的探索。其“不浪费的人类学”以多种交叉学科的长年实验，是为了人类学研究能改观实证性研究的不足，而获得哲学、人类学、诗学、艺术的关联，做整体论意义上的综观，特别是实现中国文论批判意义上的“触类旁通”。庄孔韶团队没有拘泥于学术思辨一途，是不多见的学术行动团队，在40年间用不同专业手法侧重探索言语、撰写、图像等的不可言说、言外之意，以及隐喻与直觉，为实证与诗学哲学建立了通达的桥梁，而其成果已经超越了人类学，弥补英语学术圈常常却步的不可言说研究，而这正是中国古典文论之批判性思想精粹与诗学经验的擅长。

除此以外，庄孔韶将摄影带进了医学人类学，和清华大学教授景军等一起推动了医学人类学在中国的开展，这方面也形成一个团队，并开启了“应用的影视人类学”阶段。

胸怀国际，经世致用

20世纪80年代后期，庄孔韶接过林耀华肩上的担子，任中央民族大学民族学系主任，按照世界通行的民族学人类学学科架构进行课程结构改革，中国化是他传承的学理核心思想之一，随后全国各地民族院校纷纷仿效。2004年调任中国人民大学，组建了人类学研究所，直至荣休。而后又前往浙江大学续建人类学研究所，数年后又到云南大学担任人类学首席专家至今。

透过庄孔韶的学术生涯发现，中国学术持久传承的经验有：中国本土思想系统的深度理解，大族家学与师生传续的重要性，国际学术交流之兼收并蓄，以及社会文化科技变迁中的学者自身的适应性选择等尤为重要。庄孔韶从不被动地蜂拥命题，反对学术的形式主义，而是依据师生团队的知识结构和外界建立联系和引申学术悟性，因此他的调研和写作始终是依据学术意义的感知、兴趣、公益心等随机设计。

庄孔韶的金翼山谷和虎日两个研究点的多样性成果，表明国际长时段田野点具有比一次性调研点的优势已被证实，因为长时段点不仅可以提出一次性点提不出的问题，而且为长年陆续卷入的多学科师生团队成员做单独的或组合的研究提供坚实的基础。事实证明，他的“不浪费的人类学”便是得益于可以不断调整设计，特别是团队不同学科志同道合者不断进入，导致了人类学交叉学科实验的随机性。

庄孔韶认为，中国人类学跨学科研究一定是跨系跨学院和跨校的，尤其跨校的志同道合者团队是推动重构批判性认识论和趋向更完满综观而不可多得的手段。这一数十年设计并实践的跨学科学术研究更改了人文社会学科既有模板，将中外各自擅长的实证与诗学等不同的方法融合起来，以实现中国古典文化思想精粹进入世界人类学理论之林的努力。他从文化撰写创新到诗歌歌谣、电影、戏剧、数字媒体、绘画和策展运用，呈现了跨学科和运用多模态的平台效应，使中国人文社会学科（以及生态、医学等）教学科研得以在此交汇，因为其意义不止于人类学本身，在策展过程中实现不同学科的主体性和主体间性成为必要的深化学术环境。

几十年来应邀国内外各地演讲和参加学术会议更是难以计数。他开启了多项学术议题，把人类学的种子播散到全国各地，组建起一支文科研队伍，被学界视为当代杰出的人类学民族学教育家。庄孔韶的人类学不只是象牙塔里的学问，近年在“反哺”理念支持下，他还积极帮助金翼黄村的村民向现代文旅业转向，目前金翼黄村已经成为一个观光点，让人们体验传统文化的魅力。同时在他的启发下，玉田人民开发了一系列“金翼”品牌的地方产品，实现了传统资源的现代市场价值。由此，他跟他的祖先庄起元一样实现了阳明心学所推崇的“知行合一”。

尾语

20世纪80年代，随着改革开放政策的实施与推进，中国人类学得以恢复和重建，庄孔韶和他的同道们再度具有了世界眼光。此时，国家把现代化变迁的任务交给了他们，让他们为解决中国现代化建设中出现的各种问题而贡

2019年10月24日国际影视人类学会主席向庄孔韶颁发“中国民族志纪录片学术展终身成就奖”

2019年11月与乔治·马库斯和迈克尔·费彻对话于云南民族博物馆

2022 年 12 月云南大学校院领导考察“金翼之家”

于中国油画院画室

2023 年 9 月与云南大学民族学与社会学学院研究生踢球

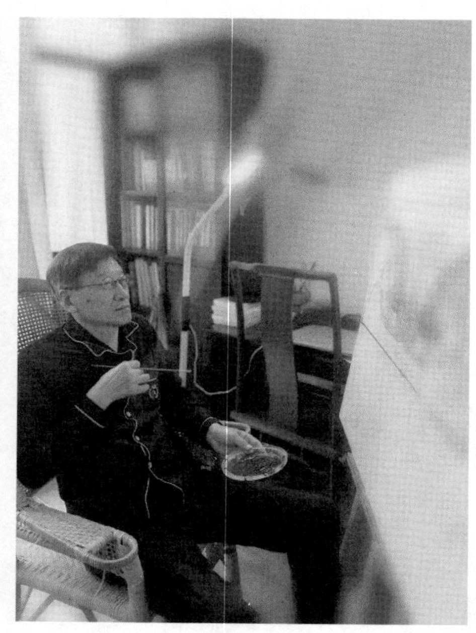

于家中创作

献智慧。因而可以说，庄孔韶的学旅人生是在从传统中国到现代中国转换、新生与发展的路途上演绎出来的，是中国当代叙事的一部分。

然而，作为生命个体的庄孔韶又是中国文化的一个集体表象，他在接受西方现代学术训练的同时，又背负着数个世纪累积的家族历史、文化传统以及相应的一套感知世界的方式。自然，出生于这个群体之中，人的眼界是很高的，把个人的生命融入国家的、民族的历史叙事之中，拥有超越个体的国家情怀自在情理之中。

庄孔韶近期作品（2019—2023）

因篇幅所限，仅列近五年作品。

（一）论文（第一作者）

[1] 庄孔韶，朝戈，林建寿，等．理论与方法的跨学科实验引申——绘画人类学二十年心得笔谈[J]．广西民族大学学报（哲学社会科学版），2023，45（3）：34-46+52.

[2] 庄孔韶．人类学的诗学探索导引[J]．民族文学研究，2022，40（4）：25-32.

[3] 庄孔韶．从生态、电影到绘画：人类学跨学科研究和适时中转[J]．西北民族研究，2021（3）：135-142.

[4] 庄孔韶．文化表征的多元方法与跨学科实验[J]．民族研究，2020（6）：55-66+140.
被《新华文摘》2021（10）全文转载，进入封面目录。

[5] 庄孔韶，乔治·马库斯，迈克尔·费彻尔．对话：文化直觉、艺术实验和合作人类学[J]．民族文学研究，2020，38（4）：166-176.

[6] 庄孔韶．过化、权力、采借与情感——中国汉人社会多点研究归纳[J]．中南民族大学学报（人文社会科学版），2020，40（3）：38-51.

[7] 庄孔韶．人类学电影和学术专著如何互释？——以《金翼山谷的冬至》和《银翅》为例[J]．广西民族研究，2019（6）：43-49.

[8] 庄孔韶．"男子汉"精神与特质从哪里来？——2018世界杯足球赛体育人类学聚焦[J]．体育学研究，2019，2（4）：88-94.
国家社科基金重点项目：中国足球振兴的文化策略研究（项目编号：17ATY004）

[9] 庄孔韶．金翼山谷冬至的传说、戏剧及电影的合璧生成研究——一个跨学科实验的人类学诗学[J]．民族文学研究，2019，37（4）：5-14.

[10] 庄孔韶，范晓君．扩大人类学的艺术研究视野——以中国当代美术教育为例[J]．思想战线，2019，45（3）：56-64.
被《中国社会科学文摘》2019（11）全文转载。

[11] 庄孔韶，张静．"并家婚"家庭策略的"双系"实践[J]．贵州民族研究，2019，40（3）：41-45.

[12] 庄孔韶．流动的人类学诗学——金翼山谷的歌谣与诗作[J]．开放时代，2019（2）：211-223+11.

（二）中英文著作

[1] 庄孔韶主编．林耀华全集（全八卷）[M]．福州：福建教育出版社，2023.

[2] 庄孔韶主编，徐鲁亚副主编．人类学的诗学探索[M]．北京：中国社会科学出版社，

2022.
[3] 庄孔韶.国学、足球与艺术：庄孔韶自选集[M].北京：中国社会科学出版社，2020.
[4] 庄孔韶等著.离别“东南”：一个汉人社会人类学的分解与组合研究[M].北京：中国社会科学出版社，2019.
[5] 庄孔韶主编.绘画人类学[M].北京：中国社会科学出版社，2019.
[6] Zhuang K S. The Silver Wings: Local Society and Cultural Changes in China (1920s–1980s)[M]. Zhonghao Xie trans. Los Angeles and Austin: Ethnographics Press, 2018.

（三）教材

[1] 庄孔韶.人类学通论[M].第四版.北京：中国人民大学出版社，2020.
[2] 庄孔韶.人类学概论[M].第四版.北京：中国人民大学出版社，2022.
普通高等教育“十一五”国家级规划教材
教育部普通高等教育精品教材
教育部高等学校社会学类专业教学指导委员会推荐教材

获奖情况

[1] 2019年10月获“中国民族志纪录片学术展终身成就奖”。

[2] 导演的《金翼山谷的冬至》(2017)获奖情况：

(1) 2019年10月，台湾国际民族志电影节入选。

(2) 2019年3月，英国皇家人类学会电影节特别展映。

(3) 2017年10月，该片在中国人类学民族学研究会宗教人类学专业委员会年会(电影节)，影视人类学先驱让·鲁什(Jean Rouch)先生百年诞辰纪念展映中荣获优秀学术展映奖。

[3] 油画《金翼山谷的婚礼：入洞房》入选2021年中国少数民族美术促进会主办“‘民族百花’——庆祝中国共产党成立100周年百幅油画精品展”。于2021年6月27日—7月9日于北京市中国油画院云上美术馆展出。

代表作

文化表征的多元方法与跨学科实验

——1979—2020年人类学团队研究的历程与思考*

庄孔韶
（云南大学）

关于文化表征的研究关系到人类学整体论的原理，涉及方法论、功能、类别综合及过程的诸种视角，被认为是“十分重要并且必要，但却不可能完全实现的美好理想”。[①]本文为此呈现的认识论指向便是讨论各种研究手段何以实现综观。其缘起不只是人类文化表征的文字撰写的限度，而是直指学术论文的局限性。为了打破学科壁垒及其撰写约束，从1978至2020年，笔者团队在因袭中国文人写作和当今人类学家“离经叛道”作品，进一步在纵横两个向度扩展研究，采纳文化表征的多元方法（写文化、拍文化、画文化、演文化、展文化等）和跨学科设计与行动实验，以求人类学学问的触类旁通，即走向人类认识论通达的综观。

一、写文化

中国古代的文人作品并没有如同近现代西洋科学主义、清晰无误的和逻辑的结论的思想影响，例如孔夫子著名的《论语》就不用确切的“定义”而是使用比喻来答疑，而且的确有些问题是具有指向和可以理喻的，不一定是确切的或唯一的答案；林语堂认为顾炎武的《日知录》三百年间所得之盛名也非来自逻辑之论证[②]。那些深受古代书写文化影响的近现代中国学者，包括人类学家、文学家的随笔和札记，如《皇权与绅权》同样喜欢用去掉论证过程

---

* 原载于《民族研究》2020年第6期。

的写法，或使用体验、直觉和逻辑混生的讲道理手法[3]；也有人在中西学术论说体例差异之间，写出以古诗、格言做引子，内含今文经学体例和“微言大义”，又符合西方论文框架的“令人愉悦的”（凯恩斯语）大著，如《孔门理财论》[4]。

科学主义主导的论文写法充斥中国大学人文学科的一个世纪以来，中国学者（有时是不得不）分别采纳了顺应的和回旋的做法。前述陈焕章中国经济史论著作不就是由特有的中国文人习惯的叙事惯习和巧用诗文推导出的实证论著吗？ 1935年林耀华先生的学士论文《义序的宗族研究》[5]使用了正文的功能主义论述写法和注解里躲藏的国学解释并置；不止于此，在他完成了哈佛学位论文后，又写了“离经叛道”的小说体的著作《金翼》(1944，1948)[6]，于是出现了真实与虚构的早期讨论，不过现在已经不成问题。有一位人类学家在非洲调研之前，惊奇地得到了林先生的人类学体《金翼》(发现不是标准论文！)，就带在了路上，也影响了她自己的田野作品。[7]当然还有一些人也这样做，有人虽不是写小说，他们的田野民族志也用了小说笔法，业已存在的多样性文化写作的努力显然影响了日后的人类学。从中国古人的“体道”和“体认”就包含了隐喻和直觉，这类作品后世数量也很大，但很难融入科学主义操盘的论文中，这是逻辑实证论以外的范畴缺失，往往民间札记随笔式的议论更为喜闻乐见。《金翼》的第二代学人完成了《金翼》的续本《银翅》[8]（1986—1989田野调查，1996桂冠出版）其中文版顺利出版和英文版一再推迟出版(2018)，似乎都和中西出版社编辑各自的“守门员”评判相关，因为这一大部头论著在论证之外，容纳了旅行文学、小说、随笔、文化的直觉与感悟的交叉手法，中国出版社似乎习惯于中国文人传统，的确，这样的学术著作的《金翼》和《银翅》在中国学界内外的销路一直不错；不过英语的出版社习惯于实证论文，所以不依照编辑的大幅修改意见就难以出版。幸好《银翅》得到熟悉中国文人写作传统的南加州大学的人类学家的出版支持，终使携带有地区性认识论惯习的思维与实践得以彰显。

关于诗作的人类学探讨，常见文化互动瞬间灵感触发的意象，而地方长久流行的歌谣则是民间群体性真情感知之精粹。[9]林耀华80年前收集的福建古田《搓圆》歌谣[10]，出现在一年一度的冬至农家厨房里，全家人围桌一边搓圆一边齐唱，实际就是展现自然轮转与民俗韵律的和谐之道，当然伴随了中国传统家族主义的政治与文化诗学。在1979—1983、1986—1994年期间开始出现人类学诗的密集创作，那是因为林耀华的第一批（1978）研究生初到云南山地和坝区民族居地，便产生了第一波田野诗作、散文和旅行随笔。他们一方面像本尼迪克特那样在文学诗刊上发表，一方面不间断刊行人类学诗集、散文集、小说和摄影集。[11]这一诗学团队到1989年以后，加入了电影摄制者、油画家、科普作者和翻译家；纪录片《端午节》[12]（1989，1992）和《金翼山谷的冬至》（2017，2018）里都有民俗歌谣和诗作的镜头，人类学诗学已经从写作表达卷入到电影镜头的美学之中。[13]

从1978年后的十余年间，我们已经形成了在一个调查点上完成文字与影视多项联合作品，比单纯发表论文和专著能够更为充分地展示那里的文化。于是笔者在1995年在北京大学的一个人类学高级研讨会上提出了"不浪费的人类学"的认知与行动理念，实际是想推动文化表现的多元方法综合实验。"不浪费的人类学"是指"人类学家个人或群体在一个田野调查点上将其调研、互动和理解的知识、体悟及情感用多种手段展示出来。著书立说以外，尚借助多种形式，如小说、随笔、散文和诗，现代影视影像手段创作；邀集地方人士的作品或口述记录，甚至编辑和同一个田野点相关的跨学科互动作品，以求从该族群社区获得多元信息和有益于文化理解与综观"。[14]笔者对"不浪费的人类学"含义的比喻是"人类学者对文化及衬在文化底色上的人性之发掘充满热忱，我们有点不满足本学科论著论文的单项收获，好像农田上功能欠缺的'康拜因'过后，还需要男女老幼打捆、脱粒、扬场，乃至用各种家什跟在后面拣麦穗一样，尽使颗粒归仓。"[15]法国人类学家Wanono教授[16]认为"不浪费的人类学"可以将其理解为"无次要材料"的人类学或"人类学资源不分主次"，应该是最为中肯的理解。

二、拍文化

中国新一代人类学家的田野调查大体在1978年以后得以恢复，林耀华后辈的田野工作率先在云南展开，三位研究生（程德祺、庄孔韶和王培英）由林耀华教授指导，黄淑娉教授带队[17]，深入西双版纳腹地的哈尼族、拉祜族和基诺族居地，考察中缅边境的干栏住房和世系群聚落开始借助绘图和摄影（主要是黑白照片和国外引进的彩卷）的方法，研究主题之一是山地民族的游耕业与家族制度。及至1986—1989年开始的回访金翼山谷（福建省古田县）则不仅使用了彩卷，还使用最新的广播级摄像机（Beta Com）拍摄了金翼山谷内外的端午节（摄影团队成员庄孔韶、杨树、张小军、王燕平、林忠诗和吴同营等)，展现他们的家庭生活、龙舟竞赛和游神活动。由于我们的电影《端午节》里的主人公是“金翼之家”后辈和邻居，就为理解林耀华的小说体人类学著作《金翼》起到了文字不可替代的作用。就像影视人类学家戴维·马杜格(David MacDougall)所言，“电影永远无法取代文字，但人类学家了解到文字表达田野经验的限制，而我们已经开始挖掘影片如何弥补这些盲点。”[18]

后来，《端午节》荣幸入围纽约玛格丽特·米德电影节，同年在华盛顿大学出版社出版(Stivan Harrell协助英译)，使得人类学著作《金翼》得到更为广泛的传播。此时，我们的团队已经对文字和图像之间的互补作用(因二者之间不可替代)有了深刻的感触。当《金翼》著作的续本《银翅》出版后，所配合的纪录片也进一步为两本专著增色。2017年笔者重返福建金翼山谷拍摄冬至节日生活，《银翅》一书中的金翼之家后辈和主人公们得以再次现身新的纪录片《金翼山谷的冬至》[19]。于是，在80年间两三代师生让《金翼》、《银翅》两本书和端午、冬至两个节日电影组成了一个地方中国文化的研究成果系列，还包括整个团队的诗学沙龙活动进入电影系列[20]。

笔者研究团队除了在福建省金翼山谷的重要调查点以外，还有上文提到的云南省宁蒗县的彝族居地。1990年代，宁蒗县的小凉山彝族头人们受困于来自金三角的鸦片和海洛因毒品危害，他们商讨沿用本民族的“虎日”盟誓戒毒仪式自救。他们在神山面前举行家支仪式和吸毒者的家庭仪式，推崇信仰尊严、道德约束、亲情教化和诚信毅力等集合的力量，先后在两组吸毒人员中实现了世界上的戒毒高成功率(64%和87%)。

“虎日”戒毒禁毒行动旨在说明，我们可以以古老的盟誓仪式所携带的文化的力量，在一定条件下战胜人类生物性的成瘾性。这一方法论成功实践的意义在于，在科学以外，还有一种可以如愿以偿的文化力量可以运用，这便提升了作为古老文化遗产的现代应用意义。这类公益与应用的实践马库斯称为公共人类学(public anthropology)，景军则使用行动人类学(action anthropology)。我们及时拍摄了2002年的虎日盟誓戒毒仪式[21]，并连续七天在当地电视台播放，促进了小凉山各地彝族头人仿效，使戒毒成果不断扩展，故而“虎日”盟誓戒毒仪式的电影被称为“应用的影视人类学”(applying visual anthropology)。因为它找到了“未经解决的新兴问题”的方法。[22]意思是说，不是所有的电影都可以应用，只有获得特定的发现机缘，人类学纪录片才能导入参与解决当今社会问题的全新的应用方向。[23]

自1999年首次拍摄《虎日》以来，已经过去了20年时间。那么，从2002年至今已有17年（1999年最早一次至今整整20年），电影里的主人公们的下落如何？笔者团队最近5年先后组织了四五次回访调研，证明了彝族人民“虎日”盟誓戒毒所携带的文化的力量的巨大意义。我们一面接续拍摄《虎日》电影里几位重要的主人公的下落，借助了电影和新媒体手法完成新作品（如触摸屏作品、“网红”作品和口述作品等），已经在云南省民族博物馆参加了展映。与以往不同的是，我们新近拍摄电影的同时，已经有20年合作经验的几位油画家也参与了写生和创作，而画作中的小凉山风光和人物里可以找到人类学论文和纪录片中的主人公。这里人类学家联合电影摄制者、画家和人类学家共同参与了云南的田野工作，产出了不同凡响的作品。直观的电影只不过是历史、哲学、文化体验在人类学主旨下的镜头展示，而多学科、多专业和新技术并置则提供了日新月异的展示效果。于是我们的学术聚焦于数字技术和新媒体对人类学文化诠释的影响上面，以及电影、网络直播如何更新高夫曼的“面对面的互动”的解读意义。

三、画文化

2019年11月30—12月1日在云南大学和云南民族博物馆召开了绘画人类学国际研讨会暨人类学绘画展，听起来像是一个通常的油画作品学术研讨，而实际上此次学术活动是以人类学绘画作品为重心的跨学科合作与行动人类学的展示，包括论文、电影、诗歌、绘画、表演、戏剧和新媒体成果[24]。而这些作品的主要田野工作依据就是1999年和2002年的“虎日”盟誓戒毒仪式后续的长时段研究，以及借鉴部分在福建金翼山谷的经验。

说到绘画和人类学结合的初起构想源于30多年前的一次体验。一次在福建的红头道士的家庭科仪，是为一位不生男孩子的村姑做法。然而当外来的人类学家、诗人、画家、电影摄制者来到现场的时候，道士宣告了科仪禁忌：不许拍照摄像，不许支画板写生！这倒不奇怪，很多人类学家都有类似的经历，如平图琵人的“梦境”图案就不许画师和不同性别的人看；[25]闾山派法事有时避外人，或允许参加却不让拍照等情形非常普遍。显然此时电影摄制人已无计可施，我们原以为画家也无能为力，殊不知画家对我说，我们和人类学家一样，也可以凭借参与观察和记忆画画。这里用的是柏格森的理念“记忆就是形象的存活”[26]，记忆既指向物质与存在，又指向表象与精神。[27]因此，在记忆中的写作与绘画“是过去与当前针对未来的‘综合体’”，[28]是一种全新的物质与精神“交接”的展示。无疑，在文化展示的多学科、多专业之间存在着难以替代的文化创作方法差异。

20年前的1999年，我们团队的人类学家和画家林建寿开始讨论如何“画文化”。那已经不是无源之水，而是水到渠成的。笔者开始静悄悄地和画家共同创意和制订绘画人类学的方案，以此作为一个出发点。于是我们诞生了第一批人类学绘画作品。如上面提到的红头师公和村姑的《祈男》[29]，而且这幅大型油画不久前用于我们的新纪录片《金翼山谷的冬至》的序幕，林耀华的晚辈，闾山派的林芳德道士在侧。其他还有作为文化冲突化解的同一时期油画《刮痧》[30]，时空快速变迁的社会文化观察油画《回访》[31]，以及跨国、跨文化、跨宗教的大型婚礼场景《波兰婚礼晚会》[32]等，均进入绘画人类学的框架之中。与此同时，人类学的绘画研究写手也对中国一个油画艺术学院群体的哲学、思维方式、艺术传承和画作流通，给予关注访谈和解析，则是以论文的形式出现的[33]。

经过近五年田野调查，从参加“虎日”戒毒仪式的戒毒者、电影中的彝族头人和男女老少的访谈与互动中，我们梳理了若干人物线索，不仅追拍了他们的记录和口述影像，还深入了解那里的民风民俗，创作了新的意义上的风景画和肖像画，那已经不是随意写生的风景画和人物肖像画。这些作品最终成了这次昆明绘画人类学会展的主要构成。

我们团队走过云南山地密林中隐约的毒品通道（欣赏胡继宁油画：《贩毒通道的风景》），并借助无人机拍摄了大面积的中缅边界山林和小路；多次探访人类学电影《虎日》中参与盟誓戒毒仪式和族群自救的几位头人。当年积极参与和组织盟誓戒毒活动的彝族头人嘉日姆几，如今已经变成了小凉山出身的第一位人类学博士和大学教授，他的肖像油画展现了无比的骄傲和社会学术担当。《虎日》电影里唱念禁毒宣言诗句的莫色不都，如今成了“虎日”戒毒的义务宣讲员（欣赏蔡志鸿油画《莫色布都》和林建寿油画《禁毒宣讲会》）。十余年间，他在学校、工厂和政府机构义务禁毒戒毒宣讲260余场。一位遇到挫折后再度成功戒毒的高大男子憨鹰（欣赏林建寿油画《一个成功戒毒的彝族男人》），在画展上现身说法，对照画家的作品，憨鹰的肖像被解读为“充满内疚、磨练和携带未来希望神情的作品”，这大概就是绘画不同于照片和电影的关键之处——一种特有的复合性情感表现手法。这些作品都入选了这次人类学画展的肖像画部分，电子触摸屏可以随手看到他们参与戒毒仪式和各种公益活动。因此，我们是在人类学田野绘画实践中改造了传统肖像画和风景画的内涵，在绘画中赋予了深度的隐喻和社会文化含义。

这次绘画包括“文化的隐喻”主题。例如一幅画上有宁蒗彝族的牧羊女和三只放牧的羊[34]。宁蒗羊是中国地方优良品种，窄脸是宁蒗羊明显的特点，还有全黑、全白和花羊（白羊毛上突显黑颈、大小不等分散的黑斑毛色）。这一幅令人赏心悦目的绘画实际上包含了更为深刻的含义。原来彝族哲学中的黑意味严重，白是轻微，而各种黑白相间的花的含义是介乎于严重与轻微之间。凉山彝族的哲学生成就直接联系于他们生活中的羊群。“隐喻将熟知的和陌生的事物特点合并在一起或将熟悉的特点进行异化的合并，便能有助于激发我们的思想，为我们带来全新的视角并使我们兴趣盎然。”[35]由此，我们看到了彝族事物分类的文化惯习。通常说二元分类具有清晰的对比意味，是认识事物的一种简洁视角，但在实际社会生活中，绝对的二分常常使人无从把握。如现代社会的法院判决，严重与轻微、胜与败的二元对立，如同黑与白的分类，难以解决那些不黑不白、亦黑亦白的各种中间状态情境，直白的法庭胜诉和败诉宣判，有时容易留下不服和仇恨。彝族人的生活处在国家法习惯法内外，到处可以感受到法律和习惯法并存遇到的一些哲理与情感判定困惑。花作为黑与白的过渡，存在着模糊的空间，因此也就意味着有很多可以协商的空间，这一空间把握的重要性在于，彝族人的哲学展现了“抚平伤口”（嘉日姆几语）而不是在伤口上撒盐的初衷。如杰伊指出在特定文化环境中的观众需要关注“观察不同视觉体制下的隐性文化规则”。[36]

《虎日》电影里为盟誓戒毒者安排邻里保人，就是为了把严重的“黑”转化为“白”，即化险为夷的做法。不过这是动态的过程。而这次画展有几幅画都包含了彝族黑花白哲学的隐喻，却是静态的。然而这不就是横向学问表达的差别吗？

你知道《百乐书》[37]吗？这是两三百年前云南彝族人的占卜签书，使用矿物和动植物天然颜料彩绘而成，图文并茂，都是解释命运福祸的主题。我们选择展出了8幅，极为难得。此外，绘画展大厅还展示了当代杨梅山彝族毕摩画师用柳木碳、红毛叶、墨汁、蛋黄、矿石颜料等画的家坛霉运祈祷绘画，绘画过程伴随着画师附体、出神的仪式，画师边击鼓，边舞蹈，并颂唱祈祷诗。（彝族李世武教授解说）会展期间，专业油画家、毕摩画师和由当地成长起来的研究人员，对颜料、绘画用途和寓意做了极为有益的学术交流。我们也看到，人类学提供了参与观察的理论依据，或许在某种意义上说，地方性、主体性、族群、哲学、情感、认知、技艺、体验、隐喻与直觉或明或暗地整合于展出的画作之中。这个团队注意到学者和当地人的长久田野交流，因有本土知识专家作枢纽，形成流动的互主认知过程，参与观众、田野原生地同胞和学者各得其所，然而这不就是人类学的初衷吗？！

四、演文化

在云南省的一次地方行政会议进行之中，一位披着“查尔瓦”斗篷的彝族汉子出现了。他有礼貌地说，可以中断一下会议吗？我只用十分钟宣传一下我们宁蒗的“虎日”戒毒精神。他的宣讲可不是照本宣科，是用押韵诗句和高亢的呼唤，动员这里的行政官员积极参加到禁毒戒毒的行动中去。

会展当天，就是这位不期而至的彝族汉子再次出现在民族博物馆的展览大厅，明眼人一下子看出，他也是20年前《虎日》纪录片中用押韵的叙事诗朗诵戒毒宣言的人。他引领着小凉山宁蒗的年轻人，身着民族服装，借助传统的招魂诗形式“旧瓶装新酒”，伴随着不时兴起的高亢的“欧拉”（“回到祖地吧”的寓意），转换为呼唤误入歧途的人回来吧：

哦，回来[38]
哦，回来
回啊回头吧
你这视毒如命的瘾君子啊
回啊回头吧
少不更事的孩子啊，回来吧
站立不稳的马驹啊，回来吧
要趟水过险河的人啊，回头吧
要攀越岩壁的人啊，回头吧
坐在阴森牢狱的人啊，回头吧
口舌之错的人啊，回头吧
脚踩歧路的人啊，回头吧
你这视毒如命的瘾君子啊，回头吧
羊到山头就回来

猪到沼地就回来
游荡山头乌鸦之乡的人啊，回头吧
藏匿深谷喜鹊之乡的人啊，回来吧
逃亡到彝汉乡间的人啊，回头吧
你爸你妈在等你
你儿你女在等你
哦，回来！
哦，欧拉！

当莫色布都伸开双臂呼喊“欧拉”之时，不仅在场的彝族，还有与会的中外学者、诗人、画家、摄影师和新媒体人都在“虎日”精神的感染中异口同声呼喊“欧拉”，让吸毒者悔悟的召唤响彻整个大厅。这一族群、信仰、尊严的庄严展演，就是小凉山彝族“虎日”戒毒精神的延续！于是彝族头人、学者、“欧拉”表演者先后站出来即兴发言，其聚焦的问题是：古老的戏剧、仪式表演的民俗文艺形式如何在恰当的契机使非物质文化遗产永存。

会展的序幕只有一两分钟。来自金翼家乡的闽剧演员“母子”（李青、吴霞饰）从大厅的观众前鱼贯而出，用古田地方戏平民化直白的唱腔和泼而不露的舞姿，演绎了劝子远离毒品的过场戏。当“瘾君子”山野落魄时，演员把泉水淋在手心、绕圈拌和燕麦炒粉的做派一出现，便一下子被在场的小凉山彝族观众所识别，和演员仅仅一米间距的观众发出会心笑声，并且和场上演员眼神互动，赢得了掌声。

中国现在有不少讲究的剧院，剧目因过于和观众区隔而招致批评，因此如何和观众互动成了业界讨论的主题。殊不知古老的中国村戏和会馆戏剧就具有良好的互动关系。因是熟人社会，乡村舞台下的观众可以喝茶，和演员打招呼，跑到后台直接评论，演员和观众用地方俚语插科打诨。可以说现代社会学家戈夫曼的“面对面的互动”[39]就在乡村舞台内外，戏剧和社会“戏剧”的边界在乡间一直混然不清。“人生如戏”这一中国民间古老的经验性俗语和现代学术观察实际是相同的。

2017年我们在金翼山谷演出了古老传说剧目《猿母与孝子》[40]，就是这出戏的上述古田籍的主角飞来省民族博物馆大厅表演了“虎日”戒毒会展的戏曲序幕《无明》。这出戏把舞台剧演到森林、公路和农家院落，让不同专业的人在不同的设计场合相遇，连《金翼》人类学小说里的林家后代和邻里，都处在了戏剧与社会戏剧之中。由此人类学家在这一戏台上下的互动中，讨论人类学家、农民演员和电影人协商和设计组合式戏剧和电影联合产品《金翼山谷的冬至》的诗学及其意义，以及融入新兴的立体与全息网络的互动问题。

五、多学科并置策展的意义

这是一场以绘画人类学为重心的多学科、多专业和多方法并置的学术实践会展活动，也是30年从乡村田野"包围"城市博物馆的历程。其含义是，原本现代博物馆学推崇的多感官、多模态展示的学术转向，以人类学的视角不仅仅是讨论新技术新媒体带来的展示方法的变化，而是聚焦于先在的田野工作，而后进入博物馆的会展集锦设计。这不单是人类学的博物馆展示，而且也是团队式的多学科交叉会展设计，突出反映了现代博物馆学推崇的多感官、多模态展示的学术转向。[41]"虎日"盟誓戒毒仪式的绘画创作是这次会展的重心。那么，相关且并置的论文写作、歌谣与诗、电影和演出，以及绘画和多感官策展中究竟分别处于何种互补的角色呢？"不浪费的人类学"是如何编织起来的呢？

（一）写作与论文

中外学术著作的多样性写文化努力由来已久，归结到人类学写作，本文主要涉及叙事建构形式的多样性和认识论的反思两方面。因抵制客观性、逻辑和科学主义在人文学科（包括人类学）中的专断地位，导致了挖掘多样性文化叙事方法的热忱，包容从古代到今天的各种叙事生成形态，关心从古代到今天、从中国到伊朗的各种叙事生成形态，这正是当今人类学作品需要的，而世界上现存的几个主要文学形式已经被利用，而且还在发展。因有"驳杂的"（heteroglossia，巴赫金语）生活，才会有无数绚丽的地方性叙事风格与多层次的文化写作技巧。不过实证科学论文中逻辑和清晰无误的刻板写作要求，一旦延伸到人文学科和人类学中，便显现了认识论被限定的缺失，即把逻辑论证直接对立的文化的直觉（不是指数学和科学发明的直觉）拦截了，因此，我们倒不是仅仅为了写作时的直觉解读，而是从田野参与时就已经在自我–他者之间努力寻获文化的直觉，因为这一直是经典功能主义和唯物主义都忽视的认知方法实验与体验。

进一步说到"虎日"盟誓戒毒实践所发表的论文主要阐述如何发现"虎日"事项及其方法论意义，即以文化的力量可以在一定条件下战胜人类生物性的成瘾性，小凉山这里能够戒毒成功的内中原委，以及仪式的意义和特定组织的社会动力学分析等。在论述与诠释的论文特定功用以外，金翼山谷的“文化的直觉”一直连接着云南小凉山的文化实践，直接导致了深度的田野学术对谈作品[42]。不仅如此，“虎日”调研点的文字论述进一步的不同之处还在于，这里文字的论述自然地引申到此次会展的绘画作品中，而且隐喻和直觉紧紧关联着两种不同的创作类型（书写和绘画）。几十年间，我们首先把深度解读的论文和直观的电影镜头语言合璧，形成对应的或非对应[43]的图文互补、互证和互释的合成综观实践，现在我们又多了绘画、诗歌和表演的多重认知和实践方式。

（二）歌谣与诗作

歌谣与诗作中的韵律与情感表达，正在弥补以往人类学偏于社会结构与权力研究的缺陷，谁也不愿意人类学者都成为政治家，而我们也不愿意诗学为文学所独有，因为“人类学诗学可以在多种民族志手法中展现”，其广义的含义包括“使个体内在的生命被他人体验的艺术”。[44]如同金翼山谷的冬至歌谣，内中“大体押韵的歌谣的节奏性，象征和对比的手法，恰好是在共同的韵律中憧憬和推动饱满的家族主义，歌颂家族理想的完满性与延续性。”[45]

而20年前“虎日”盟誓戒毒仪式中，诗学则是这样呈现的：电影使我们感受到莫色布都头人朗诵戒毒宣言时的激情与韵律，这不仅仅是有感于生活颓势的急切心理，而是源于族群“文化自救”的动力。今年（2019年）会展大厅他又借招魂有节奏地呼喊，把调动吸毒者回心转意的表演移植到与会者面前：情感、性灵、直觉的人类学诗学体验，一般也要和调查对象特别熟悉之互动后方可做到。我们团队在云南的“虎日”戒毒研究基地，已经将古老的指路经文加以解读，引申到现代社会生活的召唤之中，歌谣与押韵经文（诵读）不仅是古老的文学精粹，还和绘画、影视中的美学意涵，一同进入了应用的和公益的事务中。

（三）电影和新媒体

电影以其无与伦比的直观镜头，叙述了“虎日”仪式上的人性、文化与生命抗争实况；电影擅长的动态的过程性表达，很好地展现了生命抗争与文化自救的族群行动实践，提供了各地彝族家支头人学习与应用的情感、责任与动力。历史、哲学与民族志的文字书写在深度认知以及意义的解读上空间广阔，但文字表达的限度为电影的镜头直观展示提供了机会。电影和文字文本之间的互补与互释可以有对应性作品和非对应性作品，至于如何选择视写作与拍摄的相关情形而定。可以说，上述专著和论文与电影系列作品悉心组合，转变了传统学术著述的单一性，使图文并茂的人类学学术表达更为深邃与耐人寻味。

电影加上数字新媒体的运用，能使我们置身立体网络的“无死角”互动状态，而传统的电影展映也有了新的发展，使传统的电影展映如虎添翼。近年来，我们在新媒体众多手段中选择“网红”手法，颇有“播客”网络发布的意味，然而我们是学术的“播客”，与一般播客对应的是网上无限的“游客”不同。和播客对应的是网上无限的“游客”（笔者的比喻），虽说如此，尽管如此，网上“游客”也可以获得主体地位[46]，播客则更要及时应对。这种互为主体性或许就是新媒体网络的最大互动特点。现在已经可以讨论论文专著、电影/新媒体整合，以及网上立体互动和互释的重要意义。我们此次在民族博物馆的会展如同一个诗学、电影、戏剧、表演、绘画“面对面的互动”的学术交流“集市”，而现场实施网上直播则使全球网友进入一个“虎日”专题的立体和全息的、无限巨大的时空“集市”。当人们开始批评网上直播缺少真正“面对面的”人际感知、身体与情感互动（高夫曼式的互动）的缺陷时，我们团队起初已经设计了这次会展的平面（面对面）和立体（全息网络）兼顾的方案，这种行动的人类学实质上也是被筹划的人类学。

（四）绘画

处于这次会展的主要位置，意在表达镜头和文字陈述都难于超越的绘画创作力——来自绘画人类学的互动和认知人类学的理解；神山、信仰和族群自救的感召力；以及绘画人类学需要的田野体验。基于此的绘画创作《一个成功戒毒的彝族男人》的主人公包含着充满希望，又蕴含着内疚的眼神，这幅油画做到了呈现人物肖像的复合神情。（这一点是电影做不到的，也是没有田野参与观察的肖像画家做不到的。我曾看到网上介绍的一幅藏族人家庭绘画，却长着一副中原汉人的脸庞，这就是缺少参与观察所致；习惯于内地城市生活的画家，初到小凉山，因高原日照强烈，那里年女老幼的脸色变体多元，乃至调色都出现困难，也是因为缺少对云南高原环境、生计、彝族生活方式的体验所致，我们团队的画家深有体会。）

这次策展提供的“文化的隐喻”的绘画主题也是人类学的重要关注之一，这更是要在长时段的、充分的人类学田野调查之后才有可能寻获而进入创作。例如《牧羊女——黑花白的哲学》，神鬼人三界的《寻路》画作，以及《波兰婚礼晚会》场景中的颜色的文化隐喻都是不可多得的发现。所以我们总结人类学的隐喻的识别与觉解，绘画和撰写各具特征，难以替代，但各得其所。电影擅长呈现接续性的动态镜头，有时在绘画创作中则擅长呈现其包含多重内涵的静态复合作品，反映在肖像、群像、风景画中，还有绘画擅长的象征、隐喻、超现和直觉的理解，也不可多得。

我们不能要求画家一定要做深度的人类学田野调查后再作画，但如果我们运用了人类学擅长理解人群的理论方法，那么“深描”之后的绘画内涵会更为深厚，尤其是那些短期调查难于摸清和理解的内容。听说画家一般不很情愿画陌生人（付费街头肖像画家除外），那是因为短时难于把握对陌生人的理解。正因为如此，不少知名画家专注于固定模特（如妻子、情人和邻居等），那一定是因为有对人物个体的深刻理解，下足了功夫和情感发露所致。而人类学长时间的田野调查并非只专注个人，而是从对地方族群的整体性理解之后，再对其个体或群体居地的风景、人物写生或创作。而此时的风景画、肖像画和群像则好似被文化互动的筛子来回过滤的作品。

所以人类学的画家要用画笔来理解和诠释人与物，通常更喜欢群像活动画作以及引申性的和隐喻性的画作。人类学的整体性观察包括环境、居址、组织、民俗、信仰、隐喻和直觉中的艺术创作活动。似乎伦勃朗的《浪子回头》是很“人类学”的一幅画，而北宋张择端的《清明上河图》则包含着众多的人类学与民俗学关心的主题呀！可以认为是最早的油画《阿尔诺芬尼夫妇像》（扬·凡·艾克画，1434年）就描写了鞋子脱下放在一旁的宗教性隐喻[47]；像《波兰婚礼晚会》，更是在克拉科夫融入环境、组织、互动、象征的整体性绘画构想，其中跨文化的颜色表达已经难能可贵。然而，从其中互动的丰富故事里发现类似于黑花白羊哲学入画，则没有长期田野经历是一定难以做到的（也如同《祈男》里包含的家族主义与孝道理念）。因此可以说，人类学的田野工作可以使绘画作品的内涵达到更为深邃的程度，这也许就是人类学入画的学术魅力，倘若是解读跨文化的理解深度，也总是会在绘画表达中得到识别。

（五）表演与戏剧

我们在金翼之家演出的冬至传奇闽剧，是为了表达两千年孝道传承的流动的诗学韵律，[48]而展览大厅的“虎日”戏剧引子《无名》和表演剧《回来》所携带的社会性召唤，都因地方戏班导演、演员和人类学家磋商，以及会展导演、彝族头人、群众和"欧拉"宣讲人的屡次协商后移植到会展大厅。这就是新人类学的设计与相遇（encounter），显然是传统民族志访谈和参与观察所不能促成的。在这里我们团队从几十年的跨学科诉求和"不浪费的人类学"理念，实现了可以自然的"相遇"和需要被设计的"相遇"，并在不同的文化源头出发，在一定意义上和后现代实验合流。

这里参与设计的表演和戏剧同传统人类学独立的写作和戏剧人类学诠释的不同还在于，"我们的兴趣既在于思考民族志学者如何与公众相遇，也在于公众之间如何相遇。不仅如此，我们的目的是设计富有成效的相遇，这样的相遇处于运动模式，在合作者和公众之间互为主体的发展的发现的运动模式，"于是"我们离开了传统的表现的民族志，联系了一种具体化的、多模态的、共同建构的民族志体验。"[49]这在表演、戏剧，还有电影，以及那些需要团队内外合作的新人类学实验中最为明显。

（六）博物馆多感官策展

此次人类学绘画展映，是对20年前小凉山彝族"虎日"盟誓戒毒仪式进一步理解和推广，而展出的风景画、肖像群像画和创作画都是围绕以文化的力量战胜毒品危害的过程性成果。我们的人类学电影摄制和新媒体利用，都是为了跨学科、跨专业和跨方法横向合作的支援和帮衬。于是绘画的表达自然配合了电影和新媒体的表达，别忘了，还有民族诗学和信仰的动力与感染。

我们的摄制运用了传统记录和数字多媒体网上技术的"双轨制"的拍摄，特别是后者，我们已经不是单一面对面的互动电影过程，而是因拍摄过程直接进入网络，于是造成前所未有的、同拍摄现场"面对面互动"并存的网上"无死角"的人际时空互动网络景观。这样我们就有机会在新的电影与新媒体实验中讨论后戈夫曼的新型互动的问题。

当我们来到博物馆展现上述多学科、多专业和多方法并置的成果，显然会注意到来自20世纪末兴起的博物馆的多感官转向（the sensorial turn）[50]，观众希望进一步获得参与感，而博物馆无疑会寻求多感官的策划，将我们的创作成果从静止的罗列转为多种形式的互动体验，新技术支持了网上直播与互动，画中人和影（片）中人应邀露面，现场识别画作隐喻并和文字的直觉法加以比对，论文言说和展品解说并举，以及参与式表演与戏剧等。长时段，多点调查，交叉学科、专业与方法（各得其所），合作研究与整合性设计成为本团队的人类学研究特征，并依然兼顾“独狼”式的传统人类学方法。随后，各路研究成果分别或联合发表，并集中到博物馆展示和交流，这一漫长的过程的学术小结如下。

六、小结

学术论文规矩和文字表征的局限性，决定了“不浪费的人类学”的理念与行动的存在意义。人类认识论和哲学综观的愿景推动了挖掘多样性叙事方法的热忱和文化表征的多元方法实验行动。不仅要关注科学与实证的物质与结构过程，也包括探索意义的存在及其认知方式。如是，中国古今文人沿袭而来的容纳了小说、旅行文学、随笔、散文、歌谣和诗的直觉与感悟的手法已经显露了它的混生及其综观意义。各种文体中呈现的田野工作瞬间灵感触发的意象，不只存在于自我或他人本身，而是存在于文化的互动之中，使个体和群体内在的生命得以被他人体验，因此人类学学术论文的文化撰写，不只是文化“实证”的过程，也是流动的认知与诗学过程。

为了打破学科壁垒及其撰写约束，从1978至2020年，我们的团队在因袭中国古人图文之趣和当今人类学家“离经叛道”作品，进一步跃出文字表达系统。由于文字与图像系统功能之不可替代性，决定了图文互补之佳选，因此也成为当今跨学科和跨专业研究（如影视人类学）的理念与实践依据。

人类学研究的目的不单是为了解决应用问题而寻找一个清晰无误答案的问题意识导向，更多的是处于文化认识论的体验与过程研究之中，这正是人文学科和实证问题导向学科的重要差别。由于可清晰解说的某些“问题”的答案有时是被修剪过的，它们去除了非理性、情感和不可言说的内涵，而且先在理念、隐喻和文化的直觉也被排除与忽视。于是我们保持文字写作的探索，又从文字的限度中走出来。我们既需要科学的“事实的世界”，又不能省掉“意义的世界”。[51]我们进一步从写文化、拍文化延伸到画文化、演文化和展文化的过程性合作研究，就是为了寻找知识交叉产生的互补魅力，就是为了达成“触类旁通”的认知佳境，尽管绝对的综观可能永远也达不到。

现在探讨写作、电影、戏剧、表演、诗学、新媒体之间的不可替代性及其整合的学术意义已开始得到重视，而今日博物馆也获得了多样性文化表征与合作调研、设计、策展和知识兼收并蓄的理想空间地位。例如以往人类学写文化忽视的隐喻与直觉法，被引申到绘画特有的复合型思维实践之中；押韵的歌谣和“欧拉”即兴唱诗不只是情感与灵感突至的文学意义，已经被电影传达到非遗乃至公益的应用人类学中，这的确是始料未及的；那些被舞台限定的戏剧和表演，因数字技术得以和电影合璧而生成，扩大了各自专业理解的新生面；而电影的线性记录特点和绘画尝试中的复合型思维在并置中相互比美。于是，数十年长时段和多学科的团队田野回访成果超越了传统“独狼”式的参与观察写作，而且，我们团队也体会到了跨越学科、专业和方法之无限互补的时空存在，因而各自已难分伯仲主次（人类学资源无主次），这也正是“不浪费的人类学”实验的预想结果。

人类写文化和文化表征的探索由来已久，中国古代文人更有书画神韵的累世体验综观，均早于后现代的理论反思。不过如今跨学科的合作研究趋向和正在进行的后现代人类学反思实验已经部分合流，尽管各有各的学术导向。

注释

① J. Peacock, *Holism: Impossible but Necessary: The Anthropological Lens*, Harsh Light, Soft Focus, New York: Camb.

② 林语堂:《吾国吾民》,远景出版社1977年版,第79页。

③ 吴晗、费孝通等:《皇权与绅权》,上海观察社1948年版。

④ 1911年陈焕章博士论文《孔门理财学》的英文著作出处：Chen, Huan-chang, The Economic Principles of Confucius and His School (Vol.1&2), New York: Columbia University, Longmans, Green & Company, Agents, 1911；著名经济学家梅纳德·凯恩斯曾热情评论和推荐陈焕章的《孔门理财学》，注意到他在论述中穿插古诗和格言的独有的写法，认为是一本“令人愉悦的书”。该评论原载《经济学杂志》(*Economics Journal*),1912年12月号。

⑤ 林耀华:《义序的宗族研究》,生活·读书·新知三联书店2000年版(1935燕京大学)。

⑥ 林耀华:《金翼》,生活·读书·新知三联书店1989年版(1944,1947/1948)。

⑦ Elenore Smith Bowen, *Return to Laughter*, New York: Anchor Books, 1964.

⑧ 庄孔韶:《银翅：中国的地方社会与文化变迁：1920—1990》，桂冠书局1996年版，生活·读书·新知三联书店2000年版，英文版(2018)；阮云星:《义序再访：今日宗族乡村》(日文版)，中国研究所,《中国研究月报》1997年；兰林友:《义序与中国宗族研究范式》,《中央民族大学学报》2001年第3期；周泓、徐鲁亚整理:《〈银翅〉：中日学者恳谈》,2002年；杜靖:《百年汉人宗族研究的基本范式》,《民族研究》2010年第1期。

⑨ 庄孔韶:《流动的人类学诗学》《开放时代》2019年第2期。

⑩ 林耀华:《闽村通讯》，载林耀华《从书斋到田野》，中央民族大学出版社2000年版，第287—290页。

⑪ 例如，庄孔韶：《北美花间》(中文)，华盛顿大学人类学系，1992；Zhuang Kongshao 1992: *Valentine's Day* (Translated by Steve Harrell)；庄孔韶：独行者丛书五本，含诗集《自我与临摹》、摄影集、小说、随笔、旅行文学等，湖北教育出版社2000年版。徐鲁亚：《敦煌诗刊》2002年第1期；周泓、黄剑波：《人类学视野下的文学人类学》《广西民族学院学报》2003年9月号、11月号；周泓：《人类学诗论》,《云南民族大学学报》2003年第5期，周泓：《人类学诗》,《敦煌诗刊》2002年第1期；徐鲁亚：《神话与传说——论人类学的文化撰写》，博士学位论文，中央民族大学，2003年；徐鲁亚：《马林诺斯基与英国小说家约瑟夫·康拉德》，载《林耀华先生纪念文集》，民族出版社2005年版；徐鲁亚：《〈黑暗的心灵〉与〈西太平洋的航海者〉之比较》,《中国青年政治学院》2007年第5期；徐鲁亚：《远方的梦》,《北方作家》2008年第6期；以及［美］伊万·布莱迪编，徐鲁亚等译：《人类学诗学》，中国人民大学出版社2010年版。张有春随笔集：《田野四辑》，江苏文艺出版社2016年版；张有春：《情感与人类学关系的三个维度》,《思想战线》2018年第5期；王宏印：《朱墨诗集》，世界图书出版西安有限公司2011年版。

⑫ 庄孔韶导演：《端午节》，英文版1992；中文版2000。

⑬ 我们诗学沙龙的小电影《冬至的人类学诗学》应邀在2014年法国数字人类学年会的首席讲座；以及英国皇家人类学会为《银翅》英文版和人类学电影《金翼山谷的冬至》举行了专题展映(2019年3月)。

⑭ Zhuang Kongshao, 2010, "Non-waste Anthropology", Cultural Dimensions of Visual Ethnography: US-China Dialogues, Visual Anthropology Centre, USC; in "*Perspectives on Visual Culture from China: Methodology, Analysis and Filmic Representations*", Visual Anthropology Center of USC, and Intellectual Property Publishing House, 2012.

⑮ 庄孔韶：《行旅悟道——人类学的思路与表现实践》，北京大学出版社2009年版，第369—370页。

⑯ Wanono Gautier Nadine, De La Maison des ailesd'or au Solstice d'hiver: unegénéalogiecréative au service de l'anthropologievisuelle, *Journal des Anthropologues*, No. 156-157, 2019, p.299.(此出处由张敬京译自Nadine教授的论文《从〈金翼〉到〈冬至〉》，原载法国《人类学家》2019年第156—157期，特此致谢！)

⑰ 此间专题作品主要有：黄淑娉：《拉祜族的家庭制度及其变迁》,《新亚学术集刊》1986年第6期；程德祺：《父系宗族公社》,《中央民族学院学报》1981年第1期；庄孔韶：《基诺族"大房子"诸类型剖析》,《中央民族大学学报》1981年第2期。

⑱ ［澳］大卫·马杜格(David MacDougall)：《迈向跨文化电影》(大卫·马杜格的影像实践)，麦田出版社2006年版，第260页。(翻译请查阅他的书*Transcultural Cinema*, 1998，普林斯顿大学出版社)

⑲ 庄孔韶导演：《金翼山谷的冬至》，2017—2018年，摄制团队有王海飞、龚诗尧、宋雷鸣、吴同营等跨学科成员20余人。

⑳ 即诗学人类学沙龙小纪录电影《冬至的人类学诗学》，2014年摄制，团队成员庄孔韶、徐鲁亚、张景君等8人。

㉑ 庄孔韶导演：《虎日》，2002、2006年。团队包括嘉日姆几、王华、姚润、徐鲁亚、富晓星、和继军、莫色布都、雷亮中、和柳等20余人。详见庄孔韶主编：《人类学概论》影视教材

部分，中国人民大学出版社2006年版，第42—343页。

㉒ R.C. Smith and T. Otto, *Cultures of Future: Emergence and Intervention in Design Anthropology*, in R.C. Smith, K.T. Cangkilde, M.G. Kjaersgaard, T. Otto, J. Halse, and T. Binder (eds.), *Design Anthropological Future*, London: Bloomsbury, 2016, pp.19-36.

㉓ 庄孔韶、杨洪林、富晓星：《小凉山彝族"虎日"民间戒毒行动和人类学的应用实践》，《广西民族大学学报》2005年第2期；庄孔韶：《"虎日"的人类学发现与实践——兼论〈虎日〉影视人类学片的应用新方向》，《广西民族研究》2005年第2期。

㉔ 会议主办为云南大学民社学院"双一流"文化人类学方向团队和云南民族博物馆，跨学科联合团队成员包括人类学、博物馆学、绘画、电影、戏剧、新媒体等专业人员50余人。

㉕ [美]弗雷德·R. 迈尔斯：《表述文化：土著丙烯画的话语生产》，载[美]乔治·E. 马尔库斯、弗雷德·R. 迈尔斯编，阿嘎左诗、梁永佳译：《文化交流：重塑艺术与人类学》，广西师范大学出版社2010年版，第75页。

㉖ [法]柏格森，肖聿译：《材料与记忆》，华夏出版社1999年版，第116页。

㉗ 周冬莹：《影像与时间：德勒兹的影像理论与柏格森、尼采的时间哲学》，中国电影出版社2012年版，第104页。

㉘ [法]柏格森，肖聿译：《材料与记忆》，华夏出版社1999年版，第105页。

㉙ 林建寿油画：《祈男》（庄孔韶创意）（2001）。

㉚ 林建寿油画：《刮痧》（庄孔韶创意）（2001）。

㉛ 林建寿油画：《回访》（庄孔韶创意）（2001）。

㉜ 林建寿油画：《波兰婚礼晚会》（2017）。

㉝ 庄孔韶：《绘画人类学的学理、解读与实践——一个研究团队的行动实验（1999—2017）》，《思想战线》2018年第3期；范晓君：《中国学院派写实油画家的人类学研究》，博士学位论文，浙江大学，2019年。

㉞ 庄孔韶：《牧羊女——黑花白的哲学》（综合类材料）。

㉟ [英]维克多·特纳著，刘珩、石毅译：《戏剧、场景及隐喻——人类社会的象征性行为》第一章，民族出版社2007年版，第21页，请查阅Victor Turner, 1975, *Dramas, Fields, and Metaphors: Symbolic Action in Human Society*, Cornell University Press。

㊱ M. Jay, *Downcast Eyes: The Denigration of Vision in Twentieth-Century French Thought*, Berkeley, Los Angeles, C.A., and London: University of California Press, 1993, p.9；以及嘉日姆几（杨洪林）：《尊严，利益？——云南小凉山彝汉纠纷解决方式的人类学研究》，云南大学出版社2014年版。

㊲ 普学旺、樊秀丽、[日]藤川信夫：《百乐书影印译注：汉彝对照》，云南民族出版社2012年版。

㊳ 毛建忠诗，马鑫国汉译：《欧拉（回来）》，本文收录片段。

㊴ [美]戈夫曼（E. Goffman）著，冯刚译：《日常生活中的自我呈现》，北京大学出版社2008年版，第12页。

㊵ 请观看人类学纪录片《金翼山谷的冬至》（2018）中的地方闽剧，原定2020—2021年度的专题会展包括乡村戏剧演出。

㊶ 云南民族博物馆张金文、赵菲、高力青、韩丽萍等承担了此次人类学多学科联展的设

计、实施与协调的重要工作，一并致谢。

㊷ 庄孔韶和马库斯、费彻尔2019年1月28日在滇池对谈录：《文化直觉、艺术解读与合作人类学》，和柳翻译整理。

㊸ 庄孔韶：《人类学电影和学术专著如何互释》，《广西民族研究》2019年第6期。

㊹ Rita Dove, *What does Poetry Do News for Us?* Virginia University Alumni, January/February, 1994, pp. 22–27.

㊺ 庄孔韶：《流动的人类学诗学——金翼山谷的歌谣与诗作》，《开放时代》2019年第2期。

㊻ [美] 费·金斯伯格、里拉·阿布-卢赫德、布莱恩·拉金编，丁惠民译：《媒体世界——人类学的新领域》，商务印书馆2015年版，第460页。

㊼ [英] 史蒂芬·法辛编，杨凌峰译：《艺术通史》，北京联合出版社2019年版。

㊽ 庄孔韶：《流动的人类学诗学》，《开放时代》2019年第2期；庄孔韶：《金翼山谷冬至的传说、戏剧及电影的合璧生成研究》，《民族文学研究》2019年第4期。

㊾ Luke Cantarella, Christine Hegel and George E. Marcus: *Ethnography by Design—Scenographic Experiments in Fieldwork*, Bloomsbury Academic, London, 2019. p. 4.

㊿ D. Howes: Sensual Relations: Engaging the Senses in Culture and Social Theory [M]. Ann Arbor, MI: The University of Michigan Press, 2003.

(51) 参见牟宗三：《道德的理想主义》，其中《关于文化与中国文化》一章，学生书局，1992年修订。

第八章

▼

水乡私语书同文

2024年人类学终身成就奖获奖者

——潘悟云

人类学生涯回顾

早年苦难

1943年3月，我出生在温州瑞安莘塍镇九里村。1947年，我父亲当了九里乡的代理乡长，因为胜利公债一事，被国民党当局拘留，家产也全被变卖、充公，我们一家只能挤在残破的娘娘宫里度日，父亲气急攻心一度吐血，直至奄奄一息，母亲四处举债，凑钱给父亲治病。那时家中已经断炊，为了活命，13岁的三姐只好做了童养媳，周岁的妹妹只能被送到育婴堂，最终还是没能保住小命。

上天垂怜，父亲的病治好了，为了生计他去温州找工作。离家那天，在九里的码头上，他们依依惜别，父亲一脸悲凉和决绝："如今我这个身份想找份工太难了，这一次我要能找到工作还好，如果找不到我没脸再回来，你我不再相见。"母亲哭着对他说："你如果有三长两短，两个儿子还小，我们怎么办？"不久，母亲也去温州做了保姆，赚钱寄回瑞安养活我和弟弟。当时我五岁，弟弟三岁，两人相依为命。每天上午我带弟弟走到九里湾的桥头，桥下是从温州到瑞安的塘河，一趟趟的小火轮从河边蜿蜒而过，我们两个小小身影无数次徘徊："爸爸妈妈，你们什么时候回来呀？"

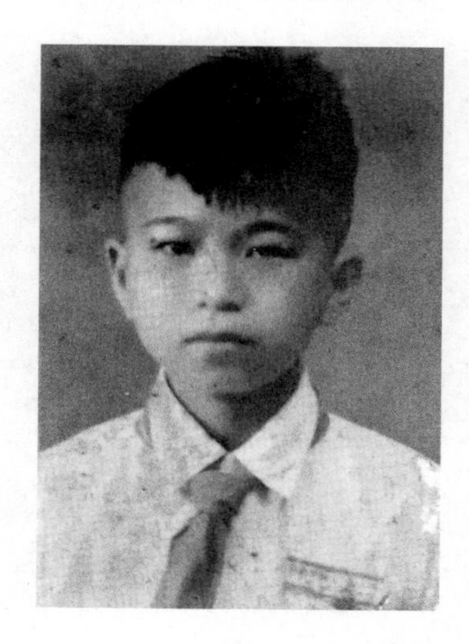

10岁时的我

如此挨过一段时日，我这小萝卜头终究还是病了。一位村民把我送到外婆家中，外婆一见到我就哭了。当时没有条件请医生，但经过外婆的精心调理，我竟然奇迹般地活下来了。

万幸,父亲总算找到了工作。

1949年后,父亲因为1947年那几个月“代理乡长”的经历,被戴上“历史反革命”的帽子。因为这顶帽子,我们四个兄弟全部无缘高考。1959年国家提出“教育要革命、学制要缩短”的口号,温州四中希望我提前一年参加高考。班级的全部任课教师参加讨论,他们几乎全部反对让我提前高考,老师们激动地说:“万一潘悟云因为提前高考成绩不够理想,影响他被北大、清华这些顶级学校录取,这会是国家的一个损失”。那时对我寄予厚望的老师们还不知道,因为父亲“历史反革命”的身份,我的命运早已被注定,“此人高考不得录取”的印章,封死了我进入大学的通道。

为了养活家人,高中毕业的我只能去做苦工:拉板车,抬煤渣,挑泥土……在7月的泥涂上赤身挖船坞,背上长满了水泡,晚上只能趴着睡觉;煤球厂一箩筐三百斤的煤渣,磨出一肩的瘀血;严冬腊月造船厂的江边,风如刀割,赤脚在水中作业。最初那段岁月,我无比伤心,陷入迷茫,眼睁睁看着昔日的同学走进大学课堂,既羡慕又哀伤。踌躇数日后我没有气馁,暗暗给自己鼓劲:我读书这么好,不应自暴自弃,不能上大学,还可以自学呀。于是,工地休息的间隙,我开始背诵古文,做高等数学习题集。那时候既看不到出路,更没想到日后会从事研究工作,只在心里铆着一股劲儿,只要能够得着的书就反复看、认真学,哲学、文学、美学、历史、地理、音乐……各种门类的书全都看,相信总有一天会用到这些知识。

引路名师

没有机会上大学的我,每每看到大学老师,心中就无比羡慕和尊敬。可能是求学的渴望打动了上苍,冥冥中老天自有安排,我无疑还是幸运的。在做苦工的时候神奇地碰上了温州老师;后来读研究生的时候,碰上了上海老师;在美国、欧洲也都碰上了海外老师。没有这些名师引路,就不可能有我今天的成长。

温州恩师

1969年,我在温州锅炉厂有了一份正式工作。按照厂里的要求,更换了好几个工种,做过钳工、车工、刨床工等,唯一不变的就是利用空余时间看书。正当我漫无边际、逢书必看的时候,一个即将改变我命运的人出现了,他就是郑张尚芳。

那时，郑张尚芳还是温州渔械厂的一名普通磨工，但在语言学界已经被人奉为奇才，被一些大家赏识。有一次，他对我说，高本汉的《中国音韵学研究》是中国科学院语言所的吕叔湘先生借给他的。吕先生来信说，图书馆要开始整理书籍了，这本书必须先归还所里，待图书馆藏书全都整理好之后才能再借。郑张尚芳对我说，这本书他每天都得用，特别是后面的方言字汇部分，更是离不开，希望我把这部分抄下来。当时我对国际音标一知半解，但依旧遵嘱认真抄好，一式两份。拿到抄本的时候，郑张尚芳非常感动，认为我能耐得住寂寞、甘坐冷板凳，是一个能做学问的人。于是他便开始教我学习音韵学。

看了高本汉的《中国音韵学研究》，我们对高本汉敬重之余，难掩心中遗憾："汉语是中国人的母语，为什么汉语语音史的奠基人会是一个只学过几年汉语的瑞典人呢？既然中古音的构拟已经由高本汉大体完成，那么中国人应该在上古音研究方面做出自己的贡献，否则中国人无以面对海外学术界。"1960年，郑张尚芳已给上古汉语构拟了七元音系统。1969年，他让我用董同龢的方法做成上古音韵表稿。我在做表稿的时候，发现同一位置上有时溢出，有时却空置，郑张尚芳就此进行改动。自1969年到1979年，这十年时间，成就了国际汉语史界著名的上古六元音系统。我不禁兴奋地说："尚芳，再努力十年，我们俩都要登上国际讲台。"十年后，我们果然实现了这一梦想。

1970年与郑张尚芳合照

回溯往事，1969—1979年，是我学术人生最为艰难的十年，也是充满发现与创造的十年，郑张尚芳先生一直引领我、鼓励我，我们相互陪伴、共同进步。一旦有突破，我们会马上跑到对方家中，分享发现的快乐，我的创造力就这样被一点点激发出来。一晃40年过去了，我们为自己的工作感到欣慰。不仅上古音研究为国际所认可，而且我们正为新的历史语言学诞生做出可喜的贡献。

上海导师

1979年，我幸运地考上了复旦大学的研究生，我的研究生导师是吴文祺先生和濮之珍先生。大革命失败以后，吴先生受郑振铎先生之邀到燕京大学任语言学教授。当时燕京大学校长是陆志韦教授，他们两个既是好朋友，又都是音韵学家。陆志韦先生正在写《古音说略》，一直与吴先生互相切磋。入学不久，学校便选中我们的《汉越语与切韵唇音字》一文，作为国庆的庆礼。有一天中文系领导去家中拜访吴先生，吴先生非常高兴地说，这么多年来第一次看到这么好的音韵学论文，他一口气仔细看完论文，酣畅淋漓浑然不觉已是凌晨五点。原来，中文系曾给我和朱晓农做语言学史方向的研究生课题，吴先生说，你们的时间如果用来写语言学史未免有些太可惜，你们的课题应当以音韵学为内容。回来的路上，我们很是兴奋："与其写他人的研究历史，不如做好自己的研究，将来自会有人写我们的研究史"。后来，我们经常到吴先生家里做客，每次都受到先生的教益和启发。

1979 年在复旦大学的研究生校友，
自左至右朱晓农、陈重业、潘悟云、余志鸿、杨剑桥

1980 年于复旦大学计算系讨论汉字输入法

1981 年桥本万太郎教授在复旦大学讲学，自左至右有沈亚明、游汝杰、张洪明、潘悟云、沈钟伟、余志鸿、汤志祥

研究生期间我的学问有幸得到系主任胡裕树和许宝华先生的认可，他们准备让我留校任教并负责语言实验室的建设。不料民政部新规已婚研究生必须分配至妻子户籍所在地，那我只能分配到温州，无法留在复旦大学。吴先生是市人大常委会委员，得知这一消息之后，立即在市人代会上为我力争留沪，同时还发文到当年8月7日的《解放日报》。

然而，这并没能起到什么作用。于是，中文系的领导又提出一个方案，让我延迟半年毕业，准备用半年时间来解决留校的问题。但还是晚了一步，人事处已经偷偷把我分配到温州师范专科学校。不仅中文系的领导很伤心，大洋彼岸的梅祖麟先生听闻以后，甚至愤然断绝了跟复旦大学的学术联系。

此后，吴文祺先生持信让我与杭州大学的姜亮夫先生联系，杭州大学半年之后才能成立新的研究所，姜先生很希望我能加盟助力，也因无法从温州调档只能抱憾。胡裕树先生还曾向浙江大学的中文系主任推荐我，但是系主任很快去世了。胡先生等虽然不是我的导师，但是他们一直把我当作自己的学生百般呵护，令我感激不尽。

1992年，上海师范大学张斌先生即将退休，学科必须聘请新的博士生导师。胡裕树先生与张先生是老朋友，当即建议张先生把我引进上海师范大学。张先生也不是我的导师，但是先生惜才，他对我关怀备至。在他的指引、关照下，我评上了博导，建立了语言研究所，当上了所长，在上海市教委做了E-研究院的首席研究员。每当有人阻挠学科建设，我难免着急上火，张先生总是温和地、不疾不徐地安慰劝导我说："我送你一句话，'有容乃大，无欲则刚'。"张先生不仅是我的老师，更像是家中长者，对我不断地关照、提携、指引。

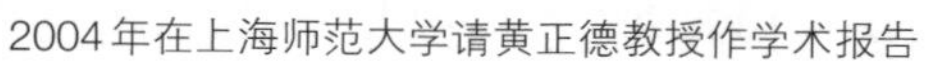

2004年在上海师范大学请黄正德教授作学术报告

2008年在上海师范大学语音实验室请Ohala教授作学术报告

海外师长

我曾撰写《汉越语与切韵唇音字》一文，那时学界还无法解释咍韵与灰韵对立的原因。一次偶然的机会，我在*lingua*上读到了王士元先生关于词汇扩散的文章，从中领悟到当时国内语言学界尚未引入的词汇扩散理论。于是，我写了封信给他，提出了我的疑问，并表达了对词汇扩散理论的浓厚兴趣。

王士元先生迅速回了信，希望我能告知我的学术背景，并且提出想要与我合作共同撰写文章。我激动万分，作为一个硕士生，能够得到如此知名语言学家的赏识，无疑是莫大的荣幸。此后，我与王士元先生的通信更加频繁，几乎每隔一段时间就能收到他的来信。我们深入讨论词汇扩散等理论问题，并且在*JCL*上发表研究成果。

1986年与梅祖麟教授一起在加州奥克兰中国语言与方言国际研讨会上

1986年，王士元先生邀请我参加在奥克兰召开的中国语言和方言会议，中国出席会议的都是朱德熙、王均、林焘等著名语言学家。

研究生第一年，我向导师交了学期作业《中古汉语方言中的鱼和虞》，后发表于《语文论丛》，文中我批评了罗常培先生的《切韵鱼虞的音值及其所据方言考》，也涉及张琨先生。那次会议上我第一次见到张先生，心有怵怵焉。会议茶歇，张先生特意找到我："你是潘先生吗？我拜读了你在《语文论丛》上的文章。"我的脑子嗡地一声，瞬间慌了神，忙说："张先生，当时我年轻，不知天高地厚。"没想到他马上说："不要这么说，罗先生当时也是年轻，你对他的批评是对的。"寥寥数语，让人动容。之后，我每次到加州大学伯克利分校访问，他总是请我这个年轻后学吃饭，给我许多指点与勉励。张琨先生不仅给我学问上的指导，而且教我怎样对待批评。有一次，他在伯克利请王辅世、戴庆厦吃饭，还特地邀请了我。席间他还向王、戴两位先生介绍我："潘先生很有学问，将来不可限量。"张琨先生的推荐，给我带来了人生的一次重要转折。1994年，学校推荐我到国家

1986 年于加州大学伯克利分校
POLA 语言实验室访学

1989 年于加州大学伯克利分校与梅祖麟、
沈钟伟、连金发在 POLA 语言实验室访学

评审组申请博导资格。很多评审专家对我并不熟悉，戴庆厦先生在会上提起张琨先生在伯克利对我的赞誉。因为张琨先生的权威性推荐和戴庆厦先生的鼎力支持，我最终评上了博导。从那以后，我培养了一百多位博士生和博士后，这是我教育事业上的一次重要转折。

在 1986 年的会议上，我还结识了罗杰瑞先生。当时，他的精彩报告和地道的闽语发音折服了所有与会者，大家都忍不住赞叹罗先生是语言学天才。后来，在西雅图，在上海，在北京，我曾多次受教于他。特别是他的《早期汉语的咽化》，是我和白一平研究上古音系统的重要来源。

1998 年在美国加州大学伯克利分校
赵元任纪念馆访问

2002 年在西雅图华盛顿大学讲学，
与罗杰瑞教授合影

我写过一篇有关声调的文章，被送给梅祖麟、桥本万太郎和陈渊泉等几位专家评审，得到了他们的一致肯定。特别是梅祖麟先生，他仔细阅读了我的文章，主动来信给予了高度评价，第一句话就是“这是一篇很精彩的文章”，随后详细阐述了他的看法。从那以后，他一直关心我的学术进展。2000年，他在香港中文大学语言学新年年会上发言，提出白一平和潘悟云属于音韵学的主流派，无意中大大压低了王力先生的学术地位。由此引爆了连续十年的批梅打潘运动。我一直抱疚在心，如果没有梅先生对我的关爱，就不会有人对他进行打压。

1986年的会议上，我还见到了蒲立本先生。蒲先生的汉语说得不太好，我的英语也是哑巴英语，我们俩交流起来很困难，几个小时的交流只能靠笔记完成。蒲立本的《上古汉语的辅音系统》涉及大量西域译名，以及欧美许多汉学文献，特别是那些梵文和亚洲的古代文献，我连地名、人名都不认识。后来有幸遇到华东师范大学的徐文堪先生，他研究历史交通史，对中亚、西亚的地名很了解。我们二人合译，花费数年，终于把蒲立本先生的《上古汉语的辅音系统》呈给国内学术界。汉语上古音的突破性进展是蒲立本先生的一大功劳，我的辅音构拟很多都得益于蒲先生。

1997年在奥斯陆大学与何莫邪教授讨论汉语上古音

1983年在复旦大学与包拟古教授讨论《原始汉语与汉藏语》的翻译，照片中有张洪明、杨剑剑、严修

梅祖麟先生曾在通信中向我介绍包拟古先生的上古汉语六元音系统。当我看到包拟古先生的那本书，惊讶地发现，他的六元音系统与我的构拟竟然惊人相似。这让我非常激动。我把信转给了郑张尚芳。为此，包拟古先生到上海跟我见了面，

让我翻译他的《原始汉语与汉藏语》。白一平是包拟古的学生，他们的上古汉语体系反映西方此类研究的最高水平。如果说，包拟古先生通过汉藏比较提出他的上古音构拟，白一平先生则通过他的数学家的头脑，使之成为一个体系：由他们师生俩共同命名的包—白体系。

2002 年参加李方桂 100 年诞辰研讨会，从左到右：白一平、郑张尚芳、许思来莱、潘悟云、何莫邪合影

近年来，白一平先生和沙加尔先生合作，在以往基础上对上古汉语研究提出许多新的想法，成为白一平—沙加尔体系。2002 年在巴黎，沙加尔先生向我提出上古音构拟的几点共识，并于 2005 年 11 月 13 日至 18 日，由白一平（William H. Baxter，美国密歇根大学）、金力（复旦大学生命科学学院）、沙加尔（Laurent Sagart，法国科学研究院）和潘悟云（上海师范大学）共同发起了“上古汉语构拟国际学术研讨会”，这也是历史上规模最大的上古汉语构拟国际研讨会之一，邀请了来自包括港澳台在内的中国各地以及美国、法国、加拿大、英国、荷兰、泰国的五十余位学者进行研讨，这些学者的讨论和研究逐渐成为汉语上古音构拟的主流。

2003 年在布拉格与白一平教授等讨论了一个月的上古音构拟

朱子云：问渠哪得清如许，为有源头活水来，在我看来王士元先生就是中国语言学的创新源泉。从 1979 年到现在，我的多向音变、汉语方言计算机处理系统、汉语历史音韵学、汉藏语的演化语素、音节理论，它们都是在汲取王先生的灵泉。特别是王先生提出演化语言学的时候，我提出了演化音系学：包括演化音系学的基本单位，基本单位的性质和分类，音变链及其语言演化的表现，等等。

正是因为有了老师们的辛勤汗水和创造灵泉的灌溉，才结成了我自己的学术成果。

上古汉语

汉语约有六千年的历史，承载着汉民族的悠久文化。语音是语言的物质外壳，也是语言研究的基础。因为汉字缺乏表音性，人们很难通过汉字本身了解汉语历史上的读音。所以，从明清之交到现在，音韵学一直是汉语史研究最困难的部分。

音韵学的一次次突破性进展，都是方法论的突破。

语音史研究中最基础的部分是韵母演变，特别是主元音撑起来的框架。

比较古老语言的文字系统以声母辅音为字母本位，每个辅音都有相同的主元音，实际上就带有主元音框架。

阿拉伯语字母全部为辅音字母；元音通过字母上方或下方的符号来表示，通常省略。梵文的字母也代表辅音，元音符号分别写在字母的前面、后面、上面或者是下面，短音的a为无标记。东南亚的许多语言文字，如藏文、柬埔寨文、缅甸文、泰文大体上都采用梵文的书写方法。

这些古老的文字系统说明，主元音比较简单，辅音最复杂，上古汉语也是如此。任何复杂系统的分析都是从其中的子系统开始。先把最简单的子系统确定好以后，才能从未知到已知，从不确定到确定。这个最先确定的子系统，就是研究的基础部分。韵母系统就是上古音研究的基本框架，基本框架确定以后，才能一步步地确定声母和声调。所以，音韵学的研究都是从韵母开始的。

清儒通过《诗经》押韵，得到上古的每个韵母，为了与中古音相区别，叫作韵部。如《诗经》有以下押韵：華家（桃夭）—蘇華都且（山有扶蘇）—華塗居書（出车）— 且辜幠（巧言）—莫除居瞿（蟋蟀）—狐烏邪且（北风）—祖屠壺魚蒲车且胥（韩奕）—華夫（皇皇者华），这些互相押韵的字合在一起“華家蘇都且塗居書辜幠莫除瞿狐烏邪祖屠壺魚蒲胥夫”，叫作鱼部。顾炎武系联出10个韵部，江永得出13个韵部。段玉裁发现同一个声符都有同一个韵部，如声符“古”，有模韵字“棒嫴姑辜酤蛄鴣沽鹽古罟鹽詁沽酤故沽固稒痼錮棝涸棒枯苦苦葫鶘瑚湖鶘猢醐糊葫蝴胡瓳怙祜岵酤婟楛婟”，鱼韵字有“居据裾琚椐鶋崌涺腒鋸倨踞椐椐腒”，他们合起来都是上古的鱼部字。段玉裁得出17个上古韵部，王念孙得出22个上古韵部。乾嘉学派的本质就是从文献学进入到语文学，这是第一代音韵学。

但是，到了语文学阶段，音韵学的发展近乎停滞了。新的研究方法应运而生。高本汉等人把历史比较法引进中国，对中古音做了成功的构拟。第一代音韵学，古音只做音类的归纳，不做音值的构拟。到高本汉这个时期，音韵学家才去了解古代的实际读音，并对高本汉的方法做了详细的探讨，甚至许多非语言学的学者都会对音韵学发生兴趣，出现了音韵学的嘉年华。这就是第二代音韵学。这一时期，从韵母系统中分析得出主元音，王力《同源字典》在上古30个韵部的基础上分析出6个元音：a、e、ə、u、o、i。

高本汉和许多音韵学家在中古音的基础上继续上古音构拟的工作，但是成绩不佳。一直到结构主义的内部拟测法提出，总算有了进展。董同龢[①]的《上古音韵表稿》用这种方法进行上古音的研究，发现传统的元（月）部在谐声关系上可以分为两支，一支是an（at），一支是ɛn（ɛt）。郑张尚芳据此也把歌部分为对应的两支。

雅洪托夫[②]的《上古汉语的唇化元音》指出，传统的歌、月、元韵部中，还有带圆唇主元音的一类韵部。跟董同龢的韵部合在一起，就构成带主元音a、e、o的三类韵部。1960年12月5日郑张尚芳写给李荣的信中，详细地记录了给王力的《汉语史稿商榷书》，其中的歌、月、元分部与雅洪托夫完全一致。郑张尚芳还指出传统的脂部分为两个韵部，合起来总共有7个元音。

1969年，我跟从郑张尚芳学习汉语音韵学，用董同龢的方法，把《广韵声系》的材料做成上古音韵表稿。年底，我们发现声韵配合出现问题，有的空缺很多，有的则在位置上溢出，这说明构拟有些错误。这些问题促使郑张尚芳重新调整和进一步思考，发现宵部有其特殊的音变行为。于是郑张尚芳把宵（藥）部剔除出去，剩下的韵部就只有6个元音。这就是汉语史上著名的六元音系统，各自产生于中国（郑张尚芳-潘悟云，1969）、美国（包拟古-白一平，1972）和苏联的（斯塔罗斯金，1989）。

上古六元音系统构拟好以后，剩下的主要构拟内容就是辅音。20世纪70年代初西方语言学家有以下突破性进展。

雅洪托夫的《古汉语的复辅音声母》指出，二等钝音、莊组和B类在上古都带有 *-r- 介音。这是上古音研究最重要的突破，为上古辅音丛的演变找到了原点，也画出了上古声韵配合的主要框架。

蒲立本凭借其丰富的古代文献借词知识，特别是从上古汉语的辅音找到了极其重要的材料与工具。其中最重要的是中古的来母来自上古的 *r- ，以母来自上古的*l-。因为*r-和*l-是辅音丛中的主要辅音，他跟雅洪托夫的相互构拟是上古辅音丛的最重要根据。

我在他们和包拟古的基础上，在上古辅音的方面也有多个研究成果。

90年代以后，我在雅洪托夫和蒲立本的基础上对上古辅音的研究做出以下贡献。

对雅洪托夫复辅音构拟的修正

雅洪托夫对二等和B类的声母构拟为复辅音*Cr-，我改拟为复杂辅音*C^{r}-。复杂辅音中的主要调音是声母中的主要成分，在演变的过程中会保留下来，次要调音则演变为介音性质的近音。根据复杂辅音的音变规则，二等字“家”会变成k^{r}a>k^{ɯ}a̠>k^{i}a，如果根据复辅音的演变规则，它要变成kra̠>ʈa>tʂa[3]。

李方桂、白一平把二等中的知组构拟为*tr-、*t^{h}r-、*dr-，我则改拟为上古的*T-类单辅音>ʈ-，*Kr-类和*Kl-类的复辅音声母>ʈ。

一个半音节

包拟古根据藏缅语同源词更可能是浊塞音加-r-的复辅音这一点，把它们构拟为*g-r-、*d-r-、*b-r->r-，不过他采用的这种拟音只是一种特别的标写法，其中的连字符并不表示第一个成分必定是前缀，也不能说明它们与二等的拟音*Cr-在语音上有什么区别。Matisoff[4]把次要音节和主要音节组成的词叫作一个半音节的词（sesquisyllabic word），次要音节短而弱，容易失落，王敬骝[5]等许多南亚语学者，都发现一个半音节存在这种音系的演变过程。我根据有些来母字具有类似音变规则，前面的塞音失落剩下主辅音r-，把上古读音构拟为*g.r-、*d.r-、*b.r->r-，其中塞音是次要音节，加“.”表示次要音节[6]。

复辅音

上古汉语有许多复辅音。

蒲立本提出一个音变例子，唐*g-laŋ>dɑŋ。包拟古提出一系列的音变规则*k-l->t-、*k^{h}-l->t^{h}-、*g-l->d-，*p-l->t-、*p^{h}-l->t^{h}-、*b-l->d-。其中横杠只是一种标记。我把包拟古上述构拟确定为复辅音：Cl>T，T取C的发音方法和l的发音部位。如“跳”*k^{h}le̠ws>t^{h}eu，可比较汉越语跳k^{h}ieu[3]。

冠音加辅音丛

有些音韵学家的构拟偶尔出现三辅音组。我提出三辅音组就是冠音加辅音丛，同时通过语音解释，提出三辅音组的应用规则③。

冠音和辅音丛合在一起，每一个音段的时长会缩短。冠音往往是无音位价值的元音，第三个辅音是近音，第二个辅音是阻音，响度和长度最小，所以更加容易失落。如“硬” *ŋg^{r}a̱ŋ>ŋ^{r}a̱ŋ，“明” *mk^{wr}aŋ>*k^{wr}aŋ，龐 bg^{r}o̱ŋ>b^{r}o̱ŋ。三辅音的这种音变不仅出现于上古汉语，还出现于藏文。西藏《贤者喜筵》肿胀 sraŋs，16 世纪该书作者有的地方写作 skraŋs，说明在历史上第二个塞音失落。

上古的小舌音

我首次提出上古汉语存在小舌音，到中古汉语发生了以下音变⑥。

上古汉语	*q-	*q^{h}-	非三等 *ɢ-	三等 *ɢ-/*ɢl-	三等 *ɢ^{w}- 前高元音	三等 *ɢ^{w}- 非前高元音	三等 *ɢ^{r}-
中古汉语	影母 ʔ-	晓母 h-	匣母 ɦ-	以母 j-		云母 ɦ^{j}-	

清鼻音

董同龢、李方桂、郑张尚芳的构拟中都有清鼻音。我根据上古声母的格局，确定上古清鼻音到中古音发生了以下音变⑦。

上古	中古	上古	中古	上古	中古
清鼻音	送气清阻音	送气清鼻音	擦音	常态鼻音	鼻音
m̥>	p^{h} 滂母	m̥^{h}>	h 曉母	m>	m 明母
n̥>	t^{h} 透母	n̥^{h}>	h 曉母	n>	n 泥母
ŋ̊>	k^{h} 溪母	ŋ̊^{h}>	h 曉母	ŋ>	ŋ 疑母
m̥^{j}>	tɕ^{h} 昌母	m̥^{hj}>	ɕ 書母	m^{j}>	ȵ 日母
n̥^{j}>	tɕ^{h} 昌母	n̥^{hj}>	ɕ 書母	n^{j}>	ȵ 日母
ŋ̊^{j}>	tɕ^{h} 昌母	ŋ̊^{hj}>	ɕ 書母	ŋ^{j}>	ȵ 日母

来自sC-的上古声母

蒲立本Pulleyblank 于1962—1963年提出*sC-型辅音：*st->ts-、*st^{h}->ts^{h}-、*sd->dz-、*sk^{h}->ts^{h}-。李方桂[8]也有类似构拟并且更加系统：*st->s-，*st^{h}->ts^{h}-，*sd->dz-，*sk（w）->s（w）-，*sk^{h}（w）->ts^{h}（w）-，*sg->dz-。

我在此基础上提出[9]：

*sk-> st->ts	*sk^{h}-> st^{h} >ts^{h} >ts^{h}-	*sg-> sd- ->dz-
*sp-> st->ts	*sp^{h}-> st^{h} >ts^{h} >ts^{h}-	*sb-> sd- ->dz-

塞擦音是塞音的一种，与除阻的快慢有关。快除阻就是爆音，慢除阻就是塞擦音[10]。T在前面s的影响下，成阻的时候插上s的音姿，变成慢除阻，就成了塞擦音，之后前面的s失落。

上文已经讨论过音变：kl->t-，k^{h}l->t^{h}-，gl->d-,pl->t-，p^{h}l->t^{h}-，bl->d-，我们会更多地碰到下面的音变：

* skl-> st-> ts-	* sk^{h}l-> st^{h}-> ts^{h}-	* sgl-> sd-> dz-
* spl-> st-> ts-	* sp^{h}l-> st^{h}-> ts^{h}-	* spl-> sd-> dz-

上古汉语的塞韵尾

中古汉语的塞韵尾都是清的不爆破音，俞敏[11]和郑张尚芳[12][13]提出上古入声韵尾是*-b、*-d、*-ɡ，我通过汉藏语的历史比较与古汉语的借词，把上古汉语构拟为爆破的浊塞韵尾[14]。

上古汉语的音节结构

我在2024年的国家重大项目中提出汉语的音节结构：

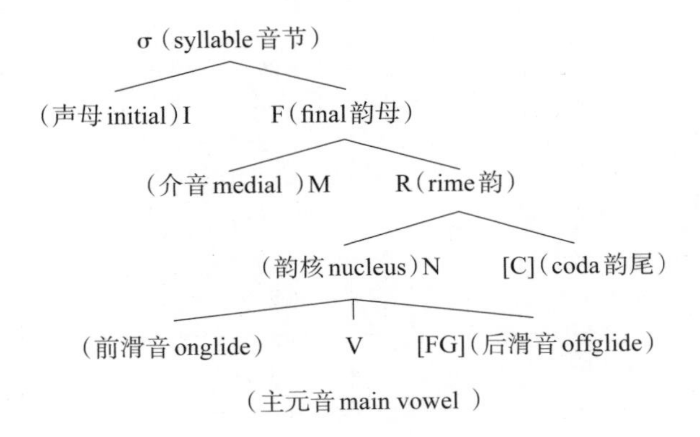

同时，提出原始汉藏语声母结构如下：

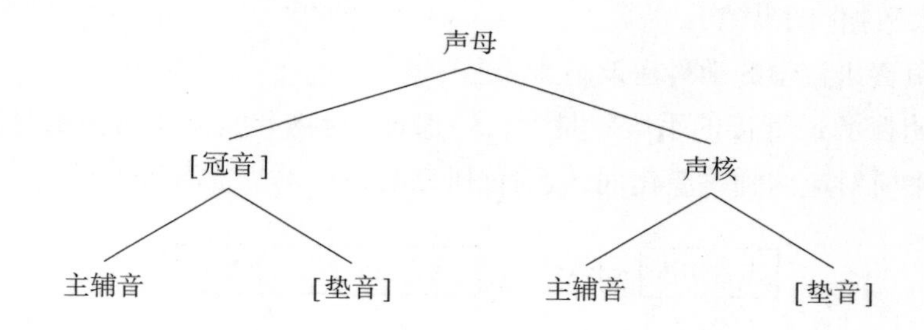

文明探源

文明探源，主要有三个窗口：遗传、语言、考古。

2015年，我在复旦大学金力、林尚义的领导下，成立了中华文明数据中心。

十九大以来，国家从中华民族、中华文明的高度提出一系列的政策措施。复旦大学在中华文明数据中心的基础上，明确提出以遗传、语言与考古为主要研究内容。

复旦大学最早提出东亚人群起源于非洲的学说，认为他们有着共同的遗传来源。

历史地理是探寻中华文明研究的重要手段。复旦大学在这些方面做了诸多尝试，不乏重磅成果。比如，谭其骧的《中国历史地图集》。

复旦大学的出土文献与古文字研究中心，是在裘锡圭教授的领导下建立的，在国内外具有无可争议的领先地位。

复旦大学东亚语言研究中心，已经采集了大量汉语方言和民族语材料，可以服务于多种研究，其中最重要的，就是从海量材料中提炼演化语素，用演化语素进行语言演化研究。

复旦大学提出“中华早期文明跨学科研究计划”，包括以下7个学科，括号中为该学科校内相关机构：

1. 语言与古文字研究（现代语言学研究院、出土文献与古文字研究中心）
2. 考古与科技考古（文物与博物馆学系）
3. 人类学（以人类遗传学与人类学系为主）
4. 民族学和民俗学（以民族研究中心和哲学学院为主）
5. 中华民族形成史（历史系）

6. 计算机科学与古籍保护（计算机科学技术学院、中华古籍保护研究院）

7. 大数据（图书馆）

我负责语言学的学科建设。

文明探源最直接的手段就是考古。遗址、遗物等实体所对应的是概念，在语言中所对应的则是演化词，做演化研究的最小单位是演化语素。

文明产生于文化长河中的某个阶段，这一阶段属于早期文化，考古就是通过实体探寻早期文化。但是，在文明探源过程中作为主要手段的考古仍会有局限性。第一，考古材料的发现和获取有极大的偶然性，需要“老天爷赏饭吃”。而语言材料随时可得，只要具备历史语言学的基本知识，随时可以在现存的语言中去挖掘“语言化石”。第二，考古研究有赖于极大的财力、物力和人力的支出作为支撑，每次地下挖掘都需要有政府的巨额资助。相较而言，历史语言学的研究成本简直微乎其微。第三，用通常的考古手段不一定能得到考古实体的时间序列，用语言学的材料却可以确定时间先后甚至传播路径。例如，伴随古代交通贸易的发达，很多地方都可以找到丝绸的实物，不可能有丝绸就可以还原丝绸的传播路径。这时语言学便能发挥它的积极作用。人们生活中用到的实物，总归需要有个名称来称呼它，较早出现丝绸的地方一般相应存在早期的丝绸名称。从语言学的角度来看，汉语“丝”出现于朝鲜、满、蒙古、波斯、阿拉伯、叙利亚、希腊、罗马、俄罗斯和西欧，这些地方就是通过语言学所能考定的丝绸之路。语言学中有一门新的学科叫“演化语言学”，最早出现的词通过历史演化，在各个地方演化成不同的读音，叫作演化词。

演化词跟普通的词不一样的地方，是它们具有时间元素，可以通过时间元素来推断考古实体的演化过程。有时候实体传播的时间很难确定，但是恰好各语言的读音却可以根据自然音变规则来推断时间的先后。什么叫自然音变？它是可以用发音与感知的原因加以解释的音变。比如：一个演化语素在A语言的读音是ɢ-，B语言的读音是g-，根据自然音变规则ɢ会变为g，而g不会变为ɢ，由此可以推测A语言的读音比B语言更为古老，可以用语言的演变时间来推断考古实体出现的时间。下文举稻米的语素为例。

中华五千年文明的标志是良渚文化[15]，良渚是稻作文化所产生的文明。

水稻人工栽培有许多遗址，如印度柯尔迪华（Koldihwa）遗址，泰国农诺他（NonNokTha）遗址，中国的河姆渡遗址，等等。世界上关于水稻种植起源一直存在争议，可举其中有重要科学依据的文章，Jeanmaire Molina等[16]从遗传角度指出水稻在长江流域首次种植。Fabio Silva等[17]从400个稻米遗址的地理特征作回归分析，认为水稻种植和传播中心主要集中在长江中游和长江下游。我和龙国贻通过演化语素分析的方法，不仅同样得到长江流域的结论，而且还进一步认为是从长江下游传播到中游。侗台人群居住于古代的长江下游，苗瑶人群居住于古代的长江中游。

我们把侗台人群的稻米演化语素全列出来：ɣaːu ɣau ɣɑu gau haːu hau hɑu ɦau ɦɑu ɦɯ hu khau khəu khɜu ŋau taːu tau xau ʔu ʔou ʔəu ʔɔu ʔɐu ʔău ʔau qou qəu qet qɐu qau ou kou kəu kɐu kau ɦu hou ho həu hɐu ɣəu əu ˀau ɐu au ɗɔu tau 。

依据Labov[18]的元音三大变化原理，“稻米”一词的韵母在侗台语中发生了以下音变：

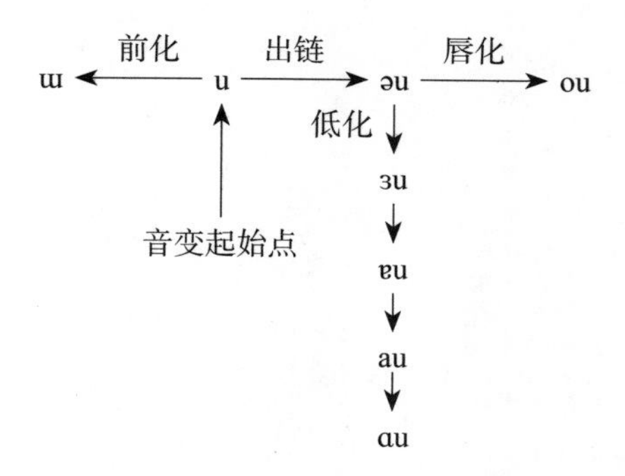

根据自然音变通则，梳理得出侗台语的声母音变链：

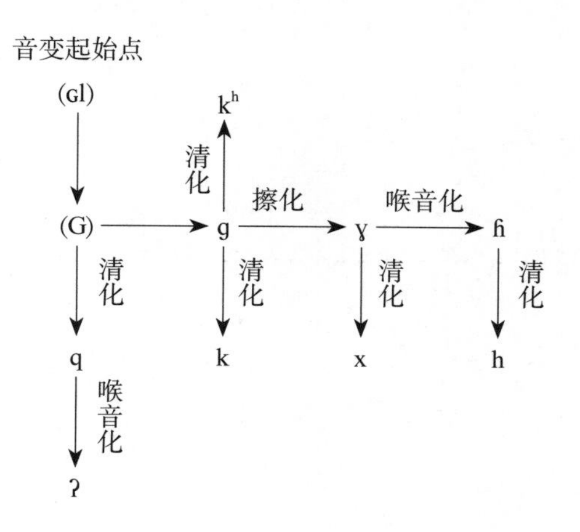

声母的起始音变是ɢl。侗台语的调类是4，来自原始侗台语的喉塞韵尾-ʔ[⑤]，整个音节是ɢluʔ。

“稻”的汉语中古读音是定母豪韵上声，上古音可拟成ɡlu̯ʔ。长江中下游稻米的原始遗址，比黄河流域更早。从历史上看，汉族的谷物里，最重要的是稷，即小米，而不是稻。稻米作为汉族的主食在汉以后才出现。可见，中原地区的稻米是从侗台族群传过去的，侗台人群的稻最早是ɢluʔ>ɡluʔ。

我们再梳理得出苗瑶语中“稻米”有以下演化语素：

bau bjau ɓjau blau bu ɓu ɓʁu bo mbjo mble mjau mjou mɯ mpje mpla mplæ mple mplei mplɛ mpli ndli na ne nei nɯ ntle ntɬe ntli nu pʰɔu pʰɔuʔ pja ple

我们同样根据语音演变的原理，提取苗瑶语韵母中的韵核的音变链：

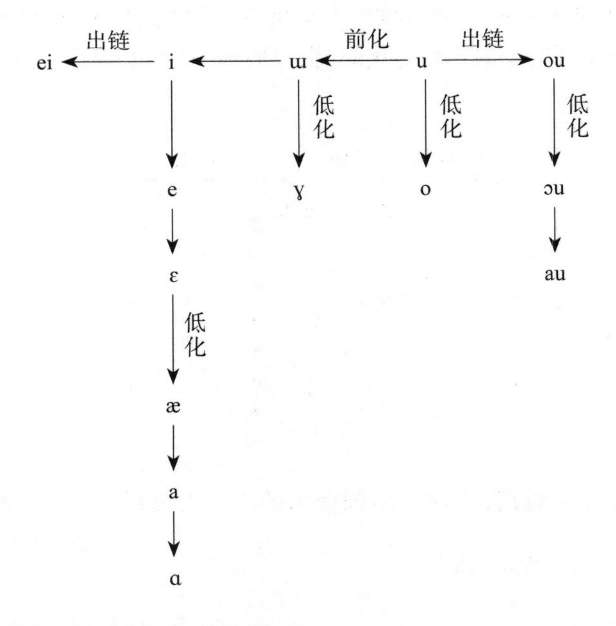

苗瑶语的声母主辅音的音变链：

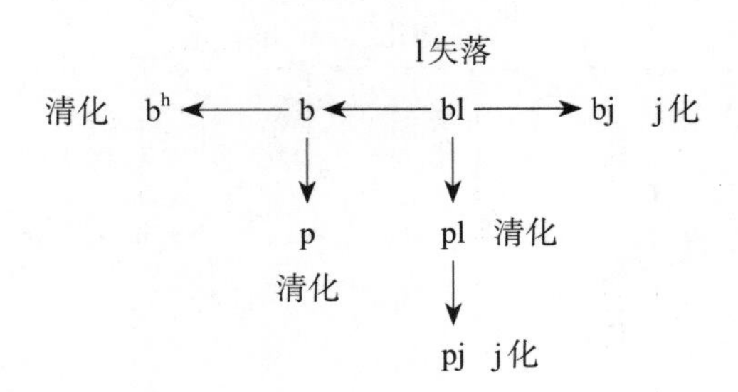

苗瑶人群"稻"的声母起始点是bl-，韵母起始点是u，合起来就是blu。许多语言中的自然音变有glu>blu，塞音g在圆唇元音的同化下变成圆唇的b，许多闽语都有这种音变，如：清溪、揭西 、秀篆 、长汀、宁化、陆川 、香港的"姑"（娘娘）pu<ku。

通过"稻"这个词的演变先后，可以得出以下演化路径：侗台→汉、南亚→苗瑶、南岛、藏缅、泰米尔。东亚语言中稻米的语素集中地出现于侗台地区，这也能说明它的发源地。

从上述稻米的例子可见，从语素演化推演语言演化，再从语言演化推演人类演化，这是从语言学研究人类学的一种基本方法。

根据一个演化语素在不同亲属语中的读音和自然音变规则，可以推演求出音变链，即语言演化的详细过程。语言间的语音对应关系，严格说来就是不同语言之间不同语素具有相同的音变链。一个音节分声母、韵母、声调，有着不同的音变链，例如：下表7个方言有8个语素，都是歌韵开口字，它们之间都有语音对应关系，每个语素都有相同的韵母音变链。

语言点	音变链	歌	可	蛾	何	罗	多	驼	左
高丽	a	ka	ka	n	ha	na	ta	ta	tɕa
汉音	a	ka	ka	ga	ka	ra	ta	ta	sa
安南	a	ka	kha	ŋa	ka	la	ɗa	ɗa	ta
广州	o	ko	ho	ŋo	ho	lo	to	tho	tso
汕头	o	ko	kho	ŋo	ho	lo	to	tho	tso
福州	o	ko	kho	ŋo	ha	lo	to	tho	tɕo
温州	u	ku	kho	ŋ	ɦu	lu	tu	du	tsu

又如：下表是9个苗瑶语、4个语素，每个语言的语素之间都有语音对应关系，每一个语素都有相同的声母音变链。

语言点	音变链	廪	犁	菜	石
罗香	g	gam	gai	gau	gaŋ

（续　表）

语言点	音变链	廪	犁	菜	石
览金	gj	gjam	gjai	gjau	gjaŋ
长坪	ɣ	ɣam	ɣai	ɣau	ɣaŋ
大坪江	l	lam	lai	lai	laŋ
南岗	j	jum	—	ju	jɔŋ
砂坪	z	—	zau	—	zo
大坪	dz	dzum	—	dzu	dzɔŋ
摆托	v	vɦoŋ	vu	—	vɦɔŋ
吉卫	ʐ	ʐe		ʐɯ	ʐɑŋ

表中苗瑶语中的“廪”，吉卫话读ʐe，罗香话读gam，即使是同族人都难以判断它们是不是同一个演化语素。但是，我们从语音对应关系得出相同的音变链g、gj、ɟ、ɣ、l、j、ð、dz、ʐ、z、$z^{ɦ}$、$ʐ^{ɦ}$，吉卫的ʐ-与罗香的g-都是来自同一条音变链中的声母。

表中不同语言和语素之间的不同读音，可以通过计算机统计方法求出音变链的对应关系，由此可以用来确定它们是不是演化语素。但是，因为读音的不规则，统计会有错误。上表中出现横杠的语言就是音变链中的不规则对应。这就是我们后续要用AI等技术手段解决的新课题。

2000年武夷山东南方言比较研讨会报告

2001年在台湾师范大学演讲

2002 年在华盛顿大学访学，与余霭芹教授和学生合影

2002 年接受复旦大学兼职教授聘书

2002 年在华中理工大学召开电子文献会议

2002 年在日本爱知大学演讲

2003 年捷克查理大学访学

2005 年瑞典社会科学研究院做三个月访学

2006 年在南开大学讲学，与邢公畹教授、
校友沈钟伟合影

2007 年在哈尔滨参加汉藏语言学第 40 次年会，
与马提索夫教授合影

2007 年在纽约哥伦比亚大学作学术报告

2008 年与邢向东跟学生一起在咸阳作田野调查

2009 年在五指山参加黎语调查，与发音人的合影

2012 年在汉阳大学访学，与严翼相教授合影

第五十届国际汉藏语言暨语言学会议，左起潘悟云、李壬癸、孙宏开、郑张尚芳、戴庆厦、马提索夫

注释

① 董同龢.上古音韵表稿.史语所集刊,1948.

② 雅洪托夫.古汉语的复辅音声母.汉语史论文集.北京：北京大学出版社,1960：42–52.

③ 潘悟云.上古汉语的复杂辅音与复辅音声母.中国民族语言学报,2017,1：55–61.

④ Matisoff J A. Tonogenesis in Southeast Asia//Larry Hyman(ed.). Consonant types and tone. Southern California Occasional Papers in Linguistics1, 1973, 5: 73–95.

⑤ 王敬骝.傣语声调考.民族语之研究丛刊,1983.

⑥ 潘悟云.汉藏语中的次要音节.第30届国际汉藏语会议论文，中国语言学的新拓展.香港：香港城市大学出版社,1999：125–147.

⑦ 潘悟云.上古汉语鼻音考［J］.民族语文,2018,4：3–9.

⑧ 李方桂.上古音研究.清华学报，1960.新9卷1、2期合刊.本文参用商务印书馆1980年再版本.

⑨ 潘悟云.汉、藏语历史比较中的几个声母问题.语言研究集刊,1987,1：10–36.

⑩ 朱晓农.语音学.北京：商务印书馆,2010.

⑪ 俞敏,中国语言学论文选.东京：光生馆,1984.

⑫ 郑张尚芳.上古韵母系统和四等、介音、声调的发源问题.温州师范学院学报，1987，4：67–90.

⑬ 郑张尚芳.上古入声韵尾的清浊问题.语言研究，1990年第1期.

⑭ Pan W Y. The voiced and released stop codas of old Cinese. Journal of Chinese. Linguistics, 2023, 51(1): 1–21.

⑮ 王巍.中华文明探源研究主要成果及启示.求是，2022年第14期..

⑯ Jeanmaire M. Molecular evidence for a single evolutionary origin of domesticated rice. PNAS, 2011: 8351–8356.

⑰ Fabio S. Modeling the geographical origin of rice cultivation in Asia using the rice

archaeological database.Journal PLOS One, 2015.
⑱ Labov. Principles of linguistic change. Blackwell Publishers, 1994.

主要论著

[1] 潘悟云.朱晓农.汉越语与《切韵》唇音字[J].中华文史论丛,语言文字研究专辑,1982: 323-356.
[2] 潘悟云."轻清、重浊"释——罗常培《释轻重》《释清浊》补注[J].社会科学战线,1983,2: 324-328.
[3] 潘悟云.中古汉语方言中的鱼和虞[J].语文论丛(第二辑),1983: 78-85.
[4] 潘悟云.汉语词典的审音原则[J].辞书研究,1984,5: 75-80.
[5] 潘悟云.非喻四归定说[J].温州师专学报(社会科学版),1984,1: 114-125.
[6] 潘悟云.词汇扩散理论评介[J].温州师专学报(社会科学版),1985,3: 53-62.
[7] 许宝华,潘悟云.不规则音变的潜语音条件——兼论见系和精组声母从非腭音到腭音的演变[J].语言研究,1985,1: 25-37.
[8] 潘悟云.章、昌、禅母古读考[J].温州师专学报(社会科学版),1985,1: 93-111.
[9] 潘悟云.吴语的语法、词汇特征[J].温州师专学报(社会科学版),1986,3: 19-25.
[10] 郑张尚芳,潘悟云.国际音标拉丁字母代用方案征求意见稿[J].温州师专学报(社会科学版),1986,3: 93-94.
[11] 潘悟云.吴语的语音特征[J].温州师专学报(社会科学版),1986,2: 1-7.
[12] 潘悟云.谐声现象的重新解释[J].温州师范学院学报(社会科学版),1987,4: 57-66.
[13] 潘悟云.越南语中的上古汉语借词层[J].温州师范学院学报(社会科学版),1987,3: 38-47.
[14] 潘悟云.高本汉以后汉语音韵学的进展[J].温州师范学院学报(哲学社会科学版),1988,2: 35-51.
[15] 潘悟云.温州方言的指代词[J].温州师范学院学报(哲学社会科学版),1989,2: 13-22.
[16] 潘悟云.中古汉语擦音的上古来源[J].温州师范学院学报(哲学社会科学版),1990,4: 1-9.
[17] Pan W Y. An Introduction to the Wu Dialects[J]. Journal of Chinese Linguistics, 1990: 57-273.
[18] 潘悟云.上古汉语和藏语元音系统的历史比较[J].语言研究增刊,1991: 127-135.
[19] 潘悟云.上古汉语使动词的屈折形式[J].温州师范学院学报(哲学社会科学版),1991,2: 48-57.
[20] 潘悟云.苍南蛮话[J].温州师范学院学报(哲学社会科学版),1992,4: 85-96.
[21] 潘悟云.上古收-p、-m诸部[J].温州师范学院学报(哲学社会科学版),1992,1: 1-12.
[22] 潘悟云.连调和信息量[J].温州师范学院学报(哲学社会科学版),1993,4: 2-12.

[23] 潘悟云."囡"所反映的吴语历史层次[J].语言研究,1995,1:146-155.
[24] 潘悟云.温处方言和闽语[M].中国东南部方言比较研究丛书(第一辑).上海:上海教育出版社,1995:100-121.
[25] 潘悟云.汉语方言史与历史比较法[M].中西学术(第一辑),1995:370-385.
[26] 包拟古.原始汉语与汉藏语[M].潘悟云,译.北京:中华书局,1995.
[27] 潘悟云.温州方言的体和貌[M].中国东南部方言比较研究丛书(第二辑),1996:254-284.
[28] 潘悟云.上古阴声韵部不带塞韵尾的内部证据[M].中西学术(2).上海:复旦大学出版社,1996:524-539.
[29] 潘悟云.喉音考[J].民族语文,1997,5:10-24.
[30] 潘悟云.温州方言的动词谓语句.动词谓语句[M].广州:暨南大学出版社,1997:58-75.
[31] 潘悟云.上古汉语的韵尾[M].中西学术第二辑,1997.
[32] 高本汉.修订汉文典[M].潘悟云,译.上海:上海辞书出版社,1997.
[33] 汉、藏语历史比较中的几个声母问题[J].语言研究集刊,1987(1):10-36.
[34] 潘悟云.三等腭介音的来源[C].李新魁教授纪念文集.北京:中华书局,1998:29-38.
[35] 潘悟云.温州音档[M].上海:上海教育出版社,1999.
[36] 潘悟云.汉藏语中的次要音节[C]//第30届国际汉藏语会议论文,中国语言学的新拓展.香港:香港城市大学出版社,1999:125-147.
[37] 陈忠敏,潘悟云.论吴语的人称代词[M]//代词.广州:暨南大学出版社,1999:1-24.
[38] 潘悟云.陶寰.吴语的指代词[M]//代词.广州:暨南大学出版社,1999:25-67.
[39] 潘悟云.上古汉语元音系统构拟述评[C]//汉语现状与历史的研究——首届汉语语言学国际研讨会文集.北京:中国社会科学出版社,1999:410-428.
[40] 潘悟云.汉语历史音韵学[M].上海:上海教育出版社,2000.
[41] 潘悟云.缅甸文元音的转写[J].民族语文,2000,2:17-21.
[42] 潘悟云.温州方言的介词[M].中国东南部方言比较研究丛书(第四辑).广州:暨南大学出版社,2000.
[43] 潘悟云.汉语音韵研究概述[M].汉藏语同源词研究(一).南宁:广西教育出版社,2000:117-308.
[44] 蒲立本.上古汉语的辅音系统[M].潘悟云,译.北京:中华书局,2000.
[45] 潘悟云.反切行为与反切原则[J].中国语文,2001,2:99-111+191.
[46] 潘悟云.上古指代词的强调式和弱化式.语言问题再认识[M].上海:上海教育出版社,2001.
[47] 潘悟云.避忌讳与古音韵考证[J].中国语文研究,香港中文大学,2001.
[48] 潘悟云.著名中年语言学家自选集・潘悟云卷[M].合肥:安徽教育出版社,2002.
[49] 潘悟云.汉语否定词考源——兼论虚词考本字的基本方法[J].中国语文,2002,4:302-309+381.

[50] 潘悟云.流音考：东方语言与文化[M].上海：上海东方出版中心，2002：118–146.

[51] 潘悟云.吴闽语中的音韵特征词——三等读入二等的音韵特征词[C].声韵论丛(第12辑)，2002：175–188.

[52] 潘悟云.吴语中麻韵与鱼韵的历史层次.闽语研究及其与周边方言的关系[M].香港：香港中文大学出版社，2002：47–64.

[53] 李辉，潘悟云，文波，等.客家人起源的遗传学分析[J].遗传学报，2003，9：873–880.

[54] 潘悟云.汉语南方方言的特征及其人文背景[J].语言研究，2004，4：89–95.

[55] 潘悟云.语言接触与汉语南方方言的形成[C]//语言接触论集.上海：上海教育出版社，2004：298–318.

[56] 潘悟云.汉语方言的历史层次及其类型[C]//乐在其中——王士元教授七十华诞庆祝文集.天津：南开大学出版社，2004：59–67.

[57] 潘悟云.20世纪的中国社会科学·语言学卷[M].上海：上海人民出版社，2005.

[58] 潘悟云.汉语方言学与音韵学研究方向的前瞻[J].暨南学报(哲学社会科学版)，2005，5：104–107.

[59] 潘悟云.客家话的性质——兼论南方汉语方言的形成历史[J].语言研究集刊，2005：18–29+393.

[60] 潘悟云.字书派与材料派——汉语语音史观之一[C]//音史新论——庆祝邵荣芬先生八十寿辰学术论文集.北京：学苑出版社，2005：368–375.

[61] 潘悟云.上古汉语的流音与清流音[C]//汉藏语研究——龚煌城先生七秩寿庆论文集.台湾“中央研究院”语言学研究所《语言学暨语言学》专刊外编之四，2005.

[62] 潘悟云.关于东亚语言的谱系分类的争论[M]//20世纪的中国社会科学·语言学卷.上海：上海人民出版社，2005：332–336.

[63] 潘悟云.竞争性音变与历史层次[J].东方语言学，2006：152–165.

[64] 李龙，潘悟云.国际音标输入法及其实现[J].语言研究，2006，3：67–70.

[65] 潘悟云.汉语的音节描写[J].语言科学，2006，2：39–43.

[66] 潘悟云.朝鲜语中的上古汉语借词[J].民族语文，2006，1：3–11.

[67] 潘悟云.从几个词语讨论苗瑶语与汉藏语的关系[J].语言研究，2007，2：1–9.

[68] 潘悟云.上古汉语的韵尾*-l与*-r[J].民族语文，2007，1：9–17.

[69] 潘悟云.汉藏二族，血肉相连——生物学与语言学的视角[C]//2008年度上海市社会科学界第六届学术年会文集(哲学·历史·文学学科卷).上海社会科学界联合会，2008：7.

[70] 潘悟云.藏文的ɕ-与ʑ-[J].民族语文，2008，4：3–8.

[71] 潘悟云.吴语韵母系统主体层次的一致性[J].东方语言学，2008，1：132–137.

[72] 潘悟云.吴语鱼韵的历史层次[J].东方语言学，2009，2：151–165.

[73] 潘悟云.吴语形成的历史背景——兼论汉语南部方言的形成模式[J].方言，2009，31(3)：193–203.

[74] 潘悟云.吴语鱼韵的历史层次[J].东方语言学，2009，1：90–103.

[75] 潘悟云.历史层次分析的若干理论问题[J].语言研究，2010，30(2)：1–15.

[76] 潘悟云.从地理视时还原历史真时[J].民族语文，2010，1：3–12.

[77] Pan W Y. Compteting sound change and historical strata, Linguistics in China, Vol.1, Singapore: World Publishing Corporation, 2010: 293-313.
[78] 潘悟云.汉藏语与澳泰语中的“死”[J].民族语文,2011,6: 3-8.
[79] 潘悟云.面向经验科学的第三代音韵学[J].语言研究,2011,31(1): 59-63.
[80] 潘悟云,江荻,麦耘.有关计算机数据处理的记音规范建议[J].民族语文,2012,5: 3-7.
[81] 潘悟云.音韵论集[M].上海: 上海教育出版社,2013.
[82] 潘悟云.汉语元音的音变规则[J].语言研究集刊,2013,1: 133-140.
[83] 潘悟云.东亚语言中的“土”与“地”[J].民族语文,2013,5: 3-12.
[84] 潘悟云,张洪明.汉语中古音[J].语言研究,2013,33(2): 1-7.
[85] 潘悟云.汉语历史音韵学[M].韩国首尔学古房,2014.
[86] 龙国治,潘悟云.壮族族称考[J].广西民族大学学报(哲学社会科学版),2014,36(6): 110-112.
[87] 潘悟云.对三等来源的再认识[J].中国语文,2014,6: 531-540+576.
[88] 章杰鑫,潘悟云.古籍数字化技术的新思路[J].语言研究,2014,34(1): 124-126.
[89] 王奕桦,潘悟云.韩语字母“·”所表记的音值及其音变路径[J].民族语文,2015,6: 59-65.
[90] 潘悟云.方言考本字“觅轨法”[J].方言,2015,4: 289-294.
[91] Pan W Y. Middle Chinese phonology and Qieyun[M]//The Oxford Handbook of Chinese Linguistics, 2015: 80-90.
[92] 潘悟云.再论方言考本字“觅轨法”——以现代韵母为u的滞后层为例[J].语文研究,2016,4: 9-11.
[93] 潘悟云.侗台语中的几个地支名[J].民族语文,2016,5: 3-11.
[94] Pan W Y. On some theoretical issues about historical strata analysis. Macrolinguistics, 2016, 4(5): 11-38.
[95] 潘悟云.释地名中的阳[C]//丁邦新、张洪年、邓思颖、钱志安主编.汉语研究的新貌: 方言、语法与文献——献给余霭芹教授.香港: 香港中文大学出版社,2016.
[96] 潘悟云.上古汉语的复杂辅音与复辅音声母[J].中国民族语言学报(第一辑),2017: 55-61.
[97] 白一平,潘悟云.上古音对谈实录[J].语言研究集刊,2018,2: 394-416.
[98] 潘悟云.丝绸传播路径与年代考[J].语言研究集刊,2018,2: 478-481+662.
[99] 潘悟云.滞二等现象——考本字的一个特殊视角[J].方言,2018,40(3): 257-262.
[100] 潘悟云.上古汉语鼻音考[J].民族语文,2018,4: 3-9.
[101] 潘悟云.汉藏语的使动态——兼评OldChinese[J].汉语史与汉藏语研究,2018,1: 1-26.
[102] 潘悟云,王奕桦,葛佳琦.中韩两国古代文化交流的印证——语言学视角[J].广西师范大学学报(哲学社会科学版),2018,54(1): 76-83.
[103] 王轶之,张梦翰,潘悟云.汉语史上开闭音节中元音演变速度差异的原因——感知实验的视角[J].中国语文,2018,1: 96-105+128.

[104] 潘悟云.汉藏语的使动态[J].汉语史与汉藏语研究(第三辑).北京：中国社会科学出版社,2018：1–26.

[105] 潘悟云.对白—沙体系的评论[J].饶宗颐国学院院刊,2018(5)：403–415.

[106] 潘悟云.汉语音韵学与文字学的互动[J].饶宗颐国学院院刊,2019,6：11–28.

[107] 潘悟云.南方汉语中的"毒"字[C]//汉语与汉藏语前沿研究——丁邦新先生八秩寿庆论文集.北京：社会科学文献出版社,2018：751–756.

[108] 潘悟云.东亚语言声调起源的内因与外因[J].韵律语法研究,2019,2：1–17.

[109] Pan W Y. Phylogenetic evidence for Sino-Tibetan origin in northern China in the Late Neolithic[J]. Nature, 2019, 569(7754): 112–115.

[110] 潘悟云.上古音构拟[J].出土文献,2020,2：127–135+158.

[111] 潘悟云.地理虚时与音变链[J].方言,2020,42(2)：142–147.

[112] 潘悟云.语言借用和历史比较[J].语言战略研究,2022,7(4)：1.

[113] 潘悟云.同源语素与音变链[C]//中国语言学研究第一辑.北京：社会科学文献出版社,2022：1–9

[114] 潘悟云,刘航宇.喉音再考[J].民族语文,2023,5：3–9.

[115] Pan W Y. The voiced and released stop codas of old chinese[J]. Journal of Chinese Linguistics, 2023, 51(1): 1–21.

[116] 潘悟云.汉语古音手册[M].上海：中西书局,2024.

[117] 潘悟云.重述六元音系统——怀念王力先生、梅祖麟先生和郑张尚芳[J].辞书研究,2024,4：86–88.

[118] 陈鹏,潘悟云.上古汉语的第一人称[J].古汉语研究,2024,3：115–125+128.

[119] 潘悟云.知组考[J].中国语文,2004,5：515–519.

代表作

THE VOICED AND RELEASED STOP CODAS OF OLD CHINESE

Pan Wuyun[1], Zheng Zining[2]
([1]*Shanghai Normal University*
[2]*Shenzhen Ghostvalley AI Lab*)

ABSTRACT

This paper puts forward the property of voicing and releasing in coda stops in Old Chinese is mainly based on the historical comparison between Chinese and Tibetan Languages and Chinese loan words in other languages. For example, in ancient Tibetan the stop codas were transcript to voiced letters. The majority of bound function words in Classical Tibetan has two variants based on whether the preceding codas are voiced or not. The L - (- R) coda in the ancient Chinese northern languages and early Sanskrit-Chinese phonetic transcription comes from the - D coda. Japanese Old Chinese loan words have voiced stop codas. Modern Chinese dialects also have voiced stop codas. In addition, stop codas in Tibetan Xigaze, Balti, Lajiao and other dialects, <Xi Fan Yi Yu> and Qiang languages all have the nature of releasing. The 'broken tone' in Chinese is caused by a released glottal stop.

KEYWORDS

Old Chinese Stop codas Release Voicing Sound pattern Historical comparison

1. STOP CODAS IN OLD CHINESE ARE VOICED

The stop codas of entering tone (入声 *rùshēng*) in Middle Chinese manifested as -p, -t, -k, not -b, -d, -g based on Sino-Japanese phonetic transcription. For example, If Japanese people at that time heard the character 各 pronounced kag, they would hypothetically transcribed it as kagu instead of kaku, the correspondence in reality. Therefore Karlgren reconstructed the stop codas of rùshēng as voiceless *-p, *-t, *-k, not voiced *-b, *-d, *-g, which has become the consensus of Chinese historical phonology.

However, -p、-t、-k codas in Middle Chinese do not mean necessarily that they should also be -p、-t、-k in Old Chinese. Yu (1984) and Zhengzhang (1987, 1990) both brought out the possibility that the stop codas of Old Chinese could be *-b, *-d, *-g. Comparison between the stop codas of Chinese with those of Tibetan and other languages could reach the conclusion that they were indeed voiced.

1.1 Top Codas In Classical Tibetan Were Voiced

Tibetan scripts contains voiced letters བ(b), ད(d), ག(g) as well as voiceless ones པ (b), ཏ (d), ཀ (g). Stop codas of Tibetan are written using the voiced letters only. Tibetan script was modelled on an Indian Brahmic script. When writing Sanskrit stop codas, Brahmic scripts use voiceless letters. If the stop codas of Tibetan language were voiceless when Tibetan script was invented, they should have been represented by voiceless letter as in Sanskrit. Therefore the stop codas of the Tibetan language should be voiced as represented by the letters.

There remains questions on the voicing of Tibetan stop codas. One of the principle reasons doubting the validity of the voiced nature of the Tibetan stop codas is that the stop codas in almost all other Tibeto-Burman languages are voiceless. More sound change rules, however, shows that the stop codas in Classical Tibetan must be voiced.

One example given by Zhang (1992) shows the voiced nature of stop codas in Classical Tibetan through phonological evolutionary rules. The majority of bound function words in Classical Tibetan have two variants based on whether the preceding codas are voiced or not. Preceding voiced codas -m、-n、-ŋ、-l、-r precondition variant A, and voiceless coda -s preconditions variant B. The three stop codas of Classical Tibetan are succeeded by both variant A and variant

B. If they were voiceless, it would be difficult to explain why some succeeding function words belonged to variant A. If they were originally voiced, however, we could raise the following explanation: the devoicing of voiced stop codas happened successively and some of them devoiced under certain phonological conditions which made succeeding function words variant B while other stop codas had remained voiced, preconditioning variant A. Zhang(1992) proves that the appearance of variant B following some stop codas can be attributed to the devoicing incurred by phonological conditions with rigorous material. The manifestation of bound function words is enough proof that the stop codas of Old Tibetan were voiced.

Voiced stop codas in Classical Tibetan have devoiced in large part in Balti however remain voiced in some Balti words (Huang, 2007):

Table 1 A summary of Balti lexicons with voiced stop codas in Huang (2007)

	Needle	Hide	Eight	Guide	Sheep
Classical Tibetan	khab	jib	brgjad	khrid	lug
Balti	khab ^{53}	ʔib^{53}	bgjad^{53}	khjid^{53}	lug^{2}1

Balti is a living attestation to the continuance of voiced stop codas in Classical Tibetan.

However, the stop codas in Classical Tibetan being voiced do not guarantee that the stop codas in Old Chinese were voiced. We will now give evidence that support the voiced nature of Old Chinese stop codas.

1.2. Northern Neighbours Of Middle Chinese Borrowed Coda -D As -L(-R)

The stop coda –k in <Qieyun> manifests as -k among Sino-Korean words: 索 sæk 색，角 kak 각, and the stop coda -p manifests as -p: 塔 thap 탑，集 tɕip 집, but the stop coda -t manifests as -l in Sino-Korean words, e.g. 舌 səl 설, 活 hwal 활, 笔 phil 필, 割 hal 할.

Coda -t exists in Korean, e.g. grain: nat 낟, walk: kət 걷, fetch water: kit 긷, carry: 싣 sit, being first: mat 맏, bury: mut 묻, straight: kot 곧 etc. Theoretically the -t coda in <Qieyun> should have been borrowed into its most similar counterpart in Korean, i.e. -t. Why did the Koreans transcrib it as -l, a less similar sound? This illustrates that the -t coda in <Qieyun> was pronounced voiced in the Chinese dialect in northern China and it was an unreleased -d. We can try to pronounce an unreleased d, the vocal cord vibrating during the holding phase would result in a sound aurally similar to a syllabic l̩, that is the reason for the Ancient Korean treatment of transcribing Chinese -t as -l.

The Gaochang residents of Qocho during Tang Dynasty were mostly ethnic Chinese from the east, i.e. the Hexi（河 西）region. Huili（慧 立）, a monk from the Tang era transcribed <大唐大慈恩寺三藏法师传> in Uyghur script, representing the literate pronunciation system of Hexi region at that time (Lin, 2012). The Chinese stop coda -t corresponded to -r in that system, e.g. 达dɑr，喝hɑr，笔pir，佛bur，律luer. Some corresponds to -l as well, e.g. 末mɑl，喝hɑl. Luo (1933) presents data from Sino-Tibetan transcription, and the Chinese stop coda -t corresponds to -r in Tibetan transcription as well. For example, 脱thar, 杀sar, 舌ʑar, 血hjar, 涅ɴder, 察char, 厥kwar, 悦war, 密ɴbir, 实ɕir, 逸jir, 弗phur, 物bur, 鬱gur in <汉藏对音千字文>。毕pir, 蜜ɴbjir, 一ʔir, 骨kur, 佛phur, 物bur, 出chur, 別phar in <大乘中宗见解>;悉sir, 一ʔir, 佛bur, 萨sar, 末ɴbar, 八par, 发phar, 灭ɴbjer in <金刚经>; and达dar in <阿弥托经>. These materials indicate that the Middle Chinese -t was pronounced -d in the north, similar to what the Korean material shows. It is perhaps reasonable for some to propose that the coda in these Chinese dialects was not -d but -l. However, in <阿弥陀经>, one finds 跋bad, 发phad, with the codas being -d, not -r. One particular value is the transcription of character 发 as phar in <金 刚 经> but phad in <阿弥陀经>, indicating the coda must have been pronounced an unreleased -d as an -l pronunciation is highly incompatible with a -d transcription. The Middle Chinese -d has become -ɫ in Xiushui(修水), Nanfeng(南丰), Duchang(都昌) and Gaoan (高安) in Jiangxi Province, and -n in Dongshan(东山) and Jinjiang(晋江) in Fujian Province, indicating that -t was in origin a voiced unreleased -d in all probability, thus can change into -l and -n.

1.3. The Chinese Rùshēng Voiced Codas In Early Sino-Sanskrit Transcription

Yu (1948) presents the following data:

Table 2　A summary of selected Early Sino-Sanskrit transcriptions in Yu (1984)

Chinese	Sanskrit	Original Transcription	Original Word	Original Scripture
遏	ar	遏迦	arghya	摩登迦经
鬱	ud	鬱頭	udraka	中本起经
揭	gar	蔡揭	sāgara	阿弥陀经
掘	gul	鴦掘摩羅	amgulimala	撰集百缘经
涅	nir	涅槃	nirvāṇa	般舟三昧经
弗	pur	弗沙	puruṣa	摩登伽经
拔	bhad	拔陂	bhadrapāla	拔陂菩萨经
佛	bud	佛	buddha	理惑论
律	rud	阿那律	aniruddha	维摩诘经
薩	sar	薩云若	sarvajña	般舟三昧经
越	var	震越	civara	文殊师利问菩萨署经

The rùshēng characters in Chinese are commonly used to transcribe Sanskrit -p, -t, -k. The majority of the instances in the table above shows rùshēng codas transcribing -r, -l, being congruent with the evidence provided by Korean -l and Middle Northwestern Chinese -r. Some instances show rùshēng codas were corresponding Sanskrit -d and a few others -dr, e.g. 鬱 udra, 拔 bhadra, exhibiting some ambivalence between -d and -r, showing that unreleased –d is close to -r and -d simultaneously.

1.4. Rùshēng Codas Still Manifest As Voiced Consonants In Modern Chinese Dialects That Preserve Voiced Stops

Zhengzhang(1990) indicates that stop codas in the following dialects are voiced, e.g. Lianshan, Guangdong Province (Cantonese): 白 bag, 域 ɦuɑg, 特 dɑg, 族 zog, 绝 zod, 别 bed, 碟 ded, 十 zɑd/zɑb, 悦 ɦyd 乙 yd, 脱 thud, 骨 kuɐd;

Liufang & Jiangqiao, Hukou, Jiangxi Province (Gan): 直 dzig, 角 各 kɔg, 踢 ḍig, 拔 bal, 夺 lɘl, 阔 guɛl, 割 kol, 刷 sol, 骨 kuɛl, 夹 kal.

Meanwhile, Gan dialects spoken in Tongcheng (Hubei Province), Xiushui, Nanfeng, Duchang and Gaoan (all in Jiangxi Province) all have -l coda.

1.5. Old Chinese Loans In Japanese Indicate Voiced Codas In Rùshēng Syllables

麥 mugi, 直 sugu

葛 kadu，seen in "葛羅" kadura

綴 tudu, c.f. Tibetan sdud(link)

物 mono（<-modo）

頜 kubi，meaning head or neck in Modern Japanese

蛤 gama（<-gaba），synchronous change with Chinese "蛤">"虾蟆"

蓋 kabusu

Therefore, the stop codas of Old Chinese should be voiced. The devoicing of the stop codas happened successively, *-b, *-g devoiced earlier while *-d devoiced later. Apart from devoicing to -t or -ʔ, the voicing of *-d is preserved in some languages: *-d>-l. It can be attributed to their difference places of articulation. Compared with coronal consonants, the articulation of labial and velar consonants involves slower movement of speech organs and greater occlusion, resulting in higher pressure in the oral cavity. High intraoral pressure, just like high vowels, could cause glottal friction and epiglottal release. In addition to the above reason, another favourable condition for the preservation of voicing of -d coda is that the tip of the tongue can be more easily and precisely controlled, the oral cavity can be depressurized through the lateral sides of the tongue so as to maintain relatively longer duration of vibration of the vocal cord. The aforementioned evolution of the voiced stop codas can be evidenced from many materials, e.g. Classical Tibetan *-b, *-g are realized as -p, -k whereas *-d is realized as -l in Zeku dialect. Similarly, Middle Chinese -t is realized as -n in some Min dialects, which in fact evolves from this kind of unreleased -d coda, i.e. the voiced consonants, especially when succeeding high vowels, exhibit transvelar nasal coupling, producing low frequency resonance which is aurally similar to nasal consonants (Zhu & Cun, 2006), hence -d>-n.

2. THE STOP CODAS OF OLD CHINESE WERE RELEASED

The majority of stop codas in Sino-Tibetan languages are unreleased. Recently, we have found that some languages possess stop codas that are released.

In particular, some conservative Tibetan dialects do have released stop codas.

Among the northern branch of Amdo Tibetan, only the Arou and Beishan, Ledu dialects possess released stop codas. Arou, Tibetan ʔa rigis also known as Alike. The place name is transcribed阿柔(Arou) if the stop coda is unreleased and 阿力克(Alike) if it is released, representing variation of stop codas in Amdo Tibetan. 拉达克(Ladake) is the Chinese transcription of the Tibetan word la dwags. Similar to 阿力克，克 corresponds to coda -gs, illustrating that the -s in the cluster –gs is mute and -g is release (Wang, 2010).

When words in Balti (Huang, 2007) involve high vowels followed by stop consonants, there is often an epenthetic vowel, the same as the preceding vowel but pronounced lighter, which follows the coda. E.g. kjoq^{o} (wry), mbruk^{u53} (dragon),mjik^{i53} (eye)；The coda -t is often aspirated, e.g. the -t in nat^{53} (sick) is pronounced [t^{h}]. It is obvious that they show that the stop codas are released. The releasing of a stop coda is in essence the return of the speech organ from the point of restriction to its neutral position. In the majority of languages, releasing means returning to a gesture similar to schwa. The phenomenon in Balti language mentioned above, i.e. stop codas following high vowels, when released, are accompanied by a voiceless vowel similar to the preceding one is in fact the return speech gesture after the release.

The same released stop codas exist in Xigaze Tibetan (Skalbzang & Skalbzang, 2002). When preceded by high vowels in monosyllabic words, -p & -k are released and aspirated, and followed by voiceless vowels.

Table 3　A summary of Xigaze reflection of voiced stop codas in Classical Tibetan in Gesang and Gesang (2002)

	One	Six	Cloudy	West
Classical Tibetan	gtɕig	drug	thibs	nub
Xigaze Tibetan	tɕikhi̥	ʈʂhukhu̥	thiphi̥	nuphu̥

These two Tibetan varieties all show the same phenomenon, i.e. stop codas are released and aspirated when following high vowels, and they all carry epenthetic voiceless vowels, which is the speech gesture of post-release return.

The Lajiao dialect belongs to the Dbus variety of Dbus-gtsang Dialect of Tibetan language, and the releasing of stop codas is more widespread in this variety (Xu, 2020). In this variety, the realization of Classical Tibetan -g shows free variation between unreleased -k and unreleased glottal stop -ʔ. In monosyllabic words, some -k codas are released and aspirated, followed by voiceless vowels, i.e. the aforementioned return from the point of restriction to neutral position. E.g. "leopard" gzig, pronounced sikʰi̥ 132 ； "dragon" ɦbrug, pronounced ndʐukʰʉ̥ 132 ； "face mask" ɦbag, pronoucned nbakʰə̥ 132 . When the main vowel is the low vowel a, -k is more commonly realized as voiceless fricative x, e.g. "yak" g-jag, pronounced jax 53.

Classical Tibetan -b has mostly evolved into unreleased -p in Lajiao, it is released and aspirated in only a few words with high vowels and carries the speech gesture of natural return, e.g. "west" nub, pronounced nupʰʉ̥132, "book" steb, pronounced tepʰə̥ 132.

The release of stop codas in the languages above show the same characteristics which can be explained phonetically.

Firstly, stop codas of monosyllabic words are released and aspirated. Under the same pulmonary dynamic condition, the pulmonary pressure is released in shorter duration in monosyllable words with faster airflow through the vocal cord. It is therefore easier to formulate aspirated glottal fricative noise.

Secondly, releasing correlates with the height of the vowel. A high vowel leaves a narrower passage in the oral cavity, enhancing the pressure under the vocal cords. With the same aperture of the vocal cord, it is easier to cause epiglottal release. The -g coda after low vowels, to the contrary, evolves lower and fricativize to -x.

Thirdly, labial and velar stop codas involve slower movement of speech organs with greater occlusion, causing higher intraoral pressure and making stop codas more prone to loss of release. The coda -d changes to –l in some dialects (e.g. Zeku).

Fourthly, all stop codas lose release eventually. The loss of release is cross linguistically common, representing a common type of stop coda weakening.

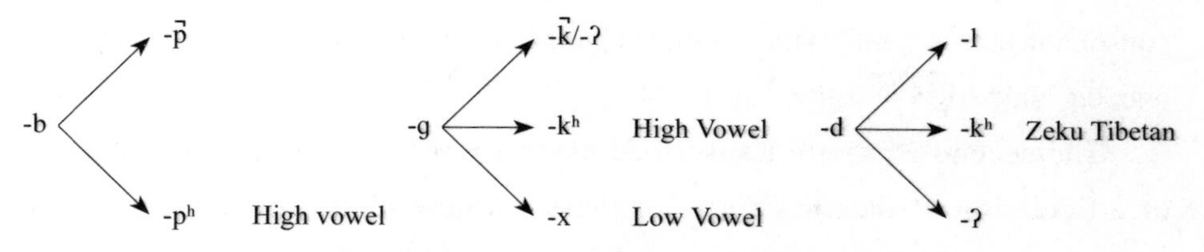

Figure 1　Proposed evolution of stop codas in Tibetan dialects

Historical documents show that the stop codas of earlier Tibetan were released. For example, <Xi Fan Yi Yu>in the Ming dynasty shows that the stop codas were released in Tibetan of the same period (Wang, 2010).

Table 4　A summary of Sino-Tibetan Transcription as recorded in Xifan yiyu as presented in Nishida (1970)

Classical Tibetan	Chinese Transcription	Classical Tibetan	Chinese Transcription
tʰog (thuner)	托	drag pa (tight)	扎罷
stegs (platform)	思楪克思	dbugs (air)	物克思
sgjogs (cannon)	思脚克思	lpags pa (skin)	失罷克思罷
drug (six)	竹	gtogs (belong)	黑奪克思
rgjab (hind)	兒甲	tʰob (get)	托
slebs (arrive)	思列卜思	srab (thin)	思剌
tsʰigs btɕad (proverb)	戚克思卜乍	nub (west)	奴
brgjad (eight)	本兒甲	ɦbjed (divided)	恩别
dred (brown bear)	折	skud (thread)	思谷

The stop codas are transcribed in two classes. The first class involves stop codas followed by another coda –s, transcribed with one Chinese character plus ‘思’, e.g. -gs is transcribed with ‘克思’, -bs is transcribed with ‘卜思’. The second class involves the whole syllable transcribed with one Chinese rùshēng character, tʰog is transcribed with ‘托’, drag is transcribed with ‘扎’. If there is a pre-consonant that is continuant, another Chinese character is added before the first one, e.g. srab '思剌', rgjab '儿甲'.

The second class are transcribed using rùshēng characters but the place of articulation of the stop codas of these Chinese characters do not strictly correspond with the ones of the Tibetan words, indicating that the stop codas in Chinese had weakened to a neutral glottal stop. The corresponding Tibetan syllables only show resemblance to the onsets and rhymes of the Chinese characters without transcribing the codas into another character, which means that the codas were not released. The stop codas in the first class were transcribed using an extra Chinese character, very different from the treatment in the second class, shows that the stop codas before the coda –s must be released.

Stop codas in Qiang language, with the exception of -p and the ones in the middle of multiple syllables are almost always released (Shen & Liu, 2010). The stop codas in the Re'ena variety of Qiang are all able to be released (Zhou, 2019). However, some of the stop codas of Qiang are secondary, being the result of weakening of vowels at end of syllables and subsequently become to apocope, coalescing with the original onsets to become a released coda. E.g. 'ear' is pronounced nə^{55}tsu^{55} in Jiulong, na^{53}pi^{53} in Ersu, and nɐ^{55}pɐ^{53} in Taoba, the first morpheme being 'ear', the second being a suffix. The original form of Mawo nə ků should be nə ku. After the weakening of the suffix, the pronunciation is weakened ku>ku̥>k^{w}, the final result being a phonetically labialized stop coda k^{w}. Apart from secondary stop codas, the majority of stop codas in Qiang are primary though.

Some Kra-dai languages possess released stop codas as well. The variey of Ha dialect of Hlai language spoken in Zhizhong, Dongfang, Hainan Province possess relatively obvious released stop codas. There are four stop codas in this variety, -p, -t, -k, -ʔ. Many of the occurrences are released with audible airflow, vibration on oscillogram and release burst on spectrogram (Shen & Liu, 2010).

We can trace the loss of release through phonetic evolution. The stop codas in most modern Tibetan dialects have lost release with the original –g, -d becoming -ʔ in some varieties such as Lhasa. <Xi Fan Yi Yu> shows that the Tibetan at that time still had released stop codas but they had mostly disappeared in the modern era. The words with released stop codas in Balti, Xigaze and Lajiao are in the process of losing release. The stop codas in Amdo Tibetan and Qiang have both released and unreleased variants. The stop codas in Re'ena Qiang can all be released. The majority of stop codas in Qiang are released with only –p and stop codas in the middle of multiple syllables showing unreleased variant. The ones in Balti, Xigze and Lajiao following non-high vowels have lost release. All the above show that the unreleased variant rose through phonetic conditions.

Lastly, we need to discuss whether the stop codas in Old Chinese were released. Up till now we have not found clear evidence of released stop coda in modern Chinese dialects and no evidence from historical documents shows it either ,but we may reach the conclusion that the stop codas in Old Chinese were indeed voiced from the following two aspects.

Firstly, since the stop codas in Old Tibetan were released and the Sino-Tibetan languages come from a common ancestor, the stop codas at some stage in Old Chinese should be released as well. The stop coda in the common ancestor of Chinese and Tibetan are either released or unreleased and we have to pay attention to the evolution of Chinese and Tibetan.In Tibetan, the evolving process of stop codas from voiced released to voiceless aspirated is strengthening from weak codas to strong codas. Subsequently, voiceless aspirated released stop codas have gradually become unreleased stop codas, fricativized or glottalized, all weakening processes with the final destination being complete loss. The stop codas in Old Chinese were voiced and they became unreleased and voiceless unaspirated in Middle Chinese, finally glottalized and dropped, as shown in the diagram below.

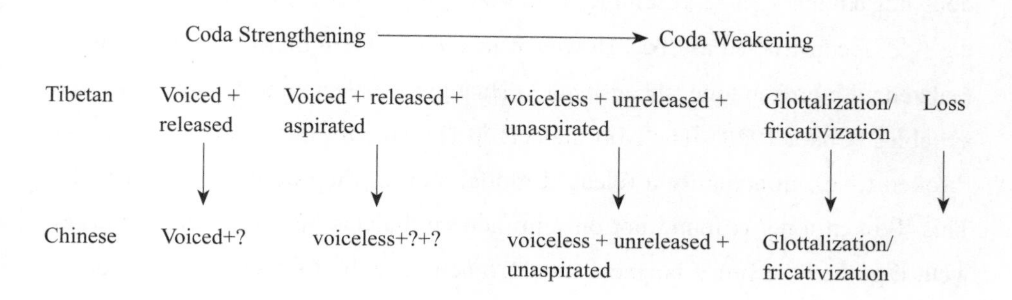

The chronological evolution of stop codas in these two languages are for most part identical with some evolution in Tibetan lacking exact counterpart in Chinese. For example, were Old Chinese stop codas released and aspirated like Tibetan? The common ancestry of the two languages and the correspondence of the majority of characteristics in each evolutionary stage suggest that these two may show similar evolution pathways and the questions marks in the evolution of Chinese can be substituted by attested characteristics in Tibetan, i.e. The stop codas in Old Chinese, like their Tibetan counterparts, were released and they became aspirated from devoicing.

Secondly, shǎngshēng (上声) in Proto-Chinese possessed released glottal stops. Haudricourt (1954) indicates that shǎngshēng in Chinese originates from glottal stop -ʔ. There are several types of glottal stops, one from creaky voice, resulting in low domain of frequency, another from head voice, resulting in high domain of frequency, another from tense voice, resulting in middle domain of frequency. Mei (1970) put forward evidence from many Chinese dialects and historical documents to prove that shǎngshēng in middle Chinese was a short and high tone so that the original shǎngshēng should come from head voice. From the perspectives of rhyming and phonological theries in Old Chinese, rùshēng forms a category of its own and the other three tones form the other class. If shǎngshēng carried a glottal stop in Old Chinese, it would make it more similar to rùshēng, in conflict with rhyming evidence. The glottal stop must have already being losing during Old Chinese and what remained in shǎngshēng was the short and high head voice. Bodman (1980) claims that shǎngshēng in Old Chinese only had glottal constriction, i.e. head voice with its short and high characteristics. As IPA does not have a sign representing head voice, we adopt Bodman's usage of '˙', e.g. '父'reconstructed as *ba˙. However, in some modern dialects, shǎngshēng is realized with broken tone, shǎngshēng in Huangyan dialect almost sounds like two syllables (Chao, 1928). The Min dialects in Hainan we investigated also shows 'broken tone', in actuality a released glottal stop in the middle of the syllable. This 'broken tone' is found not only in modern dialects but in ancient record as well. E.g. the 9th century Japanese monk Annen gave the following description of shǎngshēng: ‘The heavy version of shǎngshēng is somehow similar to the light and heavy píngshēng(平声), with the heave first and light later. It sounds peculiar and there exists differences in terms of lips and tongue as well.’ The description concerns the special case of yángshǎng, i.e. the heavy version of shǎngshēng. Annen claims that yángshǎng is the combination of yángpíng (阳平) and yīnpíng (阴平) and they show different contours, like the ngã tone in Vietnamese, starting as low even with a sudden rise to high even, showing the characteristic ‘broken tone’. Sino-Tibetan bilingual Buddhist sutras excavated in Dunhuang often have the Tibetan version transcribing Chinese shǎngshēng characters using two vowel letter (Luo, 1933). E.g. in <Qian Zi Wen>: 组 dzo‘o, 纺 pho‘o, 酒 dzu‘u, 举 ku‘u, 象 syo‘o, illustrating that the northwestern dialect at that time also had this tonal ‘peculiarity’ which sounded like two syllables to Tibetan ears. Head voice glottal stop often evolves into creaky voice glottal stops, shǎngshēng in Beijing is sometimes pronounced ‘broken tone’. The ‘broken tones’ in these dialects accomplish the evolutional chain of Chinese stop codas from released to unreleased, and then to complete loss.

Below is the diagram of Chinese stop coda evolution. The stop codas in Old Chinese were voiced and released, devoiced and unreleased by Middle Chinese. In modern dialects, some have evolved to glottal stops while others have completely lost stop codas, turning to open syllables. The –d in some dialects were voiced and unreleased in Middle Chinese which has become –r, -l, -n. Proto-Chinese had released glottal stop coda which had been preserved in some pre-modern and modern Chinese varieties, usually surfacing as ‘broken tone’. The glottal stop coda was lost in Old Chinese and caused high tone.

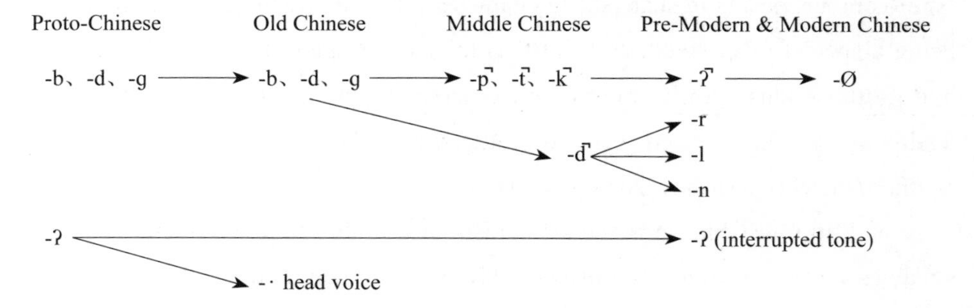

3. CONCLUSION

The pronunciation of codas is intricately related to syllabic theory and may have implication in rhythmic theory as well. This paper proposes that the stop codas in Old Chinese were voiced and released, which could solve many linguistic problems.

The codas of Sino-Tibetan languages tend to be unreleased, making them typical monosyllabic languages. However, the stop codas in Indo-European languages are typically unreleased. The vocal folds are still vibrating after the release of coda –g from its velar constriction and the subsequent return of the tongue to its neutral position, articulating a muffled vowel ə̭ or its voiceless counterpart. To Chinese 'tank' in English sounds like a disyllabic words, hence the Chinese transcription '坦克'. The nasal coda /n/ in English is unreleased, similar to their counterparts in Chinese monosyllables ending with nasal codas. The /n/ coda in French, however, is released and French syllables with coda /n/ sounds like two syllables to Chinese.

According to Chao(1928), the 'split contour' in Chinese dialects sounds as if the syllable has been broken into two. E.g. '火' is pronounced /hue^{213}/ in Ding'an, Hainan. Its pronunciation is rather peculiar: the first component is /hue^{21}/, and then after a brief break, comes /ʔe^{3}/. Some people record it as /huʔe^{213}/, /hue^{21ʔ3}/ or even two syllables. This 'split-contour' is in fact a syllable with a glottal coda, i.e. /hueʔ/. When the glottal coda is released in Ding'an, it comes with a muffled vowel /ḙ/ when the speech organs return to their natural state. Glottal codas in Chinese usually occur in rùshēng characters, therefore Chinese linguists do not consider the glottal codas in rùshēng characters as released. In particular, when 'split-contour' occurs in shǎngshēng characters, they are often relegated to merely being a special tone. Haudricourt (1954) thinks that shǎngshēng in Old Chinese had glottal codas. 'Split-contour' only occurs in shǎngshēng characters is a testimony that they carried released glottal stop codas in Proto-Chinese and the southern dialects still show some residue.

"盍" in Old Chinese is the synaeresis of "何不". The contraction of two syllables into one, while certainly possible, was not common in Old Chinese. If

"盍" was pronounced *ga̱b in Old Chinese with released stop coda –b, the coda would entail a muffled neutral vowel ə̥ in the return phase, making the actual pronunciation of ga̱b similar to ga̱bə̥. The "不" in "何不" belongs to onset category "帮" and rhyme category "之", i.e. pronounced pɯ. As a closed class word, the ɯ in '不' was often weakened to be ə̥. The pronunciation of '何不' was ga̱lpə̥, very similar to that of '盍' (ga̱bə̥). Consequently, it would be better consider '盍' not a case of synaeresis, but to be of sound borrowing. Synaereses were rare in Old Chinese while sound borrowings were commonplace.

REFERENCES

CHAO, Yuen.Ren.1928. Xiandai Wuyun de Yanjiu.(Studies in the Modern Wu Dialects). Beijing Qinghua Xuexiao Yanjiu Yuan Congshu.5.

DUAN, Yu.Cai.1928. Xiandai Wuyun de Yanjiu.(Studies in the Modern Wu Dialects). Beijing Qinghua Xuexiao Yanjiu Yuan Congshu.5.

HAUDRICOURT,André-Georges.1954.De l'origine des Tons en Vietnamien，*Journal Asiatique*.242.

HUANG, Bufan.2007. Cong Baerti Hua Kan Gu Zangyu Yuyin. (The Ancient Tibetan speech Sounds Viewed from the Balti Language).*Zhongyang Minzu Daxue Xuebao*.4.

LIN,Xunpei, X.2012. *Huihu Wen Cienzhuan Hanzi Yin Yanjiu*(*Study of the Uyghur transcription of Cienzhua*). .Ph.D. Dissertation.Shanghai Normal University.

LUO, Changpei .1933. Tangwudai Xibei Fangyin. (The Northwestern Dialects during the Tang and Wudai Period). Shanghai: *Academia Sinica.*

MEI,Tsu-lin（1970）Tones and Prosody in Middle Chinese and the Origin of the Risinng Tonne，*HJAS*.Vol.30.

NISHIDA, Tathuo .1920. 西番馆译语の研究 .(*Study of Translation of the Western Barbarian Language*). 日本：松香堂 .

SHEN, Xiangrong & Liu, Bo.2010. Hanzangyu zhong de Sewei Baopo Xianxiang(Release of Stop Codas in Sino-Tibetan languages). *Minzu Yuwen* 1.35-40.

GESANG, Jumian& Gesangy, Yangjing. 2002. Zangyu Fangyan Gaikuang.(*Sketch of Tibetan Dialects., Minzu* Chubanshe.

WANG, Shuangcheng.2010. *Anduo Zangyu Yuyin Yanjiu*.(*The Phonology of Amdo Tibetan*) .Ph. D. Dissertation.Shanghai Normal University.

XU, Shibo.2020. *Zangyu Shannan Lajiao Hua Yuyin Yanjiu*.(*Phonology of Lajiao Tibetan in Shannan*) .Ph.D. Dissertation.Shanghai Normal University.

YU, Min .1984. Zibgguo Yuyanxue Lunwen Xuan.(*Selection of Publications on Chinese Linguistics*) , 光生馆，东京 .

ZHANG, Jichuan .1982. G Zangyu Seyin Yunwei Duyin Chutan.(Pioneering Study on the Sound of Old Tibetan Stop Coda) .*Minzu Yuwen*.6.

ZHENGZHANG, Shangfang .1987. Shanggu Yunmu Xitong he Siden Jieyin Shengdiao de Fayuan Wenti.(The Rhyme System of Old Chinese and the Origin of the Four Division, Glides, and Tones).*Wenzhou Shifan Xueyuan Xuebao*.4.

——. 1990. Shanggu Rusheng Yunwei de Qingzhuo Weinti.(On the Voicing of Old Chinese Stop Codas).*Yuyan Yanjiu*.1.

ZHOU, Facheng.2019. *Reena Qiangyu Cankao Yufa*.(*A Reference Grammar of Re'ena Qiang*). Ph.D. Dissertation. Shanghai Normal University.

ZHU, Xiaonong and CUN, Xi. 2006. Shilun Qingzhuo Yinbianquan – Jian lun Wumminyu Nneibaoyin bu Chu yu Dongtai Diceng.(On the Voiceless and Voiced Sound Change Circle). *Minzu Yuwen*.3.